Methods in Neurosciences

Volume 12

Receptors: Molecular Biology, Receptor Subclasses, Localization, and Ligand Design

Methods in Neurosciences

Edited by

P. Michael Conn

Department of Pharmacology
The University of Iowa
College of Medicine
Iowa City, Iowa

Volume 12
Receptors: Molecular Biology, Receptor Subclasses, Localization, and Ligand Design

ACADEMIC PRESS, INC.
Harcourt Brace Jovanovich, Publishers
San Diego New York Boston London Sydney Tokyo Toronto

Front cover photograph: Autoradiographic distribution of the thymic β_2-adrenergic receptor. A Loats Image Analysis System (Amersham) was used to analyze differences in receptor localization. Cold colors (including blue, violet, and rose) indicate the absence or low concentration of receptor; green, yellow, orange, and red represent increasing receptor concentrations. This section from the thymus gland of a female rat shows nonspecific binding after an excess of (-) propranolol. Courtesy of Drs. Bianca Marchetti and Maria C. Morale, Department of Pharmacology, Medical School, University of Catania, Italy.

Academic Press, Inc.
1250 Sixth Avenue, San Diego, California 92101

United Kingdom Edition published by
Academic Press Limited
24–28 Oval Road, London NW1 7DX

International Standard Serial Number: 1043-9471

International Standard Book Number: 0-12-185273-3

PRINTED IN THE UNITED STATES OF AMERICA
93 94 95 96 97 98   EB   9 8 7 6 5 4 3 2 1

Table of Contents

Section I Molecular Biology (Molecular Cloning, PCR, Transcription, and Transfection)

Section II Receptor Subclasses

Section III Receptor Localization

Section IV Ligand Design

Contributors to Volume 12

Article numbers are in parentheses following the names of contributors. Affiliations listed are current.

E. X. ALBUQUERQUE (16), Department of Pharmacology and Experimental Therapeutics, University of Maryland, Baltimore, Maryland 21201

A. BALASUBRAMANIAM (23), Department of Surgery, University of Cincinnati Medical Center, Cincinnati, Ohio 45267

LU BAO (5), Department of Medicine, Harvard Medical School, Boston, Massachusetts 02115

JAMES F. BATTEY (6), Laboratory of Biological Chemistry, Developmental Therapeutics Program, Division of Cancer Treatment, National Cancer Institute, National Institutes of Health, Bethesda, Maryland 20892

STEPHEN A. BERMAN (19), Department of Gerontology, Laboratories of Molecular Neuroscience, McLean Hospital/Harvard Medical School, Belmont, Massachusetts 02178

JÉRÔME BERTHERAT (17), INSERM, Centre Paul Broca, 75014 Paris, France

D. BERTRAND (16), Department of Physiology, CMU, 1211 Geneva 4, Switzerland

EMIL BOGENMANN (10), Division of Hematology/Oncology, Children's Hospital Los Angeles, Los Angeles, California 90027

MARIE-LOUISE BOUTHENET (4), Université René Decartes, 75006 Paris, France

PATRIZIA BOVOLIN (12), Department of Animal Biology, University of Torino, 10123 Torino, Italy

NOEL J. BUCKLEY (1), Division of Physical Biochemistry, National Institute for Medical Research, London NW7 1AA, England

SHERRY BURSZTAJN (19), Department of Psychiatry, Laboratories of Molecular Neuroscience, McLean Hospital/Harvard Medical School, Belmont, Massachusetts 02178

S. A. LEWIS CARL (23), Department of Physiology and Biophysics, University of Cincinnati Medical Center, Cincinnati, Ohio 45267

NANCY E. CROSBIE (7), Worcester Foundation for Experimental Biology, Shrewsbury, Massachusetts 01545

THAN-VINH DAM (14), Neuroscience Division, Douglas Hospital Research Centre and Department of Psychiatry, McGill University, Verdun, Québec H4H 1R3, Canada

EVELINE P. C. T. DE RIJK (21), Department of Animal Physiology, University of Nijmegen, 6525 ED Nijmegen, The Netherlands

JACQUES EPELBAUM (17), INSERM U159, Centre Paul Broca, 75014 Paris, France

CAROLA EVA (2), Instituto di Farmacologia e Terapia Sperimentale, Universita di Torino, 10125 Torino, Italy

RICHARD I. FELDMAN (6), Berlex Biosciences, Inc., Alameda, California 94501

D. G. FERGUSON (23), Department of Physiology and Biophysics, University of Cincinnati Medical Center, Cincinnati, Ohio 45267

J. E. FISCHER (23), Department of Surgery, University of Cincinnati Medical Center, Cincinnati, Ohio 45267

F. JAVIER GARCIA-LADONA (18), Institute of Pathology, Department of Neuropathology, University of Basel, CH-4056 Basel, Switzerland

LEVI A. GARRAWAY (5), Department of Biological Chemistry and Molecular Pharmacology, Harvard Medical School, Boston, Massachusetts 02115

CRAIG GERARD (5), Department of Medicine, Harvard Medical School, Boston, Massachusetts 02115

NORMA P. GERARD (5), Department of Medicine, Harvard Medical School, Boston, Massachusetts 02115

JOHN R. GILBERT (11), Department of Medicine, Division of Neurology, Duke University Medical Center, Durham, North Carolina 27710

BRUNO GIROS (4), Unité de Neurobiologie et de Pharmacologie de l'INSERM, Centre Paul Broca, 75014 Paris, France

DANIEL GOLDMAN (8), Mental Health Research Institute and Department of Biological Chemistry, University of Michigan, Ann Arbor, Michigan 48109

DENNIS R. GRAYSON (12), Department of Biology, Fidia-Georgetown Institute for the Neurosciences, Georgetown University, Washington, D. C. 20007

RICHARD HARKINS (6), Berlex Biosciences, Inc., Alameda, California 94501

CHRISTINA A. HARRINGTON (1), Division of Physical Biochemistry, National Institute for Medical Research, London NW7 1AA, England

XIAO-PING HE (5), Department of Medicine, Harvard Medical School, Boston, Massachusetts 02115

BARBARA L. HEMPSTEAD (7), Division of Hematology/Oncology, Cornell University Medical School, New York, New York 10021

VIRGINIA HIEBER (8), Mental Health Research Institute, University of Michigan, Ann Arbor, Michigan 48109

DANIEL HOYER (15), CNS Department, Sandoz Pharma Ltd., CH-4056 Basel, Switzerland

CHRISTINE HULETTE (11), Department of Pathology, Duke University Medical Center, Durham, North Carolina 27710

MARYANNE E. HUNT (13), Basic Science Research, Neuropsychiatric Research Institute, Fargo, North Dakota 58103

HIDEYA IIJIMA (5), 1st Department of Medicine, Tohoku University School of Medicine, Sendai, Japan

JOHN L. KRSTENANSKY (24), Syntex Research, Division of Syntex, Inc., Palo Alto, California 94303

RALPH H. LORING (20), Department of Pharmaceutical Science, College of Pharmacy and Allied Health Services, Northeastern University, Boston, Massachusetts 02130

MARIE-PASCALE MARTRES (4), Unité de Neurobiologie et de Pharmacologie de l'INSERM, Centre Paul Broca, 75014 Paris, France

HIROSHI MATSUSHIMA (10), Department of Pediatrics, Jikei University School of Medicine, Tokyo 105, Japan

GUADALUPE MENGOD (15, 18), Department of Neurochemistry, CID, CSIC, 08034 Barcelona, Spain

DAVID S. MIDDLEMAS (9), Molecular Biology and Virology Laboratory, The Salk Institute, San Diego, California 92037

MIRTA MIHOVILOVIC (11), Department of Medicine, Division of Neurology, Duke University Medical Center, Durham, North Carolina 27710

FREDRICK J. MONSMA, JR. (3), Experimental Therapeutics Branch, National Institute of Neurological Disorders and Stroke, National Institutes of Health, Bethesda, Maryland 20892

JOSÉ M. PALACIOS (15, 18), Department of Neurochemistry, CID, CSIC, 08034 Barcelona, Spain

JEAN-LUC PAQUET (5), Marion-Merrell Dow, Strassbourg, Germany

DOROTA K. POLUHA (7), Worcester Foundation for Experimental Biology, Shrewsbury, Massachusetts 01545

PIERRE POULAIN (22), INSERM U156, 59045 Lille, France

RÉMI QUIRION (14), Department of Psychiatry, Neuroscience Division, Douglas Hospital Research Centre, McGill University, Verdun, Québec H4H 1R3, Canada

ALONZO H. ROSS (7), Worcester Foundation for Experimental Biology, Shrewsbury, Massachusetts 01545

ERIC W. ROUBOS (21), Department of Animal Physiology, University of Nijmegen, Toernooiveld, 6525 ED Nijmegen, The Netherlands

MARIA-RITA SANTI (12), Department of Biology, Fidia-Georgetown Institute for the Neurosciences, Georgetown University, Washington, D. C. 20007

DAVID W. SCHULZ (20), Pfizer Central Research, Groton, Connecticut 06340

JEAN-CHARLES SCHWARTZ (4), Unité de Neurobiologie et Pharmacologie de l'INSERM, Centre Paul Broca, 75014 Paris, France

S. SHERIFF (23), Department of Surgery, University of Cincinnati Medical Center, Cincinnati, Ohio 45267

DAVID R. SIBLEY (3), Molecular Neuropharmacology Section, Experimental Therapeutics Branch, National Institute of Neurological Disorders and Stroke, National Institutes of Health, Bethesda, Maryland 20892

PIERRE SOKOLOFF (4), Unité de Neurobiologie et de Pharmacologie de l'INSERM, Centre Paul Broca, 75014 Paris, France

ROLF SPRENGEL (2), Center for Molecular Biology, University of Heidelberg, D-6900 Heidelberg, Germany

XING SU (19), Laboratories of Molecular Neuroscience, McLean Hospital/ Harvard Medical School, Belmont, Massachusetts 02178

JAMES K. WAMSLEY (13), Department of Pharmaceutical Sciences and the Departments of Pharmacology and Neuroscience, North Dakota State University, Grand Forks, North Dakota 58201

MARYVONNE WAREMBOURG (22), INSERM U156, 59045 Lille, France

S. WONNACOTT (16), Department of Biochemistry, University of Bath, Bath BA2 7AY, England

Preface

Receptors initiate the means by which cellular regulators exert their actions on target cells. Because of the central role of cell–cell communication and signal transduction, receptors are of intrinsic interest to neuroscientists.

Receptor studies utilize both traditional methods of analysis and modern molecular techniques. Volumes 11 and 12 of this series are divided into sections describing, in a pragmatic way, "model" receptor techniques, molecular techniques, and techniques for the determination of receptor subclasses and for localization and consideration in ligand design. The chapters are written in a way that will allow readers to "export" the technology described to the study of other receptor systems in their own areas of interest. Techniques include PCR protocols, methods for the assessment of gene expression, transfection, cloning, autoradiography, *in situ* hybridization, radioligand binding, receptor solubilization and purification, and coupling to effector systems.

The goal of these volumes—and of the others in this series—is to provide in one source a view of the contemporary techniques significant to a particular branch of the neurosciences, information which will prove invaluable not only to the experienced researcher but to the student as well. Of necessity some archival material has been included, but the authors have been encouraged to present information that has not yet been published, to compare (in a way not found in other publications) different approaches to similar problems, and to provide tables that direct the reader, in a systematic fashion, to earlier literature as an efficient means to summarize data. Flow diagrams and summary charts will guide the reader through the processes described.

The nature of this series permits the presentation of methods in fine detail, revealing "tricks" and short cuts that frequently do not appear in the literature owing to space limitations. Lengthy operating instructions for common equipment will not be included except in cases of unusual application. The contributors have been given wide latitude in nomenclature and usage since they are best able to make judgments consistent with current changes.

I wish to express my appreciation to Mrs. Sue Birely for assisting in the organization and maintenance of records and to the staff of Academic Press for their efficient coordination of production. Appreciation is also expressed to the contributors, particularly for meeting their deadlines for the prompt and timely publication of this volume.

P. MICHAEL CONN

Methods in Neurosciences

Section I

Molecular Biology (Molecular Cloning, PCR, Transcription, and Transfection)

[1] Methods for Studying the Control of Receptor Gene Expression: Promoter Mapping and Analysis of Transcription Regulation

Christina A. Harrington and Noel J. Buckley

Introduction

The process that leads to the expression of a particular receptor molecule at the cell surface is complex, with the potential for regulation to occur at multiple levels (i.e., transcription, RNA processing, and translation). Activation (or derepression) of a receptor gene in a cell and initiation of transcription are the first steps in the cascade leading to the insertion of a mature protein in the membrane. Once RNA synthesis has been initiated, transcription of the gene may then continue for the life of the cell, it may be transient, or it may be modulated over the cell's lifetime, responding to intracellular and extracellular cues. Transcription is controlled by a repertoire of cis regulatory elements in the genomic DNA and trans-acting protein factors that bind to these elements to regulate gene expression.

In this chapter, we describe methods we are using to look at control mechanisms that operate at the transcriptional level to regulate the expression of neurotransmitter receptor genes. Analysis of transcriptional mechanisms is a broad and technique-rich area in modern molecular biology, and full consideration of the range of technical approaches is beyond the scope of a single chapter. We limit ourselves to a discussion of strategies and methodologies that we have chosen to isolate receptor genes and to begin the characterization of the genetic elements involved in the spatiotemporal control of receptor gene expression. Our goal is to produce an overview of the approaches we are taking to analyzing the transcriptional regulation of receptor genes. We describe alternative experimental approaches when appropriate and have included reference sources that provide thorough discussions of each area covered.

The chapter is divided into three sections: (i) strategy and methods for isolating genes and genomic flanking regions, (ii) methods for transcript mapping and identifying promoters, and (iii) *in vitro* and *in vivo* methods for analyzing transcription signals. The methods we describe presume that a cDNA for the receptor of interest is already available.

Isolation of Receptor Gene and Promoter

In this section we describe strategy and techniques for isolating the region of the genome that encodes the receptor and controls the expression of receptor mRNA. The DNA sequences involved in controlling gene expression can fall within the transcribed region of the gene itself and up to tens of kilobases upstream or downstream of the gene. In general the gene promoter is defined as that portion of the genome found adjacent to the 5' end of the RNA-encoding region that promotes basal expression of the RNA transcript. If the receptor is encoded by a single exon, isolation of the gene and its promoter is relatively straightforward. In this case, any genomic DNA clone that contains significant sequence upstream from the predicted transcription initiation site will in all likelihood contain the basic promoter elements, and, probably, other control elements as well. Genes composed of more than one exon can make promoter isolation more difficult, particularly if the introns separating the exons are large. In this situation, the more complete the cDNA is that one starts with, the more likely it is that the gene promoter can be isolated directly from a genomic library. Methods for obtaining full-length or extended cDNAs are dealt with in a later section.

Genomic Library Selection

Historically, the most popular vectors for genomic libraries have been derivatives of the bacterial phage λ. In the recombinant phage a portion of the λ DNA is replaced with genomic DNA. Genomic cloning vectors have been engineered to accommodate 9–23 kb of foreign DNA, and many of the available vectors have additional features that facilitate library construction and recombinant mapping. Some of the commercially available vectors we have found useful and easy to work with include λGEM (Promega, Madison, WI), λFIX (Stratagene, La Jolla, CA), and EMBL3 and EMBL4 (Promega; Stratagene; Clontech, Palo Alto, CA).

When multiple exons or large introns are anticipated, libraries containing larger genomic inserts are useful and will minimize the amount of recombinant "walking" (i.e., moving sequentially along the gene by cloning adjacent regions) that needs to be done. Cosmids are cloning vectors that combine aspects of plasmid and bacteriophage systems. The presence of the phage *cos* site allows *in vitro* packaging of large genomic DNA inserts into phage heads and high-efficiency transduction of recombinants into bacteria. Following transduction, cosmids replicate in bacteria as large plasmids. Several cosmid vectors are now available that can stably accommodate up to around 45 kb of foreign DNA. (We have had success with the pWE15 cosmid vector.) One drawback to using cosmid vectors in place of λ vectors is that the recombinant copy number in a colony is lower than in an equivalent λ plaque. Thus, for a given hybridization probe, a positive radioactive signal will be less intense for a cosmid recombinant colony than for a λ plaque. Nonetheless, if library screening procedures are performed carefully to maximize probe specific activity and to minimize background, the lower signal intensity of cosmid recombinants should not be a problem.

Genomic libraries from a wide number of species are available in both λ and cosmid vectors. Commercial sources include Stratagene, Promega, and Clontech. If it is necessary to construct your own library, detailed procedures for genomic cloning can be found in *Methods in Enzymology,* Volume 152 (1–3), or in Sambrook *et al.* (4).

Library Screening

Library Plating

For library screening, we generally plate 4–5 genome equivalents of cosmid or λ recombinants. For mammalian genomic cosmid libraries, this works out to approximately 4.0×10^5 colonies (assuming an average insert size of 35 kb). This number of colonies can be grown on two to three 22×22 cm bioassay dishes that can accommodate up to 250,000 colonies each (although we usually plate at a density of 200,000 colonies per plate, or less). Assuming an average insert size of 15 kb for λ recombinants, 8×10^5 plaques would correspond to 4 genome equivalents. Libraries in phage λ are plated at a density of 40,000–50,000 plaques per 15-cm petri dish (these can also be plated on the 22×22 cm square dishes, if desired). Our procedure for plating λ recombinants is essentially as described by Boulter and Gardner (5) using the bacterial host recommended by the supplier of the λ vector.

Cosmid libraries are plated directly onto membranes that are recommended for colony hybridization and that will survive handling and long-term storage at −80°C. We have had considerable success with GeneScreen Plus [New England Nuclear (NEN), Boston, MA]. Our procedure for cell plating and making replica filters is taken from Grosveld *et al.* (6, 7); a similar protocol is described in Sambrook *et al.* (4). (Any standard procedure for plating and replicating plasmid-containing bacterial libraries should be adequate.)

Probe Selection and Labeling

The probe(s) selected for screening a genomic library can be either double-stranded cDNA fragments or synthetic, single-stranded oligomers corresponding to cDNA sequence. *In vitro* synthesized complementary RNA (cRNA) probes and DNA fragments generated by polymerase chain reaction (PCR) amplification can also be used. If the gene contains multiple exons, it may be necessary to screen with sequences representing the entire cDNA, or at least both ends of the cDNA. If a complete cDNA is not available, some gene mapping (see below) will probably be required to determine whether genomic walking is required after the first round of positive recombinants are isolated. In this case, terminal sequences in isolated genomic clones are used to select adjacent genomic regions until the entire gene has been isolated.

When possible, we find it useful to confirm clone identification by screening successively with two different probes, generated from different regions of the cDNA or by different labeling methods (e.g., random primer labeled cDNA versus 3'-tailed oligomers or riboprobes). Successive screening can minimize the number of nonpositive colonies or plaques that are isolated.

Double-stranded DNA fragments are labeled by the random primer method (8, 9) to high specific activity ($>1 \times 10^9$ dpm per μg). If recombinant cDNA clones are used, the portion used for library screening must be separated from any regions of the vector DNA that would cross-hybridize with the λ or cosmid vectors carrying the genomic DNA inserts. Oligomer probes are labeled by 3' tailing with terminal deoxynucleotide transferase (10) to a specific activity of greater than 4×10^7 dpm/pmol.

Hybridization and Clone Isolation

Library filters are incubated with a probe and subsequently washed according to standard hybridization procedures (e.g., see Ref. 4, pp. 9.47–9.55). Stringent wash conditions determined according to the length and sequence (or average base composition) of the probe are required to minimize background. For a discussion of how to determine hybridization temperatures and wash conditions, see Wahl *et al.* (11). Appropriate wash conditions should be confirmed empirically by including a control filter in the hybridization which contains both positive and negative DNA controls.

Washed filters are exposed to Kodak (Rochester, NY) X-OMAT AR film sandwiched between two intensifying screens. Exposures of 3–7 days are usually required to detect positive signals. Positive recombinants are selected and individually isolated by a second round of screening. Cosmid or λ DNA is then prepared from recombinant clones for further analysis.

Gene Mapping and Promoter Indentification

Mapping the receptor gene involved (1) generating a physical map of the genomic DNA with restriction enzymes, (2) locating and positioning the introns (if any) which are spliced out of the mature mRNA, and (3) determining the 5′ and 3′ terminals of the transcript. The principles of physical gene mapping using partial and multiple restriction enzyme digests are described by Danna (12). Many of the commercially available vectors are engineered to contain bacteriophage RNA polymerase promoters flanking the genomic DNA insert. These promoter sequences allow specific labeling of the ends of the recombinant DNA and facilitate mapping of the insert. Commercial kits are available for this purpose. Methods for mapping genomic inserts in vectors containing phage promoters are described by Evans and Wahl (13).

After physical mapping of the genomic DNA, a transcript map of the gene is determined. This involves defining the boundaries of the transcribed region, determining whether introns are present, and, if they are, positioning the exons in the gene. We do not describe exon/intron and 3′ terminal mapping here, although several of the techniques described below are used for this purpose. We focus on 5′ terminal mapping and locating the promoter.

Generally, the first step in identifying the promoter for a gene is to determine the site of transcription initiation (which corresponds to the 5′ terminus of the messenger RNA). In many cases, this is a precise base located approximately 30 bp downstream from a TATA sequence (14). However, there are many examples of genes which do not initiate transcription at a precise location, and even a few which have two initiation sites that are widely displaced and generate mRNAs with different 5′ exons (reviewed in Refs. 15 and 16). At present the characteristics of G-protein-coupled receptor promoters are just beginning to be elucidated. However, sequence analysis of opsin (17, 18), substance P (19), and β_2-adrenergic (20) receptor genomic clones has demonstrated the presence of consensus TATA sequences upstream from discrete transcription initiation sites in these genes. Other conserved sequence motifs that may be involved in regulation of these genes have also been described (19–21).

There are several approaches that can be taken to identify the promoter region and map the transcription initiation site. The extent of cDNA sequence that is available and the structural complexity of the gene (i.e., its intron/exon structure) will determine the order and success with which the various techniques are applied. The five techniques generally used for promoter mapping are (1) full-length cDNA analysis, (2) nuclear run-on assays, (3) nuclease protection mapping, (4) primer extension assays, and (5) transient transfection assays. We describe each technique in the order listed, but the actual order of application depends on the particular situation. Appropriate use and the pros and cons of each procedure are discussed in the individual sections. In general, the sensitivity and resolution of all of the above assays are determined by the levels of specific RNA expression in the cell lines or tissues that are utilized in the assay. Therefore, tissues or lines expressing high levels of receptor RNA should be determined prior to initiating gene mapping and promoter analysis.

Full-length cDNA Analysis

Generating Full-length cDNAs

The ready availability of many high-quality cDNA libraries, either commercially or privately, in many cases obviates the need for cDNA library construction. However, isolation of full-length clones (or at least clones containing as much 5' information as possible) greatly facilitates subsequent promoter mapping studies. Standard methods for producing cDNA can be found in Sambrook *et al.* (4). Whichever protocol is chosen, the preeminent considerations are the quality of the mRNA, the use of high-quality reagents, and the exercise of care at all stages. Beyond these considerations there are a few nuances which may be helpful: (i) inclusion of a mixture of oligo(dT) and random primers during the reverse transcription may be useful, especially when the cDNA is long; (ii) use of methylmercuric hydroxide to denature the mRNA prior to reverse transcription may help in destroying any secondary structure that inhibits transcription ("5'-STRETCH" libraries are available from Invitrogen, San Diego, CA, that employ this procedure); and (iii) vector-primed synthesis of cDNA frequently generates a large proportion of full-length clones (22).

Generating and Sequencing cDNA 5' Ends

PCR-based techniques can greatly aid in the generation and analysis of cDNA clones corresponding to the 5' ends of mRNA. RACE (rapid amplification of cDNA ends, Ref. 23) or anchored PCR (24) is particularly useful in transcribing cDNA from small amounts of total RNA (see Fig. 1). The products can then be cloned, if necessary, but usually obtaining sequence information is sufficient. This being the case, PCR-based sequencing procedures (25) can be used to sequence the RACE product directly; this offers the advantage that an adequate sequencing ladder can be generated from as little as 100 ng of crude PCR product. The use of nested primers to sequentially generate (i) first-strand cDNA, (ii), RACE product, and (iii) sequencing ladder greatly increases the specificity of each step.

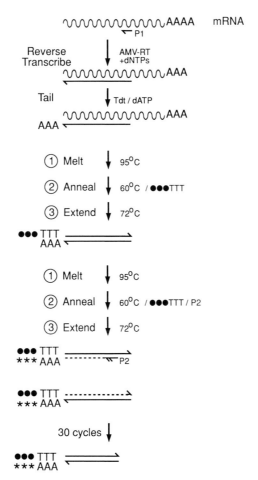

FIG. 1   Schematic diagram of the RACE or anchored PCR procedure. Total RNA or mRNA is reverse transcribed using avian myeloblastosis virus reverse transcriptase (AMV-RT) and a primer (P1) that anneals near the 5′ end of the transcript. The 3′ end of the cDNA is tailed using terminal deoxynucleotidyltransferase (Tdt) and dATP. The RNA/cDNA hybrid is then denatured, and the cDNA is used as a template for a PCR to synthesize the complementary DNA strand using an oligo(dT)-linker primer (. . . TTT). A further 30 cycles is then used to amplify the 5′ end of the cDNA using the same linker primer and a nested sequence-specific primer. Use of a nested primer to drive the PCR instead of the primer used in the reverse transcription increases the yield of specific reaction product. The resultant PCR product may then be cloned or sequenced.

Using cDNA Sequence to Map Exon Locations and to Map 5' Ends

Labeled portions of the cDNA can be used to determine which regions of the cloned gene encode the corresponding mRNA exon(s). Synthetic oligonucleotides derived from the cDNA sequence can be used to sequence genomic DNA and to map exon junctions precisely. The only exon of direct importance to mapping of the promoter is the first exon. If full-length cDNA information is available, then a probe corresponding to the 5' end of the cDNA can be constructed and used to isolate a genomic clone containing the transcriptional initiation site and/or to identify restriction fragments of genomic clones containing exon 1. Precise mapping of the site can then be achieved using primer extension and nuclease protection assays (see later).

Nuclear Run-On Assays

In the process of mapping the gene, it is necessary to determine which portions of the genomic recombinant(s) are transcribed into RNA. The results of a nuclear run-on analysis can be used to determine the approximate boundaries of the RNA transcript. The transcript boundaries can be mapped more precisely with nuclease protection mapping or primer extension assays described below, but, in the first instance, it is often useful to identify the transcribed regions and the general location of the ends of the gene with the nuclear run-on assay, particularly if a full-length cDNA is not available.

The principle underlying the use of the nuclear run-on assay is the uniform labeling of the nascent gene transcript as it is being synthesized. Isolated nuclei are incubated in the presence of radioactive nucleoside triphosphates and allowed to continue transcription that was initiated *in vivo*. In this process, labeled RNA molecules are generated that correspond stoichiometrically to the entire transcribed gene (see Fig. 2). Both exon and intron regions are labeled to the same extent.

The labeled RNA is then purified and annealed to cloned genomic DNA fragments. Recombinant DNA is immobilized on membranes either as a Southern blot of restriction enzyme digests or as dot/slot blots of subcloned or isolated restriction fragments. Genomic DNA that is homologous to an RNA transcript will anneal to the labeled, nascent RNA. DNA fragments which do not hybridize are considered to lie outside transcriptionally active regions. Occasionally, this technique is complicated by the presence of repetitive DNA sequences in the genomic DNA clones. Many of these sequences are transcribed *in vivo,* and labeled run-on transcripts from different areas of the genome will hybridize to a homologous sequence in the clone. The presence of repetitive sequences in the genomic clone can be determined by

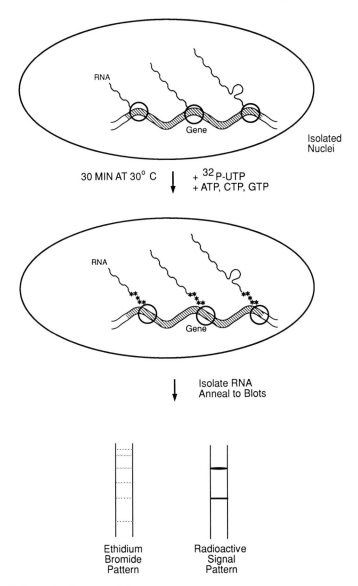

FIG. 2   Schematic diagram of the nuclear run-on assay. Circles indicate RNA polymerase II transcribing genomic DNA. Single wavy lines correspond to nascent RNA.
Asterisks indicate the incorporation of radioactive uridine monophosphate into nascent RNA.

labeling the clone, or fragments of it, and hybridizing it to a Southern blot of total genomic DNA. If a repetitive sequence is present in the clone, hybridization to multiple bands or a smear of radioactivity will be observed. If a repetitive sequence is present in the genomic recombinant, its location can often be distinguished from the transcribed regions of the gene by the use of appropriate restriction fragments.

A nuclear run-on protocol we have used successfully to map genes expressed in PC12 cells is described in Harrington *et al.* (26). Methods for isolating nuclei from fresh tissue and other cell lines will vary. A more complete discussion of the nuclear run-on assay is presented by White and LaGamma (27) (see also Refs. 28 and 29).

Nuclease Protection Mapping

Nuclease protection assays are used to define transcribed sequences and to map exon positions and the 5' and 3' ends of the RNA transcript. In the protection mapping procedure, labeled fragments of the cloned gene are hybridized to RNA prepared from a tissue or cell expressing the receptor gene (Fig. 3). After annealing, the RNA/genomic probe mixture is digested with a nuclease selected to digest all probe sequences which are not protected by being in a hybrid molecule with the receptor RNA. Only sequences complementary to RNA transcribed from the cloned fragment will be present in the hybrid. The nuclease-treated mixture is then analyzed to determine the size of any probe sequences that have been protected from nuclease digestion. This analysis generally involves denaturing the RNA/probe hybrid and separating the labeled nucleic acid fragments on a denaturing acrylamide gel.

Berk and Sharp (30) first described a procedure for nuclease protection mapping in 1977. Today there are several technical approaches that can be employed for protection mapping. In general, they differ in the nature of the probe and the single-strand-specific nuclease employed. Commonly used probes include (1) end-labeled DNA restriction fragments (2) internally (uniformly) labeled single-stranded DNA and (3) uniformly labeled cRNA transcripts. Although the actual nuclease protection protocol is similar in each case (i.e., anneal probe with cellular RNA, digest single-stranded probe not protected in hybrid molecule, and analyze protected products on gel), the information obtained with probe (1) is distinct from that obtained with probes (2) and (3). Basically, end-labeled probes will only be detected if the restriction site used to generate the labeled fragment end is included in the transcript. When uniformly labeled probes are used, any portion of the probe that corresponds to a transcribed region will be detected. Differences in band

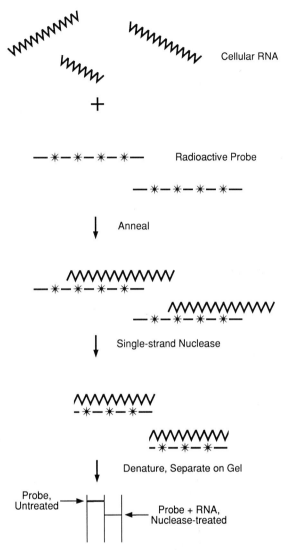

FIG. 3 Schmatic diagram of the nuclease protection assay. Cellular RNA is annealed with a labeled, single-stranded probe (see text for details). The hybrid molecules are then treated with single-strand-specific nuclease to digest any of the probe not protected by being in hybrid formation. The protected probe fragments are then sized by electrophoresis through a denaturing gel.

intensity can be used to distinguish protected exon sequences in mRNA from intronic regions of the transcribed gene fragment present in nuclear RNA.

Additional information can be obtained by using single-strand nucleases with different specificities. For example, the size and location of small introns in a transcribed region can be determined by comparing the results of a digest with endonuclease S1 with that of exonuclease VII. S1 nuclease will digest all single-stranded DNA, whereas exonuclease VII will only digest DNA processively from single-stranded ends. Therefore, a small intron in a transcribed fragment will not be protected from S1 nuclease, but it will be protected from exonuclease VII when the fragment is annealed to mRNA. When the protected fragments are run on a gel, the exonuclease VII digestion products will correspond to the full length of the transcribed region (i.e., exon plus intron), whereas the S1-digested products of probe that annealed with mRNA (exon sequence only) will correspond to protected exon-encoding regions.

A thorough discussion of nuclease protection mapping is presented by Calzone *et al.* (31) and by Sambrook *et al.* (4). Further discussion of nuclease protection assays and the use of both end-labeled cDNAs and uniformly labeled cRNAs can be found in Krause *et al.* (32).

Primer Extension Assays

Primer extension assays are used to map precisely the 5′ end of the RNA transcript. An oligomer complementary to a region near the 5′ end of the receptor RNA transcript is radioactively labeled with polynucleotide kinase, annealed to RNA from a tissue or cell line expressing the gene, and then used to prime reverse transcription of the RNA to generate a DNA copy that extends to the end of the message (Fig. 4). The size of the DNA copy strand is determined by analyzing the extended DNA on a denaturing acrylamide gel. The exact base (or bases) where transcription begins is determined by running the primer-extended product alongside a DNA sequence ladder generated from the genomic clone using the same end-labeled oligomer as a primer in a dideoxynucleotide sequencing reaction.

Best results are usually achieved when the complementary oligomer is within a hundred or so bases of the 5′ end of the mRNA. Secondary structure in the RNA can interfere with the reverse transcription process and produce premature stops during cDNA synthesis. The longer the region to be reverse transcribed, the greater the probability of running into these premature stops. Use of a tissue or cell line with relatively high levels of receptor RNA is also important for successful results. If the specific RNA levels are not at

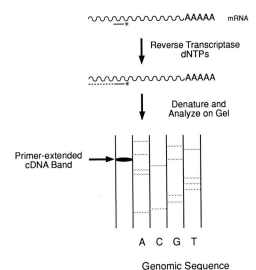

FIG. 4 Schematic diagram of the primer extension assay. Straight line with asterisk indicates 5′ end-labeled oligomer primer complementary to receptor mRNA which is annealed to total or poly(A)$^+$ RNA. Dashed line corresponds to primer-extended cDNA synthesized by reverse transcriptase. The primer-extended cDNA is sized on a denaturing acrylamide gel. The exact base (or bases) corresponding to the 5′ end of the message can be determined by running the cDNA band alongside a sequence ladder generated from genomic DNA using the same end-labeled oligomer.

the level of approximately 0.05% or greater of messenger RNA ($>10^{-5}$ of total RNA), it will probably be necessary to use purified mRNA [poly(A)$^+$ RNA] instead of total RNA for these experiments. Further discussion and protocols for the primer extension assay can be found in Calzone *et al.* (31).

Transient Transfection Assays

Transient transfection assays can be used to detect the presence of promoter activity in regions of genomic DNA. In the basic method, a fragment of genomic DNA thought to have promoter activity is inserted in a vector upstream of a promoterless reporter gene. The hybrid construct is then introduced into cultured cells that normally express the receptor gene under analysis and, thus, are expected to contain factors required for initiation of transcription from the receptor promoter. If the genomic fragment contains sufficient information for promoting transcription, the reporter gene will be expressed in the recipient cell. Transfection assays and their uses are discussed in more detail in the following section.

Measurement of Promoter Activity

To demonstrate that a particular clone or construct contains elements that are necessary and sufficient to promote transcription, it is necessary to introduce the DNA into a suitable expression system and to monitor the expression of the gene. The task then becomes one of distinguishing the activity of the transfected gene from the background activity of the endogenous gene.

There are two fundamentally different approaches to monitoring gene expression; one relies on the transfection of genomic clones into cell lines and the use of specific molecular probes to distinguish between endogenous and transfected gene products; the other uses fusion genes of the putative promoter elements joined to a gene that encodes a "reporter gene." The reporter gene encodes a novel enzymatic activity which can be detected if the gene is activated (see later). Each approach has its own pros and cons; the choice is largely dependent on the length of the construct, the size of the gene, the level of expression, and the availability of antibodies and probes to detect and distinguish among the endogenous and transfected gene products.

Foreign DNA may be transfected into cell lines or introduced into the germ line of transgenic mice. The former is relatively straightforward but provides little information on the tissue and stage specificity of gene expression, whereas the latter is very laborious (and thus cannot easily be used to analyze large numbers of constructs) but does yield exquisite information on the specific cellular expression driven by the construct.

The remainder of this chapter provides a brief overview of the methodologies used to study the elements involved in driving neurotransmitter receptor gene expression. Details of actual protocols can be found in any of the standard texts, including Sambrook *et al.* (4), Rosenthal (33), and Hogan *et al.* (34).

Monitoring the Expression of Transfected Genes

Although as little as a few hundred bases of 5′ flanking DNA is frequently sufficient to confer cell- and differentiation-dependent expression of a linked reporter gene (see later) in transient expression systems, it is nevertheless advisable, where possible, to compare such core promoter activity with that attainable using the whole gene. The latter approach can provide important information, because it makes no assumption about the position of regulatory elements in the gene. Because the vast majority of promoter analyses have been performed using transient expression assays, we consider this in more detail.

Transient Expression Assays

Transient expression usually involves the introduction of plasmid or cosmid DNA into appropriate host cell lines growing exponentially. During transient transfection, DNA is taken up by the cells and translocated to the nucleus. For about 2 to 3 days, multiple copies of the plasmid act as transcription units in the nucleus before they are degraded, and, therefore, assays are usually conducted within 3 days of transfection.

Transfection may be carried out by any of the routine procedures, including $CaPO_4$ precipitation, DEAE-dextran-mediated transfection, lipofection, protoplast fusion, microinjection, electroporation, and retroviral infection. Reviews and details of these procedures can be found in a number of texts (4, 35–37). It is also worth mentioning the development of a relatively recent lipopolyamine-mediated gene transfer technique that allows the expression of transfected genes in primary neurons (38, 39). As yet this latter method has not been applied to promoter analysis, although its potential is obvious.

By far the most commonly used procedures are $CaPO_4$ precipitation and electroporation. The $CaPO_4$ precipitation method has the advantages of ease and affordability, but it has the drawback of variability. In our hands, this variability can be greatly reduced by following the modifications suggested by Chen and Okayama (40), even though these modifications were originally introduced to increase the frequency of stable transformation. Another factor is the refractoriness of many neuroblastomas to $CaPO_4$-mediated transfection. In addition, the linear range of transfection efficiency is limited to about 1–20 μg DNA/10 ml/10^6 cells. Electroporation is quickly succeeding $CaPO_4$ as the method of choice, since it is rapid, consistent, and offers a much greater linear range of transfection efficiencies. Its main drawbacks are the need to optimize electroporation conditions for each individual cell line, the requirement for a relatively large number of cells, and its higher cost. However, few cell lines are completely refractory to electroporation, and it is to be recommended wherever possible.

Use of Stable Transformants

A small fraction of transfectants will incorporate a few copies (<10) of the transfected DNA into their chromosomal DNA and become stably transformed. By cotransfection with a selectable marker such as *neo*^r (*neo*^r encodes aminoglycoside phosphotransferase that confers resistance to the antibiotic neomycin) and selection in the presence of neomycin (or G418), clones of stable transformants may be isolated. This is a time-consuming process and takes several weeks to establish continuous lines but, nevertheless, is worthwhile because stable transformants can be useful in resolving whether ambiguities in analyzing transient transfection assays are resolvable by incorporating the DNA into the chromosome. Problems that can arise when using transient transfection assays stem from the fact that, although plasmids translocated to the nucleus may be efficiently used as transcriptional units, they are nonetheless devoid of any influence that neighboring chromatin may exert. Furthermore, because the plasmids are present in high copy number, the concentration of transcriptional factors may be rate limiting, possibly leading to aberrant expression. To a large extent, establishment of stably transformed cell lines obviates these problems.

Measurement of Promoter Activity

If the gene is small enough to be contained within a single cosmid clone, then the host cells may be transfected directly with the cosmid DNA (either supercoiled or linearized). However, it must be borne in mind that the concomitant low transfection efficiency of cosmid DNA necessitates the use of very sensitive assays to detect any transcription in transient assays. This problem can be overcome by selecting stable transformants. If the transfected gene is derived from a different species/strain than the host cell, then it may be possible to obtain species-specific antibodies or to synthesize species-specific oligonucleotides; the latter is a much easier option provided that cDNA sequence is available, because 5' and 3' untranslated regions frequently show considerable interspecies differences (it is often possible to make probes that will distinguish between transcripts derived from such closely related species as rat and mouse). Such oligonucleotides can be used as either hybridization probes or as PCR primers.

Analysis of transcripts by a coupled reverse transcription/polymerase chain reaction is convenient, fast, sensitive, and requires only low amounts of starting material. However, this assay is not without its problems when used to analyze extracts prepared from transiently transfected cells, because the transfected DNA can serve directly as a template for the PCR and, unlike chromosomal DNA, seems to be refractory to complete digestion with DNase I. Thus, it is necessary to design primers that cross an intron in order to unambiguously distinguish between PCR products derived from transfected DNA and those derived from transcribed DNA. This problem does not apply to stable transformants because the foreign DNA is incorporated into the chromosomal DNA and is susceptible to removal by DNase I digestion.

Use of Reporter Genes

If species-specific probes are not available (often, the test DNA and the host cell are derived from the same species), then modifications must be introduced into the DNA to render its gene products distinct from those of the endogenous gene. Earlier studies introduced deletions into the transfected gene so that its mRNA was distinguishable from endogenous transcripts by means of nuclease protection assays. This approach has been largely outmoded by the introduction of reporter genes. Reporter genes encode novel enzymatic activities that can be fused to gene regulatory elements or spliced, in frame, into the coding sequence of the genomic DNA. This latter use of reporter genes is largely restricted to analysis of gene expression in transgenic mice (see later).

The most common use of reporter genes is in measuring and comparing promoter and enchancer activity (the notional definition of an enchancer is a cis regulatory element that can act over considerable distances, function in an orientation- and position-independent manner, and possibly function in a tissue-specific manner). There are now numerous mammalian expression vectors available that allow the insertion of test sequences proximal to a reporter gene. The three most commonly used reporter genes are those for β-galactosidase (GAL), bacterial chloramphenicol acetyltransferase (CAT), and firefly luciferase (LUC). The choice of reporter gene depends on the question that is being asked. For quantitative comparison of weak promoters, LUC (41) offers the most sensitive assay system, and although scintillation counters can be modified to measure firefly luminescence, a luminometer is required to maximize the sensitivity. CAT (33) is more easily quantifiable and, with use of the liquid scintillation assay (4), is more sensitive, faster, and easier than the thin-layer chromatography (TLC) assay. Colorimetric analysis of GAL (4) is the least sensitive of the assays (although the histochemical detection of GAL is very sensitive; see later).

Whichever reporter gene is used, quantitation is a vexed problem because different transfection efficiencies can result from different plasmid preparations and will vary among different cell lines. Normalization is achieved by cotransfecting a different reporter gene under the control of a constitutive promoter, as those of such simian virus 40 (SV40), Rous sarcoma virus long terminal repeat (RSV LTR), cytomegalovirus (CMV), or β-actin, and comparing the ratio of the two activities for a given amount of protein. The two most convenient reporters are CAT and GAL because the assays can be performed on the same extract (4) and therefore do not need any further correction to account for different cell extraction procedures. However, the sensitivity of LUC might outweigh this consideration. This means of normalization has its problems. If a number of different promoter constructs are being tested on a cell line, then apparent variation can result because of varying quality of plasmid preparations. If the same pair of plasmids is used to transfect a range of cell lines, then discrepancies may reflect differences in the strength of the promoter of the normalizing reporter vector or the stability of the reporter gene products in each cell line.

Transgenic Mice

A comprehensive review of the techniques involved in producing transgenic mice can be found in Hogan *et al.* (34). Because the procedures involved in establishing and analyzing lines of transgenic mice are complex and laborious, the preeminent consideration must be that the construct is thoroughly characterized in cell lines prior to injection into mouse oocytes. There are two principal reasons for analyzing the expression of a receptor promoter in transgenic mice: (i) it is the only system in which the cell- and stage-specific expression of the transgene can be rigorously examined, and (ii) it is the only experimental system that exposes the transgene to the complete range of cues necessary for correct spatio + temporal expression. These are especially important considerations for the analysis of the expression of neurotransmitter receptor genes because they are expressed only on subsets of neurons and many show transient peaks of expression. It is difficult to conceive an alternative experimental system that would allow these features to be adequately analyzed. Indeed, there are several examples of neural genes that, when transfected into cell lines, show no cell specificity in their expression (whether assessed in transient transfectants or in stable transformants), yet the same construct when expressed in transgenic mice shows complete specificity (42, 43). The rationale for this apparent anomaly is that the transgene needs to be exposed to the full developmental program in order for its expression to be appropriately restricted.

Because the usual reason for expressing a gene in a transgenic mouse is to examine its spatiotemporal pattern of expression, the analyses generally rely on histochemical or immunohistochemical detection. By far the most common marker gene is GAL, because the histochemical detection is sensitive, rapid, and easy and there is generally very low background. If the expression of the whole gene is being examined, then the GAL construct can be ligated in frame into the coding region of the DNA. This approach was used by Oberdick *et al.* (44) to detect expression of the L7 protein in the cerebellum of transgenic mice. More recently, a small epitope corresponding to c-*myc* has been used as a reporter gene, with subsequent expression being detected with immunohistochemical procedures. The larger the genomic fragment, the more difficult is the reconstruction, and this approach generally works best on small fragments. One potential drawback is the possibility that the reporter gene will no longer function when inserted into the transfected DNA.

Acknowledgments

We are grateful to Muriel Steel and Ian Wood for helpful criticism in the preparation of the manuscript. This work was supported in part by the Medical Research Council. C.A.H. is supported by a grant from Glaxo.

References

1. B. G. Herrmann and A. M. Frischauf, *in* "Methods in Enzymology" (S. L. Berger and A. R. Kimmel, eds.), Vol. 152, p. 180. Academic Press, San Diego, 1987.
2. A. M. Frischauf, *in* "Methods in Enzymology" (S. L. Berger and A. R. Kimmel, eds.), Vol. 152. p. 183. Academic Press, San Diego, 1987.
3. A. G. DiLella and S. L. C. Woo, *in* "Methods in Enzymology" (S. L. Berger and A. R. Kimmel, eds.), Vol. 152. p. 199. Academic Press, San Diego, 1987.
4. J. Sambrook, E. F. Fritsch, and T. Maniatis, "Molecular Cloning: A Laboratory Manual," 2nd Edition. Cold Spring Harbor Laboratory, Cold Spring Harbor, 1989.
5. J. Boulter and P. D. Gardner, *in* "Methods in Nueroscience" (P. M. Conn, ed.), Vol. 1, p. 328. Academic Press, San Diego, 1989.
6. F. G. Grosveld, H. H. M. Dahl, E. deBoer, and R. A. Flavell, *Gene* **13**, 227 (1981).
7. F. G. Grosveld, T. Lund, E. J. Murray, A. L. Mellor, H. H. M. Dahl, and R. A. Flavell, *Nucleic Acids Res.* **10**, 6715 (1982).
8. A. P. Feinberg and B. Vogelstein, *Anal. Biochem.* **132**, 6 (1983).
9. A. P. Feinberg and B. Vogelstein, *Anal. Biochem.* **137**, 266 (1984).
10. W. H. Eschenfeldt, R. S. Puskas, and S. L. Berger, *in* "Methods in Enzymology" (S. L. Berger and A. R. Kimmel, eds.), Vol. 152, p. 337. Academic Press, San Diego, 1987.
11. G. M. Wahl, S. L. Berger, and A. R. Kimmel, *in* "Methods in Enzymology" (S. L. Berger and A. R. Kimmel, eds.), Vol. 152, p. 399. Academic Press, San Diego, 1987.
12. K. J. Danna, *in* "Methods in Enzymology" (L. Grossman and K. Moldave, eds.), Vol. 65, p. 449. Academic Press, New York, 1980.
13. G. A. Evans and G. M. Wahl, *in* "Methods in Enzymology" (S. L. Berger and A. R. Kimmel, eds.), Vol. 152, p. 604. Academic Press, San Diego, 1987.
14. R. Breathnach and P. Chambon, *Annu. Rev. Biochem.* **50**, 349 (1981).

15. S. E. Leff, M. G. Rosenfeld, and R. M. Evans, *Annu. Rev. Biochem.* **55,** 1091 (1986).
16. J. G. Sutcliffe, *Annu. Rev. Neurosci.* **11,** 157 (1988).
17. J. Nathans and D. S. Hogness, *Cell (Cambridge, Mass.)* **34,** 807 (1983).
18. J. Nathans, D. Thomas, and D. S. Hogness, *Science* **232,** 193 (1986).
19. A. D. Hershey, P. E. Dykema, and J. E. Krause, *J. Biol. Chem.* **266,** 4366 (1991).
20. S. Collins, M. A. Bolanowski, M. G. Caron, and R. J. Lefkowitz, *Annu. Rev. Physiol.* **51,** 203 (1989).
21. D. J. Zack, J. Bennet, Y. Wang, C. Davenport, B. Klaunberg, J. Gearhart, and J. Nathans, *Neuron* **6,** 187 (1991).
22. H. Okayama and P. Berg, *Mol. Cell. Biol.* **3,** 280 (1983).
23. M. A. Frohman, M. K. Dush, and G. R. Martin, *Proc. Natl. Acad. Sci. U.S.A.* **85,** 8998 (1988).
24. E. Y. Loh, J. F. Elliott, S. Cwirla, L. L. Lanier, and M. M. Davis, *Science* **243,** 217 (1989).
25. M. A. Innis, K. B. Myambo, D. H. Gelfand, and M. A. D. Brow, *Proc. Natl. Acad. Sci. U.S.A.* **85,** 9436 (1988).
26. C. A. Harrington, E. J. Lewis, D. Krzemien, and D. M. Chikaraishi, *Nucleic Acids Res.* **15,** 2363 (1987).
27. J. D. White and E. F. LaGamma, *in* "Methods in Enzymology" (P. M. Conn, ed.), Vol. 168, p. 681. Academic Press, San Diego, 1989.
28. M. Blum, *in* "Methods in Enzymology" (P. M. Conn, ed.), Vol. 168, p. 618. Academic Press, San Diego, 1989.
29. J. R. Nevins, *in* "Methods in Enzymology" (S. L. Berger and A. R. Kimmel, eds.), Vol. 152, p. 234. Academic Press, San Diego, 1987.
30. A. J. Berk and P. A. Sharp, *Cell (Cambridge, Mass.)* **12,** 721 (1977).
31. F. J. Calzone, R. J. Britten, and E. H. Davidson, *in* "Methods in Enzymology" (S. L. Berger and A. R. Kimmel, eds.), Vol. 152, p. 611. Academic Press, San Diego, 1987.
32. J. E. Krause, J. D. Cremins, M. S. Carter, E. R. Brown, and M. R. MacDonald, *in* "Methods in Enzymology" (P. M. Conn, ed.), Vol. 168, p. 634. Academic Press, San Diego, 1989.
33. N. Rosenthal, *in* "Methods in Enzymology" (S. L. Berger and A. R. Kimmel, eds.), Vol. 152, p. 704. Academic Press, San Diego, 1987.
34. B. Hogan, F. Constantini, and E. Lacy, "Manipulating the Mouse Embryo: A Laboratory Manual." Cold Spring Harbor Laboratory, Cold Spring Harbor, New York, 1986.
35. B. R. Cullen, *in* "Methods in Enzymology" (S. L. Berger and A. R. Kimmel, eds.), Vol. 152, p. 684. Academic Press, San Diego, 1987.
36. W. A. Keown, C. R. Campbell, and R. S. Kucherlapati, *in* "Methods in Enzymology" (D. V. Goeddel, ed.), Vol. 185, p. 527. Academic Press, San Diego, 1990.
37. W. F. Anderson and P. F. Dimond, eds., *BioTechniques* **6,** (1988).
38. J.-P. Behr, B. Demeneix, J.-P. Loeffler, and J. Perez-Mutul, *Proc. Natl. Acad. Sci. U.S.A.* **86,** 6982 (1989).
39. J.-P. Loeffler, F. Barthel, P. Feltz, P. Sassone-Corsi, and A. Feltz, *J. Neurochem.* **54,** 1812 (1990).

40. C. Chen and H. Okayama, *Mol. Cell. Biol.* **7,** 2745 (1987).
41. J. R. de Wet, K. V. Wood, M. DeLuca, D. R. Helinski, and S. Subrami, *Mol. Cell. Biol.* **7,** 725 (1987).
42. C. W. Wuenschell, N. Mori, and D. Anderson, *Neuron* **4,** 595 (1990).
43. K. Nakahira, K. Ikenaka, K. Wada, T. Tamura, T. Furuichi, and K. Mikoshiba, *J. Biol. Chem.* **265,** 19786 (1990).
44. J. Oberdick, R. J. Smeyne, J. R. Mann, S. Zackson, and J. I. Morgan, *Science* **234,** 223 (1990).

[2]   Cloning of G Protein-Coupled Receptors

Carola Eva and Rolf Sprengel

Introduction

A wide variety of cell surface receptors for hormones, drugs, and many neurotransmitters are coupled to guanine nucleotide regulatory proteins (G proteins) which in turn link to various intracellular effector systems. Biochemical and molecular characterization have identified these receptors as single polypeptides characterized by a common structural feature of seven transmembrane (TM) domains known to be involved in signal transduction and in binding of small ligands.

Based on partial protein purification and on the limited information obtained from the amino acid sequence of several proteolytic fragments, it was possible to isolate genes of various adrenergic receptors (1, 2) and several other members of the family of G protein-coupled receptors, such as the muscarinic cholinergic receptor (3) or the luteinizing hormone (LH) receptor (4). Comparison of primary protein structures of these receptors has revealed that all contain conserved amino acids motifs within the seven TM regions. Because the amino acid similarity among members of the same G protein-coupled receptor subfamily (i.e., muscarinic cholinergic receptors) can be as high as 80–90%, an increasing number of subtypes for each receptor family were first molecularly cloned by low-stringency hybridization of cDNA libraries using a related receptor cDNA as a probe (5–9). Subsequently, the sequence information obtained from additional G protein-coupled receptor genes suggested the development of two other strategies. These include (a) screening of cDNA libraries by low-stringency hybridization using highly degenerate oligonucleotide probes designed to recognize conserved DNA regions in the seven TM domains receptor genes (10–13) and (b) a polymerase chain reaction (PCR)-devised approach that uses degenerate primers to selectively amplify gene fragments of this receptor family (14–16) (Table I).

We describe in detail the principles and methodology of using cloned DNA and oligonucleotides as screening probes to isolate novel receptor genes. The powerful cloning technique of expression screening recently applied to find related and unrelated seven domain receptors will not be covered in this chapter because different methods and parameters characterize the rational of this cloning strategy.

Methods in Neurosciences, Volume 12

TABLE I   Different Strategies Used to Clone G Protein-Coupled
           Receptors[a]

Low-stringency hybridization		Polymerase chain reaction	Peptide-derived oligonucleotides
cDNA	Oligonucleotides		
D2	$\alpha 1_A$-Adrenergic	Substance P	LH
Human m1–m4	NPY-1	VIP	$\alpha 1_B$-Adrenergic
Neuromedin K	M3	5-HT1$_B$	α2-Adrenergic
Substance P	5-HT2		Porcine m1
FSH			

[a] A selected number of G protein-coupled receptors cloned by different strategies are given.
D2, D$_2$ dopamine receptor; FSH, follicle stimulating hormone receptor; NPY-1, neuropeptide
Y-1 receptor; m3, m3 muscarinic receptor; 5-HT2, 5-hydroxytryptamine 2 receptor; 5-HT1B,
5-hydroxytryptamine 1$_B$ receptor; LH, luteinizing hormone receptor.

Cloning by Receptor Purification

Purification of receptor proteins to isolate genes of receptors is still straightforward. Peptide sequences of the purified receptor are determined, and
oligonucleotides are synthesized which encode the deduced peptide sequence. These oligomers are used to screen a λ cDNA library in order to
identify the cDNA molecules derived from mRNA encoding the purified
receptor protein. For the screen, the oligomer is labeled at its 5′ end with
^{32}P and incubated with nitrocellulose filters containing lysed λ phage DNA
that represents the entire cDNA library. When hybridization is carried out
properly, the labeled oligonucleotide will base-pair with the complementary
nucleotide sequence; the location of the hybridizing phage DNA is visualized
by autoradiography. The positive phage is purified, and, finally, the nucleotide sequence of the inserted cDNA reveals the primary structure of the
receptor molecule.

Oligonucleotide Design

Synthetic oligonucleotide hybridization probes should be at least 20 bases
in length to provide sufficient specificity. The best results are obtained by
oligomers with more than 30 bases because, with probes of this length, one
can employ more stringent hybridization conditions to avoid isolation of
false-positive phages. Because the design of most oligonucleotides is based
on partial protein sequence, the degeneracy of the genetic code (one amino

acid can be encoded by up to six codons) necessitate the synthesis of several oligonucleotides which all encode the given peptide sequence. To keep the complexity of such an oligonucleotide pool low, peptide regions which are rich in Met, Trp, Lys, Asn, Glu, His, Gln, Asp, Tyr, Cys, and Phe should be selected. Those amino acids are determined by just one or two codons. The information given by codon usage tables may help to reduce the complexity even further.

Radioactive Labeling of Oligonucleotides

Usually, synthetic oligonucleotides are synthesized without a phosphate group at their 5′ termini and are therefore easily labeled by enzymatic transfer to the γ-^{32}P from [γ-^{32}P] ATP to the 5′ terminus of the oligonucleotide. Ten picomoles of an oligonucleotide is incubated in kinase buffer (50 mM Tris-HCl, pH 7.6; 1mM dithiothreitol; 0.1 mM EDTA, pH 7.6; 10 mM MgCl$_2$) with 10 pmol of [γ-^{32}P]ATP (3000 Ci/mmol; 10 mCi/ml) and 8 units ($\sim$1 μl) of bacteriophage T4 polynucleotide kinase in a final volume of 30 μl. Reaction is carried out for at least 30 min, then stopped by the addition of 50 μl of 20 mM EDTA. ^{32}P-Labeled oligonucleotide can be separated from the bulk of unincorporated radioactivity by different protocols. We suggest purifying the radiolabeled oligonucleotide by chromatography through Sephadex G-50 columns. The slurry of Sephadex G-50 (Pharmacia, Milwaukee, WI) is equilibrated in 10 volumes of water (see the manufacturer's instructions). The column of Sephadex G-50 can be prepared by placing a sterile, glass wool plug in a sterile Pasteur pipette, followed by filling 80% of the pipette with the slurry. The column, which should never run dry, is then washed with 500 μl to 1 ml of TE buffer (10 mM Tris-HCl, pH 7.6; 0.1 mM EDTA), followed by the addition of 100 μl of 1 M NaCl. Immediately after that, the radiolabeled sample is loaded on the top of the column and allowed to enter the column. The pipette is then filled with TE buffer and the eluate collected. After discharging the first 500 μl, 8–10 fractions of 100 μl (5 drops) should be collected. Two microliters of each fraction are counted, and the specific activity of those containing radiolabeled oligonucleotide (between fractions 4 and 8) that are not contaminated with unincorporated [γ-^{32}P]ATP is calculated. If the reaction is carried out efficiently, the specific activity of such probes can be as high as 10^8–10^9 cpm/μg of oligonucleotide.

Plating of cDNA Library

The aim of screening a cDNA library is to identify among all λ phages those carrying the putative receptor cDNA. For this purpose host cells are infected with the recombinant λ phages from the library and plated to semisolid media. Single bacteria are infected by single phages and produce progeny phages which in turn release another generation of daughter particles. The semisolid media limit the diffusion of released phages and, under these circumstances, successive rounds of infection result in a spreading zone of lysis that becomes visible as a so-called clear plaque in an otherwise turbid background of bacterial growth. Each plaque represents 10^5–10^7 progenies of a single recombinant λ phage representing a single amplified mRNA molecule. If a nitrocellulose sheet is placed on the top of the plaques, most phages of these plaques stick to the nitrocellulose; plaques containing the cDNA insert of interest can then be identified by hybridization to the complementary oligonucleotide or cDNA used for screening. To set up a successful screen, one should start with a cDNA library containing at least 1–2 × 10^6 independent phages. The average insert size should be 3500 base pairs (bp) because the minimum length of a G protein-coupled receptor is 1500 bp.

Before plating a library, 50 ml of a fresh overnight culture of the host cells should be prepared in advance. Cells should be grown in medium containing maltose because bacteria grown in maltose adsorb λ phages with higher efficiency. NZYDT medium (GIBCO BRL, Gaithersburg, MD) is suitable for growth. Fifty milliliters of the medium is transferred to a sterile 250-ml flask, inoculated with a single bacterial colony, and incubated overnight at 37°C with moderate agitation (250 cycles/7 min in a rotary shaker).

For regular screens 1–2 × 10^6 phages should be plated. We observed that the best screening results when 30 petri dishes, each containing 30,000–40,000 independent plaques, are analyzed. For each dish, 30,000–40,000 phages are mixed with 250 μl of host cells and 250 μl of 10 mM MgCl$_2$, 10 mM CaCl$_2$. After 20 min at 37°C, 10 ml of molten (47°C) top agarose (NZYDT, 0.7% agarose, Sigma, St. Louis, MO) is added, and the entire contents of the tube are poured onto the center of a 20-cm predried petri dish containing 50 ml of hardened bottom agar (NZYDT, 1.5% agar, GIBCO/BRL). Air bubbles should be avoided because they can give rise to false positives in the screen. Plates are gently swirled to ensure an even distribution of the top agar and kept for 5 min at room temperature to allow the top agarose to harden. Dry plates are then inverted and incubated at 37°C. Tiny plaques begin to appear after about 7 hr. However, to increase the number of phages per plaque, incubation should be continued until individual plaques begin to make contact with one another. A confluent lysis should not be reached in order to avoid recombination events between two different phages infecting the same cell. Chill the plates for at least 1 hr to allow the top agarose to harden before phage lifts are performed.

Immobilization of λ Plaques on Nitrocellulose Filters

For the hybridization analysis, nitrocellulose filters or nylon membranes can be used. We use nitrocellulose because, even though it is not as pliable as nylon membranes, it gives more reliable results. The nitrocellulose filters should be numbered with a ballpoint pen and placed onto the surface of the top agarose so that direct contact with the plaques is achieved. Then the filters are marked in at least three asymmetric locations with an 18-gauge needle. Using a pair of blunt-ended forceps, the filters are peeled off and placed (DNA side up) on Whatman (Maidstone, Kent, UK) 3MM paper soaked with 0.5 N NaOH and 1.5 M NaCl to allow the DNA to be denatured and become single stranded. After 5 min the filters are neutralized by putting them on a 3MM sheet soaked with 1.5 M NaCl and 1.5 M Tris-HCl, pH 7.4. Filters are then rinsed in 2 × SSC (20 × SSC is 3 M NaCl, 300 mM sodium citrate), and the DNA is fixed to the filters by baking them between a stack of filter paper for 2 hr at 80°C in a vacuum oven. Filters are now ready to be hybridized. A second set of filters can be placed on the same plates. They are marked with an 18-gauge needle using needle holes already present in the agar from the first filter lift and then processed as described above.

Hybridization to Nitrocellulose Filters Containing Replicas of λ Plaques

There are many methods available for hybridizing radioactive probes in solution to DNA immobilized on nitrocellulose membranes. We prefer hybridization reactions in formamide. The set up is simple and creates fewer evaporation problems at higher temperatures, and the hybridization protocol is less harsh on the filters than hybridization at 68°C in an aqueous solution. In general, we keep the hybridization conditions not too stringent, but we increase the stringency in the washing steps that follow. For hybridization, the baked filters are floated and submerged in prehybridization solution (30% formamide, 750 mM NaCl, 75 mM sodium citrate, 0.025% sodium pyrophosphate, 0.1% polyvinylpyrrolidone, 0.1% bovine serum albumin, 0.2% Ficoll, 50 μg/ml yeast tRNA). Filters are then transferred to the hybridization solution (12 filters/50 ml) containing 200,000 to 400,000 cpm/ml of the ^{32}P-labeled DNA probe. Hybridization is carried out for more than 12 hr and filters are washed 6 times for 5 min in SSC-containing washing solution (200 ml/12 filters).

In general, ^{32}P-labeled oligonucleotides are hybridized in 30% formamide at room temperature and ^{32}P-labeled cDNAs in 50% formamide at 42°C. The high-stringency wash for oligonucleotides is $0.2 \times SSC$ at 60°C and for cDNA $0.1 \times SSC$ at 65°C. Low-stringency washes are obtained with $2 \times SSC$ at room temperature for oligonucleotides and $2 \times SSC$ at 60°C for cDNAs. After washing, filters are partially dried, fixed on 3MM paper, covered with Saran Wrap, and exposed to X-ray film (Kodak XAR, Rochester, NY, or equivalent). If rewashing of filters is envisaged, filters should be exposed between two layers of Saran Wrap to avoid filters drying completely. The orientation of the filters relative to the X-ray film can be achieved by labeling the 3MM paper with radioactive ink or by using commercially available fluorescent markers (Stratagene, La Jolla, CA). Twenty-four hours later the films can be developed and positive plaques identified.

Picking λ Plaques

The area which correspond to the positive signals on the autoradiogram should be marked on the bottom of the petri dish. A Pasteur pipette is stabbed through the chosen area into the hard agar beneath, and the phages together with the underlying hard agar are withdrawn into the pipette. The agar piece is transferred to 1 ml of phage buffer (10 mM Tris-HCl, pH 7.6; 100 mM NaCl; 10 mM MgSO$_4$) containing 2% (v/v) chloroform to destroy host cells and then incubated at least 30 min at 37°C to allow the phages to diffuse out of the agar. This phage stock is diluted 1:1000 in the phage buffer, and 1, 5, and 20 μl of this dilution are replated (see above). After 2 hr plates with 1000 to 5000 isolated plaques/20-cm plate are selected, and nitrocellulose filters from these plates are rescreened for positive plaques using the same hybridization and washing conditions used for the first series of nitrocellulose filters. Replating of the phages has to be repeated until a positive, well-separated plaque can be identified. This procedure applies to λ gt10 cDNA libraries. For other λ vectors the average phage titer of plaques has to be determined in a pilot experiment.

Extraction of λ Phage DNA

To isolate λ phage DNA, a 50-ml small-scale liquid culture is set up. For this purpose 50 μl of a bacterial overnight culture is incubated together in a sterile culture tube with 10^6 phages (for λ gt10 phages, use about 20 μl of the 1:1000 dilution of a resuspended single plaque). The same volume of 10 mM MgCl$_2$, 10 mM CaCl$_2$ is added and incubated for 15 min at 37°C to allow phages to adsorb. Then 30 ml of NZYDT medium is added, and the culture is incubated with vigorous shaking until lysis occurs (8–12 hr). Bacterial debris is removed by centrifugation at 11,000 g for 10 min at 4°C. Twenty-five milliliters of the supernatant is recovered, and 20 μl of DNase (10 μg/ml) and 20 μl of RNase (10 μg/ml) are added to remove the DNA and RNA of the lysed bacteria. The supernatants are incubated at 37°C for 20 min. Seven microliters of PEG solution (20% polyethylene glycol 8000, 2.5 M NaCl) is added, mixed well, and incubated on ice for a least 1 hr to allow phage particles to form a precipitate. The precipitated phages are recovered by centrifugation at 10,000 g for 2 min, and the phage pellet is resuspended in 450 μl protease K buffer (10 mM Tris-HCl, pH 7.6; 15 mM NaCl). Protease K (2 μg, self-digested) is added and is allowed to digest phage proteins for 2 hr at 55°C. An extraction with a 50:50 mixture by volume of equilibrated phenol and chloroform is performed, and 400 μl of the supernatant is added to a new tube containing 1 ml cold ethanol (−20°C) mixed with 40 μl of 3 M sodium acetate, pH 5.6. Allow the mixtures to mix themselves and wait until the DNA is visible and precipitated. After ethanol is removed the air-dried phage DNA is resuspended in 100–200 μl of TE.

Characterization of the λ Clone

Subcloning in Plasmid Vectors

The cDNA insert of the recombinant λ phage has to be transferred to a bacterial plasmid vector for further analysis. Among many commercially available vectors, pBluescript plasmids (Stratagene) are convenient recipients. The cDNA can be inserted at a multiple cloning site, and pBluescript with inserted DNA fragments will form colorless colonies when plated on 5-bromo-4-chloro-3-indolyl-β-D-galactoside (X-Gal)/isopropylthio-β-D-galactoside (IPTG) plates. Vectors without insert will form blue colonies. For quick subcloning about 2 μg of the isolated λ DNA is mixed with 100 ng pBluescript vector DNA. The DNA mixture is then digested with restriction enzymes, releasing the entire cDNA from the λ DNA and cutting once in the multiple cloning site of pBluescript. Digestion should be monitored by checking linearization of the plasmid on an agarose gel. After complete digestion a phenol/chloroform extraction is performed, followed by DNA ethanol precipitation. The precipitated DNA is resolved in 20 μl of ligation buffer (500 mM Tris-HCl, pH 7.6; 1 mM EDTA; 10 mM dithiothreitol; 100 mM MgCl$_2$), then ligated in the presence of 100 μM ATP and 0.5 units of the bacteriophage T4 ligase (10 μl final volume). After 1 hr of ligation at room temperature 5 μl of the reaction is added to 200 μl of competent *Escherichia coli* DH5α cells (Stratagene), incubated on ice for 30 min, and heat-shocked 5 min at 37°C, followed by addition of 1 ml of growth medium. Cells carrying the recombinant plasmid are selected on agar plates containing 200 μg/ml ampicillin, 120 μg/ml X-Gal, and 100 μg/ml IPTG. Plasmids from white colonies are extracted and investigated by restriction enzyme analysis.

Subcloning in M13 Sequencing Vector

Sequence analysis of the isolated cDNA must be performed in order to demonstrate full length of clone and its structural homology (i.e., conserved amino acid stretches within the seven TM domains) to other members of the G protein-coupled receptor family. Analysis can be conducted directly on the plasmid construct or by further subcloning of smaller restriction fragments into vectors derived from filamentous phages. Prototypes of filamentous phage vectors are *E. coli* phage M13mp derivatives (17).

Vectors derived from filamentous phages are preferred because M13 phage-inserted DNA can be recovered in two forms: double-stranded circle replicating forms (RF) and single-stranded circles. The ready availability of single-stranded DNA is an important reason to use M13 in subcloning for sequence analysis. Base sequencing is done more conveniently if pure single-stranded DNA is available. Similarly, as in pBluescript, insertion of DNA in the polylinker region inactivates the *lacZ* α fragment. When insert-containing phages are plated with X-Gal and IPTG, they form colorless plaques; vectors that do not contain inserts form blue plaques.

The RF of M13mp is isolated by the same procedure employed for isolating plasmids, and its use as a cloning vector is exactly like that of a plasmid DNA molecule. The RF is cut by the restriction enzymes that produce sticky end compatible with those of DNA fragments isolated from plasmid digestion. The digested vector (20–25 ng) is mixed 1:3 (w/w) with the restriction fragment from plasmid DNA (which can be extracted from agarose gel as described below), and the mixture is annealed and ligated as previously described. Five microliters of the ligation reactions is added to 200 μl of rapidly thawed competent *E. coli* JM101 (Stratagene). Cells are then placed on ice for 30 min and heat-shocked at 37°C for 1 min, followed by the addition of 2 mg (20 mg/ml) IPTG, 2.4 mg X-Gal (24 mg/ml), and 10 ml of 45°C top agarose. Tubes are rapidly mixed by inversion and the contents poured onto 30-mm NZYDT plates. Plates are incubated at 37°C overnight to allow plaque formation. We recommend that self-ligation reactions of linearized phage DNA always be run in parallel in order to check for background from incomplete digestion or mutated phages.

Preparation of Single-Stranded Phage DNA from M13-Derived Vectors

Two milliliters of 2YT medium (1.6% Tripton, DIFCO, Detroit, MI; 1% yeast extract; 17 m*M* NaCl; pH 7.5) and 10 μl of an overnight culture of JM101 cells are dispensed in 10-ml culture tubes and inoculated with a colorless plaque picked with a sterile toothpick. Only isolated plaques should be picked; generally, 6–12 plaques from each plate should be grown separately. Cultures are shaken at 37°C for 5–8 hr and then centrifuged for 5 min at 10,000 g. The supernatant is incubated for 15 min at room temperature with 250 μl of PEG solution (30% polyethylene glycol 8000, 2.5 *M* NaCl), and precipitated phages are collected by 15 min of centrifugation at 10,000 g and resuspended in 200 μl of TE buffer. Phenol/chloroform extraction is performed as follows. One hundred microliters of phenol is added to resuspended phages, incubated at 50°C for 5 min, and followed by the addition of 100 μl of chloroform and 15 min of centrifugation. The upper aqueous phase (180 μl) is carefully transferred to a new tube, and DNA is precipitated for 5 min on ice after the addition of 1/10 volume of 3 *M* sodium acetate, pH 5.6, and 450 μl of cold absolute ethanol (−20°C). The DNA pellet is resuspended in 15 μl of TE buffer, and 3 μl is used for sequence analysis.

DNA Sequencing Reactions

The sequencing technique currently used in our laboratories is the Sanger method of dideoxy-mediated chain termination (18). 2′, 3′-ddNTPs differ from conventional dNTPs in that they lack a hydroxyl residue at the 3′ position of deoxyribose. They can be incorporated by DNA polymerases into a growing DNA chain through their 5′-triphosphate groups. However, the absence of a 3′-hydroxyl residue prevents the formation of a phosphodiester bond with the succeeding dNTP. Further extension of the growing DNA chain is therefore impossible. Thus, when a small amount of one dNTP is included with the four conventional dNTPs in a reaction mixture for DNA synthesis, there is competition between extension of the chain and infrequent, but specific, termination. The products of the reaction are a series of oligonucleotide chains, the length of which are determined by the distance between the terminus of the primer used to initiate DNA synthesis and the sites of premature termination. By using the different dNTPs in four separate enzymatic reactions, a population of oligonucleotides are generated that terminate at positions occupied by every A, C, G, or T in the template strand.

Sequence analysis of single-stranded DNA from M13mp derivatives is generally performed as follows. Three microliters of template DNA (0.1–1 μg) is incubated with 0.2 μg of *lac* 17-mer (GTT TTC CCA GTC ACG AC, Pharmacia) and 2× sequencing buffer (80 mM Tris-HCl, pH 7.5; 32 mM MgCl$_2$) at 55°C for 5 min (10 ml final volume). Six microliters of a labeling mix cocktail containing 1 μl of [^{35}S]dATPαS (Amersham, Arlington Heights, IL, 1200 Ci/mmol), 1 μl of 100 mM dithiothreitol, 3 units (1 μl) of T7 polymerase, and 3 μl of $-$C mix (0.5 μg dGTP and dTTP, 150 mM NaCl) are added to the cooled template–primer DNA and gently mixed by pipetting, followed by 1–5 min of incubation at room temperature. This step allows the incorporation of a small number of ^{35}S-labeled A residues into the sequence (until the first C is found) in order to label homogeneously each (longer or shorter) oligonucleotide that will be produced by the sequencing reaction. Meanwhile dddX reaction tubes can be set up by pipetting 2.5 μl of each dddX mix into four tubes, labeled G, A, T, and C (each of the dddX mixes is composed of 80 μg of all four dNTPs + 2 μg of the appropriate ddXTP). The DNA–cocktail mix is then dispensed to each of the four tubes (3.5 μl/tube) and incubated at 37°C for 5 min. Reactions are stopped by the addition of 6 μl of formamide dye (96% deionized formamide, 0.05% Bromophenol Blue, 0.05% xylene cyanole, and 10 mM EDTA), followed by denaturation at 100°C for 2–4 min. Samples are then cooled on ice and immediately loaded onto a 5% acrylamide sequencing gel that can be prepared as described (see Ref. 19).

Sequencing of supercoiled plasmid DNA can be performed, following denaturation of the double-stranded DNA, essentially by using the same protocol previously described for single-stranded DNA. Two micrograms of the plasmid DNA is incubated for 5 min with 0.2 N NaOH in a final volume of 20 μl. The incubation mix is neutralized by the addition of 10 μl of 0.2 M ammonium acetate, then the DNA is ethanol-precipitated and resuspended in 8 μl of water. Two microliters of 5× sequencing buffer and 1 μl of 5 μM sequencing primer are added to the denatured DNA and incubated at 37°C for 15 min. Six microliters of the labeling cocktail is then mixed to the cooled template–primer DNA as previously described.

Cloning of Novel Subtypes of G Protein-Coupled Receptor Families by Low-Stringency Hybridization

Cloning of G Protein-Coupled Receptors by Low-Stringency Hybridization Using cDNA Probes

Cloning of subtypes of receptors belonging to the same families has been successfully achieved by screening a cDNA library under low-stringency conditions using a cDNA encoding a related receptor as a probe. This kind of strategy has been used, for example, to isolate follicle-stimulating hormone (FSH) receptor using the LH receptor gene (9).

cDNA Probes

As DNA probes, a DNA fragment which covers not more than the coding region of the related receptor (TM region one to seven) should be selected for the cDNA library screening. Usually the DNA fragment, which should be 500–1500 bp in length, is released from vector sequences by cutting with appropriate restriction enzymes, with the digested plasmid DNA being resolved on an agarose gel. Finally the DNA fragment is isolated from the agarose gel, and 20 ng of the gel-purified fragment is radioactively labeled by incorporating α-^{32}P-labeled nucleotides in a multiprime labeling reaction (18). The specific activity of a labeled DNA fragment should reach at least 10^9 cpm/μg DNA to successfully screen 10^6 λ plaques of a cDNA library. Multiprime labeling reaction kits which guarantee high specific labeling are available from various commercial suppliers.

Screening of cDNA Library

The cDNA library generally derives from a tissue that is known, for example, from binding studies, to contain subtypes of the receptor family. The library is plated as described above, and two replica filters are pooled from each plate. Filters are then hybridized with the same cDNA probe using the standard hybridization conditions for cDNA probes. Following a 12-hr hybridization, both filter sets are washed under low-stringency conditions (50°C, 0.5× SSC), and double-positive plaques are identified after exposure of the filters to X-ray film. Then one filter set is washed again at high stringency (65°C, 0.1–02× SSC) and exposed to X-ray film to identify phage-inserted cDNA sequences, which are equal to the original probe. Plaques to be picked are those which are double positives but do not hybridize with the probe at 65°C, 0.1× SSC. These plaques should contain cDNA inserts related, but not identical, to the cDNA used for screening. Similar washing conditions should be maintained for the second and third replatings during plaque purification. Isolation and characterization of the isolated clones are then performed as described above.

Cloning of G Protein-Coupled Receptors by Low-Stringency Hybridization of cDNA Libraries Using Degenerate Oligonucleotides

Design of Oligonucleotides

Highly degenerate oligonucleotides are used for cloning G protein-coupled receptors by low-stringency hybridization of cDNA libraries. The oligonucle-

A

```
          *           * *       *   *     *   *     * * * *
bRHO    L L R T P L N Y I L L N L A L A V A D L
a1-R    L L R T P L N Y I L V N L A V I A A D L
SKR     L M R P V T S Y F I V T S L A L A L A D L
b2-R    L L Q V V T T Y F I T S L A L A C A D L

OLI-DE1  5'-CTG CAG ACA GTC ACC TAC TTT ATC ACC TCC TTG GCC TGT GCT GAT CT-3'
                     G                 C                  C
OLI-DE2  5'-CTC CAG ACA GTC AAC TAC TTT TTG AGC ATG GCC TGT GCT GAC CT-3'
                     G                 C          GT
                                                  C
OLI-DE3  5'-CAG ACA GTC AAT TAT TTT AAT TAT TTT AT-3'
                     A   CC          C       C   C

OLI-DE4              5'-ACC TTA GCT TGT GCT GAT TTT-3'
                                C   C         C   CC

rat M3-R 5'-CTC AAG ACA GTC AAC TAC TTC CTC TTA AGC CTG GCC TGT GCA GAC CTG-3'
            L   K   T   V   N   Y   F   L   L   S   L   A   C   A   D   L
```

B

```
         *   *           *       *   *       * *
bRHO     A F L I C W L P Y A G V A L P
a1-R     M F I L C W F P F F L H F F A L L I
SKP      T F A L W F Y H Y I H I F E I
b2-R     T F T L C W L P F F I V N I I C

coding pot.  L F A L C W L P L L P A L P
OLI-FC5  5'-CTC TTC GCC CTC TGC TGG CTG CCC CAT ATC AAC AAC TGC-3'
            |   |   |   |   |   |       |   |   |   |   |   |
                                           C        CTT ACC ATC TTC AAC ACT-3'
NPY-1-R  5'-GCC TTC GCG GTC TGC TGG CTG CCC CTT ACC ATC TTC AAC ACT-3'
            A   F   A   V   C   W   L   P   L   T   I   F   N   T
```

FIG. 1 Degenerate oligonucleotides identified the rat muscarinic M3 receptor (A) and the NPY-1 receptor (B) by low-stringency hybridization. (A) The amino acid sequence between TM domains 1 and 2, which is conserved among the bovine rhodopsin (bRHO), α1-adrenergic (a1-R), substance K (SKR), and β2-adrenergic (b2-R) receptors was used to design the screening oligonucleotide pools OLI-DE1, OLI-DE2, OLI-DE3, and OLI-DE4. Nucleotide mismatches between oligonucleotides OLI-DE1 to OLI-DE4 and the hybridizing M3 receptor gene fragment (M3-R) (12) are indicated by vertical bars. (B) Amino acid residues present in TM domain 7 of the bovine rhodopsin, α1-adrenergic, substance K, and β2-adrenergic receptors were used to design the screening oligomer OLI-FC5, which encodes a similar polypeptide (coding pot) (11). Nucleotide mismatches between OLI-FC5 and the hybridizing NPY-1 receptor cDNA are indicated by vertical bars. Highly conserved amino acids are depicted by stars.

otides are derived from the compilation of sequences corresponding to one consensus region located in the TM regions of different G protein-coupled receptors. Oligonucleotides derived from these consensus regions should contain a mixture of oligonucleotides with a number of degeneracies, allowing a 78% match, or better, with any of the receptors. Figure 1 shows, as an example, how the degenerate oligonucleotides that were used as probes to pull out from a rat brain cDNA library the neuropeptide Y-1 (NPY-1) receptor (OLI-FC5, Fig. 1A) (11) and the muscarinic M3 receptor (OLI-DE1 to OLI-DE4, Fig. 1B) (12) were designed. The choice of the composition was oriented arbitrarily toward a few particular receptors to avoid excessive degeneracy. Moreover, when the goal of the cloning strategy is to pick up a novel receptor belonging to a particular receptor family, the sequences used to design the oligonucleotides should be confined to those belonging to that specific family. Oligonucleotides synthesized following these criteria can be labeled as previously described, then used as probes for screening of the cDNA library.

Stringency Conditions Used for Screening of cDNA Library

Low stringency should be used for both hybridization and washing conditions. Two sets of filters should be hybridized for 12 hr at room temperature in 30% formamide. Washing should be done with $1 \times$ SSC at 42°C. The same stringency conditions for both hybridization and washing should be maintained for the second and third rounds of plaque purification.

Cloning of Novel G Protein-Coupled Receptors by Polymerase Chain Reaction

The PCR is a recently developed technique that allows a primer-directed enzymatic amplification of specific DNA sequences by using a polymerase extracted from the bacterium *Thermus aquaticus* that acts at high temperatures (20). The specificity of the PCR is based on amplification of two oligonucleotide primers that flank the DNA segment of interest and hybridization to opposite strands. The procedure involves repeated cycles of heat denaturation of the DNA, annealing of the primers to their complementary sequences, and extension of the annealed primers with DNA polymerase. The primers are oriented so that DNA synthesis proceeds across the region between the primers. The extension products of one primer can serve as a template for the other primer, so each successive cycle essentially doubles the amount of DNA synthesized in the previous cycle. Hence, each of the oligonucleotides becomes physically incorporated into one strand of the PCR products; the termini of this discrete fragment are defined by the 5' ends of the PCR primers.

Oligonucleotide Design

Oligonucleotides to prime the PCR should be designed following the same criteria described for degenerate oligonucleotides hybridization probes. Sense and antisense primers, preferably 18–24 nucleotides in length, are generally derived from comparison of consensus sequences located on two TM regions of different members of the G protein-coupled receptor family. The distance between the primers should not be greater than 1000 bp. Generally, conserved regions located in TM domains 3 and 6 or 7 have worked successfully (Fig. 2) (14–16). To facilitate the subcloning of PCR products, nucleotide sequences that contain endonuclease recognition sites are added at the 5′ termini of primer oligonucleotides.

Reverse Transcriptase–Polymerase Chain Reaction

As target for the PCR, total RNA from a tissue containing a high concentration of the receptor is first prepared. Cesium chloride density gradients (19) or guanidine thiocyanate acidic phenol/chloroform extractions (21) are convenient for this purpose. The integrity of the RNA should be checked on a formaldehyde-containing denaturing agarose gel by determining the presence of 28 and 18 ethidium bromide-stained ribosomal subunits; if most of the RNA is still intact, they should appear in a 2:1 ratio.

The complementary DNA from the isolated RNA is prepared by using oligo(dT) to prime the reverse transcriptase. One microgram of poly (A)$^+$ or 10 μg of total RNA is first incubated for 15 min at 65°C with 1μg of oligo(dT)$_{15}$ in a final volume of 10 μl to allow RNA denaturation and annealing

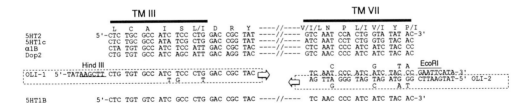

FIG. 2 Degenerate PCR primers identified a 5-hydroxytryptamine 1c (5-HT1c) receptor cDNA fragment. Oligomer pools OLI-1 and antisense oligomer OLI-2 (boxed by dotted lines) amplified a cDNA fragment encoding TM domains 3 to 7 of the 5-HT1c receptor (16). The nucleotide sequence of the primer pools OLI-1 and OLI-2 is based on a comparison of corresponding gene segments encoding TM domains 3 and 7 of the 5-HT2, 5-HT1a, α1$_B$-adrenergic, and D2 dopamine receptors. Note that EcoRI and HindIII endonuclease recognition sequences are added at the 5′ end of oligonucleotides OLI-1 and OLI-2, respectively. The primer recognition sites in the 5-HT1b receptor cDNA sequence are depicted in the bottom line.

of the oligonucleotide primer. The annealing step is followed by an incubation at 37°C for 1 hr after the addition of the reverse transcriptase buffer (50 mM Tris-HCl, pH 8.3; 75 mM KCl; 10 mM dithiothreitol; 3 mM MgCl$_2$), 1 mM each of dATP, dCTP, dGTP, and dTTP, and 35 units (~1 μl) of avian myeloblastosis reverse transcriptase (total volume 20 μl).

Synthesized cDNA is amplified by the PCR in a 50-μl reaction mixture containing 1 μl of the cDNA reaction, 1–2 units of Taq polymerase, 500 M of each dNTP, and 10 pmol of each primer together with 10 mM Tris-HCl, pH 8.7, 50 mM KCl, and 2–5 mM MgCl$_2$. Because MgCl$_2$ can differentially influence the efficiency of amplification of different transcripts, several PCR reactions using increasing Mg^{2+} concentrations should be run in parallel. PCR reactions are run in a thermocycle for 40–50 cycles; each cycle normally includes the following three steps: 0.6 min at 95°C to denature double-stranded DNA, 0.3 min at 55°C for primer annealing, and 0.6 min at 72°C for primer extension. The time for extension can be increased when fragments longer that 800–1000 bp are amplified; moreover, the annealing temperature can be decreased if primers of the reactions have a low degree of matching and no PCR products are obtained under the conditions used. Finally, the amplified DNA molecules are resolved by electrophoresis on an agarose gel, and all fragments matching the expected size are extracted for analyses.

Isolation of Amplified DNA Fragments

Extraction of the PCR-amplified DNA bands is a critical step. Extracted bands should be perfectly free of any residual agarose that may inhibit ligation to the vector. Several commercial kits are available for this purpose. As an alternative method, the piece of gel containing the amplified band can be cut, pushed through a 1-ml syringe into a tube containing 500 μl of phenol, mixed, and frozen for 5 min on an ice–ethanol bath. The melted phenol–agarose suspension is then centrifuged to spin down the gel and the supernatant processed by an additional phenol–chloroform extraction, followed by ethanol precipitation. Because the isolated fragment is flanked by restriction sites of the primers, these sites can be cleaved with the appropriate enzyme(s). The amplified DNA is then ready to be subcloned in the M13mp vector for sequencing.

Labeling of Polymerase Chain Reaction Products

Products of the PCR amplification that by sequence analysis appear to contain a partial fragment of a seven TM domain receptor, are used as probes to screen a cDNA library for a full-length clone. The same probe can also be used in parallel experiments with Nothern blot analysis (19) to establish the tissue distribution of the receptor transcript.

The double-stranded DNA to be labeled can be obtained by PCR amplification of a few nanograms of the single-stranded, insert-containing M13mp vector, using the same primers designed for isolation of the fragment. The amplified band is extracted from the gel as described above. Labeling of the DNA probe can be carried out by random oligonucleotide primers as previously described.

Stringency Conditions for Screening of cDNA Library

High stringency should be used for both hybridization and washing conditions in screening of a DNA library. Filters (two sets) should be hybridized for 12 hr at 42°C in 50% formamide. Washing should be done with 0.2× SSC at 65°C. The same stringency conditions for both hybridization and washing should be maintained in the second and third rounds of plaque purification.

Concluding Remarks

By using any of the different strategies described above, one may succeed in isolating a full-length clone (identified by sequence analysis) encoding a novel member of the G protein-coupled receptor superfamily. However, this orphan receptor has still to be pharmacologically characterized, and its functional coupling to a signal transduction system has to be demonstrated. This can be done either by injecting the synthetic mRNA (transcribed *in vitro* from the cloned receptor cDNA) into oocytes and then measuring ion channel regulation by electrophysiological techniques or by transferring the novel gene into immortalized cell lines by using the calcium phosphate transfection protocol described by Chen and Okayama (22). Transfected mammalian cells can be used for pharmacological studies, including binding assays, or for determination of second messenger systems (adenylate cyclase, phosphatidylinositol turnover).

References

1. S. Cotecchia, D. A. Schwinn, R. R. Randall, R. J. Lefkowitz, M. G. Caron, and B. K. Kobilka, *Proc. Natl. Acad. Sci. U.S.A.* **85,** 7159 (1988).
2. B. K. Kobilka, H. Matsui, T. S. Kobilka, T. L. Yang-Fen, U. Francke, M. G. Caron, R. J. Lefkowitz, and J. W. Regan, *Science* **238,** 650 (1987).

3. T. Kubo, K. Fukuda, A. Mikami, A. Maeda, H. Takahashi, M. Mishina, T. Haga, K. Haga, A. Ichiyama, K. Kangawa, M. Kijma, H. Matsuo, T. Hirose, and S. Numa, *Nature (London)* **323,** 411 (1986).

4. K. C. McFarland, R. Sprengel, H. S. Phillips, M. Khler, N. Rosemblit, K. Nikolics, D. L. Segaloff, and P. H. Seeburg, *Science* **245,** 494 (1989).

5. E. G. Peralta, A. Ashkenazi, J. W. Winslow, D. H. Smith, J. Ramachandran, and D. J. Capon, *EMBO J.* **6,** 3923 (1987).

6. J. R. Bunzow, H. H. M. Van Tol, D. K. Grandy, P. Albert, J. Salon, M. Christie, C. A. Machida, K. A. Neve, and O. Civelli, *Nature (London)* **336,** 783 (1988).

7. R. Shigemoto, Y. Yokota, K. Tsuchida, and S. Nakanishi, *J. Biol. Chem.* **265,** 623 (1990).

8. Y. Yokota, Y. Sasai, K. Tanaka, T. Fujiwara, K. Tsuchida, R. Shigemoto, A. Kakizuka, H. Ohkubo, and S. Nakanishi, *J. Biol. Chem.* **264,** 17649 (1989).

9. R. Sprengel, T. Braun, K. Nikolics, D. L. Segaloff, and P. H. Seeburg, *Mol. Endocrinol.* **4,** 525 (1990).

10. D. A. Schwinn, J. W. Lomasney, W. Lorenz, P. J. Szklut, R. T. Fremeau, T. L. Yang-Fen, M. G. Caron, R. J. Lefkowitz, and S. Cotecchia, *J. Biol. Chem.* **266,** 8183 (1990).

11. C. Eva, K. Keinanen, H. Monyer, P. Seeburg, and R. Sprengel, *FEBS Lett.* **271,** 81 (1990).

12. T. Braun, P. R. Schofield, B. D. Shivers, D. B. Pritchett, and P. H. Seeburg, *Biochem. Biophys. Res. Commun.* **149,** 125 (1987).

13. D. B. Pritchett, A. W. J. Bach, M. Wozny, O. Taleb, R. Dal Toso, J. C. Shih, and P. H. Seeburg, *EMBO J.* **7,** 4135 (1988).

14. A. D. Hershey and J. E. Krause, *Science* **247,** 958 (1990).

15. F. Libert, M. Paramentier, A. Lefort, L. Dinsart, J. Van Sande, C. Maenhaut, M.-J. Simons, J. E. Dumont, and G. Vassart, *Science* **244,** 569 (1989).

16. M. M. Voigt, D. J. Laurie, P. H. Seeburg, and A. Bach, *EMBO J.* **10,** 4017 (1991).

17. J. Messing and J. Vieira, *Gene* **19,** 269 (1982).

18. F. Sanger, S. Nicklen, and A. R. Coulson, *Proc. Natl. Acad. Sci. U.S.A.* **74,** 5463 (1977).

19. J. Sambrook, E. F. Fritsch, and T. Maniatis, *in* "Molecular Cloning: A Laboratory Manual" (C. Nolan, ed.), 2nd Ed., p. 7.19. Cold Spring Harbor Laboratory, Cold Spring Harbor, New York, 1990.

20. R. K. Saiki, D. H. Gelfand, S. Stoffel, S. J. Scharf, R. Higuchi, G. T. Horn, K. B. Mullis, and H. A. Erlich, *Science* **239,** 487 (1988).

21. P. Chomszynsky and N. Sacchi, *Anal. Biochem.* **12,** 156 (1987).

22. C. Chen and H. Okayama, *Mol. Cell. Biol.* **7,** 2745 (1987).

[3]   Molecular Cloning and Expression of a D_1 Dopamine Receptor

Frederick J. Monsma, Jr., and David R. Sibley

Introduction

Until recently, only two subtypes of dopamine receptors had been definitively shown to exist, each exhibiting a unique pharmacological profile and signal transduction mechanism. These were the D_1 receptors, which activate the enzyme adenylate cyclase in response to dopamine, and the D_2 receptors, which act to inhibit the activity of this enzyme (1). Both receptor subtypes belong to a large superfamily of neurotransmitter, hormone, and sensory (i.e., light and odorant) receptors that are coupled to their specific effector functions via guanine nucleotide regulatory proteins (G proteins) (2). Recently, the study of neurotransmitter receptors has been greatly enhanced by the application of modern molecular biological techniques. The initial cloning of prototypical members of this family, the opsins and the adrenergic receptors (2), has since led to an explosion of information concerning the sequence, structure, and membership of this family of proteins.

The cloning of novel receptors and subtypes has often been based on the prior cloning of other well-characterized members of this gene family. For example, the cloning of a β-adrenergic receptor led to the cloning of the D_2 dopamine receptor (3). Subsequently, the sequence of the D_2 dopamine receptor has been utilized to clone other members of this family of neurotransmitter receptors. In this chapter, we describe how we have utilized sequence information from previously cloned receptors to isolate a clone representing a D_1 dopamine receptor which is linked to adenylate cyclase activation (4). We have found this method, which utilized the polymerase chain reaction (PCR), to be generally applicable to identifying and isolating numerous members of the family of G-protein-coupled receptors.

Experimental Methods

General Considerations

The task of identifying and isolating clones for novel receptor subtypes can be broken down into several distinct steps: identification of a tissue or cell line enriched in a potential novel receptor subtype, preparation of high-quality mRNA from the source tissue, design and synthesis of degenerate oligonucleotide primers for PCR amplification, subcloning and sequencing of PCR products, and screening of an appropriate cDNA or genomic library to isolate a full-length clone. We discuss each of these steps in detail.

The general strategy we utilized to clone the D_1 dopamine receptor and identify other novel receptor subtypes involves the amplification of a segment of the messenger RNA utilizing the PCR and degenerate oligonucleotide primers to regions which exhibit a high degree of homology in G-protein-coupled receptors (Fig. 1). The success of this approach depends on the identification of a source of mRNA that is likely to contain the receptor subtype of interest, and it is crucial to design degenerate oligonucleotide

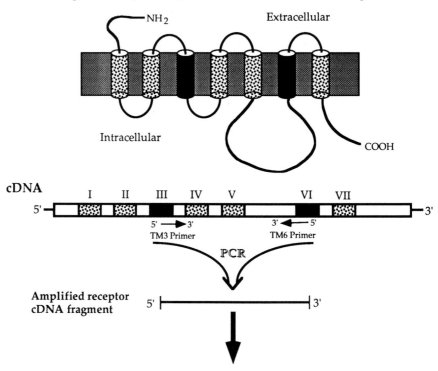

Screen cDNA library for full-length clone

Fig. 1 Diagrammatic representation of the use of the polymerase chain reaction with degenerate oligonucleotide primers to identify novel members of the G-protein-coupled receptor family.

primers with the appropriate balance of specificity and degeneracy that will enable one to amplify closely related gene products, with minimal nonspecific amplification. It should be noted that it is also possible to utilize genomic DNA as a starting template for amplification, provided the target gene does not contain long introns between the PCR primer sequences. This may be a reasonable assumption since several G-protein-coupled receptors have been shown to lack introns in the coding region of their genes. Nevertheless, we feel that mRNA obtained from an appropriate tissue or cultured cell line should be the starting material of choice owing to the enrichment of the target message as compared to genomic DNA.

mRNA Isolation

The first task is to identify a potential source of messenger RNA from which the clones of interest will be amplified. In attempting to clone the well-characterized D_1 dopamine receptor coupled to adenylyl cyclase, we chose to isolate mRNA from a murine neuroblastoma cell line, NS20Y. These cells have previously been shown to express a D_1 dopamine receptor which is pharmacologically indistinguishable from that found in the rat striatum (the classic D_1 receptor) (5) and which is coupled to activation of adenylate cyclase (6). In addition, these cells express very low to undetectable levels of D_2 dopamine and β-adrenergic receptors (F. Monsma and D. Sibley, unpublished observation, 1989), thus minimizing the greatest potential sources of "contaminating" receptors.

The method for isolation of mRNA is essentially according to Okayama *et al.* (7). NS20Y cells, cultured in Dulbecco's modified minimal essential medium (DMEM) and 10% fetal bovine serum (FBS), are washed with sterile Dulbecco's phosphate-buffered saline (0.9 mM $CaCl_2$, 2.7 mM KCl, 1.1 mM KH_2PO_4, 0.5 mM $MgCl_2$, 138 mM NaCl, 8.1 mM Na_2HPO_4) and lysed *in situ* in 5.5 M guanidine thiocyanate/25 mM sodium citrate/0.5% sodium lauryl sarcosine/0.2 M 2-mercaptoethanol, pH 7.0 (GNTC solution). DNA is sheared by three passes each through a 19-gauge and a 23-gauge needle. The GNTC lysate is divided, and approximately 20 ml is layered directly onto 17 ml of CsTFA solution (cesium trifluoroacetate/0.1 M EDTA, pH 7.0) at a final density of 1.51 g/ml. RNA is separated from DNA and cellular proteins at 25,000 rpm in an SW28 rotor (Beckman, Fullerton, CA) for 23–24 hr at 15°C. Pelleted RNA is dissolved in 0.4 ml of 4 M GNTC solution, and total RNA is precipitated in acetic acid and ethanol. After a second NaCl/ethanol precipitation, poly(A)$^+$ RNA is separated from total RNA by two rounds of oligo(dT)-cellulose chromatography (8).

TM III

Rat D2	CTG	TGT	GCC	ATC	AGC	ATT	GAC	AGG	TAC
	L	C	A	I	S	I	D	R	Y
Ham β2	CTG	TGC	GTG	ATT	GCA	GTG	GAT	CGC	TAT
	L	C	V	I	A	V	D	R	Y
Rat β2	CTG	TGC	GTG	ATA	GCA	GTG	GAT	CGC	TAC
	L	C	V	I	A	V	D	R	Y
Hum β2	CTG	TGC	GTG	ATC	GCA	GTG	GAT	CGC	TAC
	L	C	V	I	A	V	D	R	Y
Hum β1	CTG	TGT	GTC	ATT	GCC	CTG	GAC	CGC	TAC
	L	C	V	I	A	L	D	R	Y
Turk β1	TTG	TGC	GTC	ATC	GCC	ATC	GAC	CGC	TAC
	L	C	V	I	A	I	D	R	Y
Hum α2a	CTG	TGC	GCC	ATC	AGC	CTG	GAC	CGC	TAC
	L	C	A	I	S	L	D	R	Y

A

Combined DNA Sequences	CTG	TGC / T	GTC / CG	ATC / T / A	GCC / AGA	GTG / C T / A	GAC / T	CGC / A G	TAC / T

B

Degenerate Primer: 5′–CTG TG(C/T) G(TC/CG) ATC A(G/C)C AT(G/T) GA(C/T) C(GC/AG) TA–3′ **C**

TM VI

Rat D2	TTC	ATC	ATC	TGC	TGG	CTG	CCC	TTC	TTC	ATC
	F	I	I	C	W	L	P	F	F	I
Ham β2	TTC	ACC	CTC	TGC	TGG	CTG	CCC	TTC	TTC	ATT
	F	T	L	C	W	L	P	F	F	I
Rat β2	TTC	ACC	CTC	TGC	TGG	CTG	CCC	TTC	TTC	ATC
	F	T	L	C	W	L	P	F	F	I
Hum β2	TTC	ACC	CTC	TGC	TGG	CTG	CCC	TTC	TTC	ATC
	F	T	L	C	W	L	P	F	F	I
Hum β1	TTC	ACG	CTC	TGC	TGG	CTG	CCC	TTC	TTC	CTG
	F	T	L	C	W	L	P	F	F	L
Turk β1	TTC	ACC	CTC	TGC	TGG	CTC	CCT	TTC	TTC	TTG
	F	T	L	C	W	L	P	F	F	L
Hum α2a	TTC	GTG	GTG	TGC	TGG	TTC	CCC	TTC	TTC	TTC
	F	V	V	C	W	F	P	F	F	F

A

Combined DNA Sequences	TTC	ACC / GTG	CTC / G G / A	TGC	TGG	CTG / T C	CCC	TTC	TTC	ATC / T G / C T

B

Degenerate Primer: 5′–TTC ACC/GTG C(TC/A) TGC TGG C(TG/TC) CCC TTC TTC A(T)–3′ **C**

FIG. 2 Design of degenerate oligonucleotide PCR primers based on conserved sequences in the third and sixth putative transmembrane regions of the D_2 dopamine and β- and α-adrenergic receptors. Sections (A) show the nucleotide and amino acid sequences of the indicated receptors. Amino acid residues are indicated by their

Synthesis of First-Strand cDNA

Poly(A)$^+$ mRNA is converted to single-stranded cDNA for use as a template in the PCR reaction with avian myeloblastosis virus (AMV) reverse transcriptase. Approximately 4 μg of poly(A)$^+$ RNA is denatured for 10 min at 68°C in the presence of 2 μg oligo(dT)$_{15}$ primer in a total volume of 8 μl. After cooling slowly to room temperature, the annealed RNA/primer mix is added to the reverse transcriptase reaction (40 μl total volume) containing 50 mM Tris-HCl (pH 8.3), 40 mM KCl, 6 mM MgCl$_2$, 1 mM dithiothreitol, 1 mM each dATP, dCTP, dGTP, and dTTP, 40 units (U) RNasin (Promega, Madison, WI), and 44 U AMV reverse transcriptase (Promega) which had previously been diluted 1:10 and incubated on ice according to the manufacturer's instructions. The complete reverse transcription reaction is incubated for 1.5 hr at 42°C, followed by 10 min at 95°C to stop the reaction. The reaction is stored at −20°C until use in the PCR amplification.

Oligonucleotide PCR Primers

Degenerate oligonucleotide primers are designed by careful examination and comparison of the sequences of cloned α- and β- adrenergic receptors and the D$_2$ dopamine receptor (Fig. 2). Regions exhibiting a high degree of homology among the receptor sequences include the 3' end of the third transmembrane region (TM3) and the 5' end of the sixth transmembrane region (TM6). These regions offer the advantage of enabling each oligonucleotide primer to be terminated at a highly conserved amino acid which is encoded by only two codons, thus enhancing the probability that the critical 3' nucleotide would match in an unknown receptor and provide an expected product size of approximately 500–1000 bp, an optimal size for PCR amplification. In addition, amplification between TM3 and TM6 includes the third cytoplasmic loop which, in G-protein-coupled receptors, is highly variable in size, thus allowing for size selection of products for further analysis. Finally, eight nucleotides representing *Sal*I or *Hin*dIII restriction endonuclease sites plus

---

single-letter designations in bold type. Sections (B) show the nucleotides present at each position in these sequences. Sections (C) show the degenerate oligonucleotide primers which were chosen based on the sequences above. Note that for the TM6 primer the reverse complement of the sequence shown was synthesized. Because inclusion of all possible combinations of nucleotides led to an undersirable degree of degeneracy, the choice of nucleotides to include in the primers was biased first toward those represented in the D$_2$ receptor, then toward the most common nucleotide.

two random nucleotides are included on the 5' ends of the oligonucleotides to provide restriction sites to be used to subclone the PCR products for sequencing.

Oligonucleotide primers are synthesized on an ABI 381A DNA synthesizer (Applied Biosystems, Foster City, CA) with standard β-cyanoethylphosphoramidite chemistry. Following cleavage from the column and deprotection, the crude oligonucleotides are purified by denaturing gel electrophoresis on 12% polyacrylamide gels containing 8.3 M urea in 1× TBE (0.1 M Tris–borate, 2 mM EDTA, pH 8.3). The gel is placed on an intensifying screen, and the oligonucleotides are visualized by illumination with short-wavelength UV (254 nm). Bands representing full-size oligonucleotide are excised from the gel, macerated in elution buffer [0.1% sodium dodecyl sulfate (SDS), 0.5 M ammonium acetate, 10 mM magnesium acetate], and eluted overnight at 37°C with gentle agitation. The following day, acrylamide fragments are removed by filtration through glass wool, and the oligonucleotides are desalted on Sep-Pak C_{18} preparative columns (Waters, Milford, MA). Sep-Pak C_{18} columns are pretreated with 10 ml acetonitrile, 10 ml water, and 2 ml of 10 mM ammonium acetate. The filtered gel eluate is then loaded onto the column, and the flow-through is reapplied. Following 3 washes with 10 ml water, the oligonucleotide is eluted with 3 portions, 1 ml each, of 60:40 (v/v) methanol/water and the eluate dried in a Speed-Vac (Savant Instruments, Farmingdale, NY). The resulting pellet is dissolved in 200 μl TE buffer, precipitated with NaCl/ethanol, and finally resuspended in TE. The concentration is determined by absorbance at 260 nm.

Polymerase Chain Reaction

Polymerase chain reactions are carried out using reagents from the Perkin-Elmer/Cetus (Norwalk, CT) Gene Amp kit and a Perkin-Elmer thermal cycler. PCR reactions of 100 μl are set up with 4 μl of first-strand cDNA (~400 ng of RNA) as template in a reaction mixture containing 50 mM KCl, 10 mM Tris-HCl, pH 8.3, 1.5 mM MgCl$_2$, 0.001% (w/v) gelatin, 0.2 mM each dATP, dCTP, dGTP, and dTTP, and 1 μM of each primer. The reaction mix is subjected to an initial 5-min incubation at 95°C prior to addition of 2.5 U *Thermus aquaticus* (Taq) DNA polymerase. One cycle of 5 min at 55°C and 40 min at 72°C is performed to enhance the amplification of rare transcripts. This is followed by 30 cycles of 1.5 min at 95°C, 2 min at 55°C, and 4 min at 72°C, with a final extension of 15 min at 72°C.

Following PCR, 5 μl of the reaction mix is analyzed by agarose gel electrophoresis (Fig. 3). After determining the success of the amplification, the remaining 95 μl is subjected to preparative agarose gel electorphoresis, and

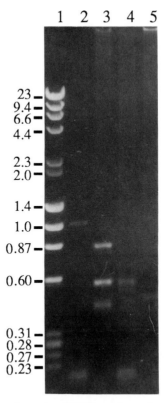

FIG. 3   Agarose gel electrophoresis of PCR products from amplification of NS20Y cDNA with degenerate primers to various transmembrane regions. Lane 1, molecular weight markers (base pairs $\times$ 10^{-3}) (λ *Hind*III digest/ϕX174 *Hae*III digest); lane 3, amplification of NS20Y cDNA with degenerate primers to TM3 and TM6.

individual bands or gel regions are excised from the gel. The DNA is removed from the gel by electroelution using a Centrilutor (Amicon, Danvers, MA) device and Centricon 30 filter units. The eluted DNA is washed twice with 2 ml of TE buffer and concentrated to approximately 50 μl by centrifugal ultrafiltration.

Subcloning of PCR Products

Gel-purified PCR products are prepared for subcloning by digestion with *Sal*I and *Hind*III (New England Biolabs, Beverly, MA). Fifty microliters of gel-purified PCR products are incubated first with *Sal*I in appropriate buffer for 4–6 hr at 37°C, followed by adjustment of the NaCl concentration to 100 μM, addition of *Hin*dIII, and overnight incubation at 37°C. (Previous experiments indicated that prolonged digestion with the appropriate restriction enzymes results in an increased efficiency of subcloning of linkered PCR products). At the same time, the vector for subcloning, pGEM 9Zf(+) (Promega), is likewise digested with *Sal*I and *Hin*dIII, although the incubation periods are limited to 1.5 hr with each enzyme. Both the vector and the PCR products are purified from the restriction reaction by ethanol precipitation and resuspended in TE buffer at a concentration of 10 ng/μl.

Ligations are carried out at a vector to insert ratio of 1:3, using 20 fmol of vector, in a final reaction volume of 20 μl, containing 50 mM Tris-HCl, pH 7.6, 10 mM MgCl$_2$, 5% (w/v) polyethylene glycol 8000, 1 mM ATP, 1 mM dithiothreitol, and 1 U T4 DNA polymerase (BRL, Gaithersburg, MD). The ligation is incubated for 4 hr at room temperature; subsequently, the reactions are diluted to 100 μl with TE buffer, and 5 ng (based on vector mass) is utilized to transform *Escherichia coli* strain DH5α. Transformed bacteria are plated onto agar plates containing 40 μg/ml X-Gal (5-bromo-4-chloro-3-indolyl-β-D-galactoside) and 0.5 mM IPTG (isopropylthio-β-D-galactoside) to allow for blue/white screening of insert-containing plasmids. Forty-eight white colonies are picked for sequence analysis. The colonies are grown overnight in 10 ml Luria–Bertani (LB) broth, 1 ml of which is utilized to prepare miniprep DNA the following day and the remainder pelleted, resuspended in 900 μl freezing medium [10 mM Tris-HCl, pH 7.5, 10 mM MgCl$_2$, 50% (v/v) glycerol], and stored at -20°C.

Miniprep DNA, for use in restriction analysis and DNA sequencing, is prepared by the alkaline/SDS method. One milliliter of overnight culture is pelleted in a 1.5-ml microcentrifuge tube and resuspended in 100 μl of solution 1 (50 mM glucose, 25 mM Tris-HCl, pH 8.0, 10 mM EDTA). Next, 200 μl of fresh solution 2 (0.2 N NaOH, 1% SDS) is added and mixed by inversion, followed by addition of 150 μl of solution 3 (3 M sodium acetate, pH 4.9) and vortexing. The resulting precipitate is removed by centrifugation at 14,000 g for 15 min at room temperature. The supernatant (~450 μl) is removed and the plasmid DNA precipitated by addition of 2 volumes of 100% ethanol, incubation at room temperature for 5 min, and centrifugation at 12,000 rpm for 10 min at room temperature. The resulting pellet is rinsed with 70% ethanol and dried under reduced pressure. The dry pellet is then dissolved in 50 μl TE buffer, and 1 μl is digested with *Sal*I and *Hin*dIII. The restriction digests are then analyzed on 1% agarose gels in 1$\times$ TAE (40 mM Tris–acetate, pH 7.8, 1 mM EDTA) to determine the insert size.

DNA Sequencing

Clones with appropriately sized inserts are sequenced by the Sanger dideoxy chain termination technique using vector-specific oligonucleotide primers to the SP6 and T7 RNA polymerase recognition sites in pGEM 9. Double-stranded plasmid DNA is utilized as template following RNase treatment and alkaline denaturation as follows. Approximately 4 μg of plasmid DNA in 100 μl TE is treated with 0.1 μg RNase A for 30 min at 37°C. The DNA is extracted with phenol/chloroform/isoamylalcohol (25:24:1, v/v) then chloroform/isoamylalcohol (24:1, v/v), precipitated with 0.4 M NaCl and 3 volumes ethanol, and dissolved in 50 μl water. Denaturation is accomplished by addition of 0.1 volume of 2 N NaOH/2 mM EDTA and incubation at 85°C for 5 min. The denatured DNA is neutralized and precipitated by addition of 0.1 volume of 3 M sodium acetate (pH 4.9) and 3 volumes of ethanol, incubation for 10 min on dry ice, and centrifugation at 12,000 rpm for 10 min at room temperature. The resulting pellet is washed with 70% ethanol and dried under vacuum.

DNA sequencing reactions are carried out with Sequenase version 2 reagents and Sequenase T7 DNA polymerase (United States Biochemical, Cleveland, OH) according to the manufacturer's protocol, with the exception that the Sequenase enzyme is diluted 1:5 prior to use instead of 1:8. The denatured DNA template is annealed to 0.5 pmol primer in 1× reaction buffer for 15 min at 37°C. The labeling reaction is allowed to proceed for 5 min at room temperature, and the termination reactions are carried out for 5 min at 37°C. The reactions are stored at −20°C and are heated for 2 min at 75°C immediately prior to loading 2 μl on a 0.4-mm sequencing gel. Sequencing gels are 6% acrylamide in 8.3 M urea and 1× TBE and are run at 60 W per gel.

Northern Blot Analysis

Following sequence analysis of various PCR-generated NS20Y fragments, a clone of interest is identified based on its homology with previously cloned G protein-coupled receptors. To obtain the full-length cDNA clone corresponding to the fragment of interest, screening of a cDNA library has to be carried out. Determination of the appropriate cDNA library to screen is accomplished by analysis of the distribution of the mRNA of interest in various brain regions and peripheral tissues. Northern blots are prepared by electrophoresis of 2 μg of denatured poly(A)$^+$ mRNA on 1% agarose containing 20 mM MOPS (pH 6.0), 5 mM sodium acetate, 1 mM EDTA, and

0.66 M formaldehyde. The RNA is electorphoretically transferred to Gene-Screen nitrocellulose paper [New England Nuclear (NEN), Boston, MA] according to the manufacturer's directions and immobilized by baking for 2 hr at 80°C *in vacuo*.

The Northern blots are probed with three 48-base oligonucleotide probes which are radiolabeled with terminal deoxynucleotidyltransferase and [α-^{32}P]dATP (Du Pont/NEN). Hybridizations are carried out for 18 hr at 37°C with 2 × 10^6 dpm/ml ^{32}P-labeled oligonucleotide probe in 4X SSPE (0.6 M NaCl, 0.04 M NaH$_2$PO$_4$, 4 mM EDTA, Na$_2$, pH 7.4), 5X Denhardt's solution, 50% formamide, 250 μg/ml yeast tRNA, 500 μg/ml sheared salmon sperm DNA, and 0.1% SDS. The labeled blots are washed in 1× SSPE (0.4% polyvinyl pyrrolidine, 0.4% bovine serum albumin, 0.4% FICOLL 400) and 0.1% SDS for 20 min 4 times at 56°C and twice at room temperature before autoradiography.

Screening of cDNA Library

Based on the results of the Northern blot analysis, which indicated a predominance of message in the striatum, a cDNA library is prepared from rat striatal mRNA in the λ-ZAP II vector. Approximately 1 × 10^6 clones from the unamplified portion of the library are screened. The probe is prepared by removal of the cloned PCR fragment from the pGEM vector by digestion with *Sal*I and *Hin*dIII followed by gel isolation and ^{32}P-labeling by nick translation. High-stringency washing of the library filters is performed with 0.15 M NaCl, 15 mM sodium citrate (pH 7.0), 0.1% SDS at 65°C. The λ phage found to hybridize to the probe are subjected to two rounds of plaque purification. *In vivo* excision and rescue of the nested pBluescript plasmids from the λ ZAP II clones is performed according to the Stratagene (La Jolla, CA) protocol.

The isolated clones are subjected to restriction enzyme analysis to determine the size of the insert and the likelihood of a clone containing the full protein coding region. One clone (pB-D1-73), approximately 3.6 kb in length, was chosen for full sequence analysis. The full sequence of this clone was obtained using both vector- and clone-specific primers, ultilizing both the full-length clone and a series of nested deletion mutants as template.

Subcloning and Transient Transfection

The positive identification of clone pB-D1-73 as a G protein-coupled receptor, and specifically as a dopamine receptor, requires analysis of its ability to bind dopaminergic (or other receptor) ligands, and to couple to some effector system. To carry out such studies, it is necessary to express the cDNA in a mammalian cell line. The entire 3.6-kb insert and polylinker region of pB-D1-73 is amplified by PCR from pBluescript with T3 and T7 primers. The reaction product is digested with *Not*I and *Kpn*I and ligated into the same restriction sites in the mammalian expression vector pCD-SRα (9) containing a modified polylinker region. Competent *E. coli* DH5α cells are transformed and the resulting colonies screened by restriction analysis. Large-scale plasmid preparations are made from a single insert-bearing clone, and the plasmid DNA is purified on CsCl gradients.

COS-7 cells are utilized for transient transfection by calcium phosphate precipitation. Plasmid DNA (30 μg) is brought to a final volume of 500 μl in 256 mM $CaCl_2$. Five hundred microliters of 1× HBS (140 mM NaCl, 0.75 mM Na_2HPO_4, 25 mM HEPES, pH 7.1) is added dropwise while vortexing, and the resulting mixture is allowed to precipitate for 30 min at room temperature. The precipitated DNA is then added dropwise to COS-7 cells in 10 ml of growth medium (DMEM, 10% fetal bovine serum, 30 μg/ml gentamycin) in 150-mm petri dishes at approximately 60% confluency. The cells are incubated for 18–24 hr at 37°C in a humidified atmosphere of 5% CO_2. The medium is then replaced with 30 ml of growth medium, and the cells are incubated for 2 days to allow expression of the transfected DNA.

Radioligand Binding to Transfected Cells

D_1 receptor binding activity is assayed in transfected cells with [^{3}H]SCH 23390. Transfected cells are harvested with 1 mM EDTA in Ca^{2+}-, Mg^{2+}-free Earle's balanced salt solution (EBSS) 72 hr after transfection with pSRα-D1-73 plasmid. The cells are washed twice in EBSS and once in binding buffer (50 mM Tris-HCl, pH 7.4, 1 mM EDTA, 5 mM KCl, 1.5 mM $CaCl_2$, 4 mM $MgCl_2$, 120 mM NaCl). The washed cells are lysed with 15 strokes of a Dounce homogenizer, and a crude membrane pellet is obtained by centrifugation at 43,000 g for 10 min. The crude membrane pellet is resuspended with a Teflon/glass homogenizer at a protein concentration of approximately 1 mg/ml, and 100 μl is utilized for receptor binding assays. For saturation binding studies, membranes are combined with 0.05 to 2nM [^{3}H]SCH 23390 in the absence or presence of 1 μM (+)-butaclamol to determine nonspecific binding in a final volume of 1 ml. For competition binding studies, the membrane suspension is combined with 0.5 nM [^{3}H]SCH 23390 and increasing concentrations of competing ligand in a final volume of 1 ml. Assays are incubated for 1 hr at room temperature and are terminated by rapid filtration through GF/C glass fiber filter paper (Whatman, Clifton, NJ) pretreated with 0.3% polyethyleneimine. The filters are washed 4 times with

4 ml of 50 m*M* Tris-HCl (pH 7.4) at 4°C. The radioactivity trapped on the filters is quantitated by liquid scintillation spectrometry in 5 ml Liquiscint cocktail (National Diagnostics, Mannville, NJ).

Results and Discussion

As indicated in Fig. 3, when cDNA derived from NS20Y mRNA is amplified by PCR with degenerate primers to the putative transmembrane regions 3 and 6, several products are produced ranging in size from approximately 0.8 to 0.4 kb. Sequence analysis of these bands has revealed that the largest band represents the D_2 dopamine receptor, while the band at approximately 0.5 kb consists of several DNA species, including a fragment which exhibits a high degree of homology with previously cloned catecholamine receptors. When the distribution of mRNA for this fragment was examined by Northern blot analysis, it was found to be localized predominantly in the striatum, with very low levels of expression also observed in the cortex and retina.

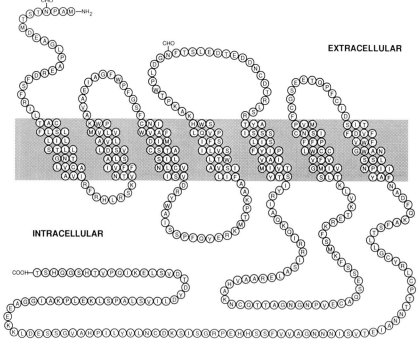

Fig. 4   Proposed membrane topography of the rat D_1 receptor based on the deduced amino acid sequence and homology with β-adrenergic receptors and rhodopsin.

This distribution is very similar to that of the D_1 dopamine receptor as demonstrated by receptor binding and autoradiography studies.

Isolation and sequence analysis of a full-length cDNA clone revealed an open reading frame of 1461 bases yielding a protein of 487 amino acid residues, although preliminary data from expression of this cDNA suggest that the first Met residue in this sequence is not used as a translational start (F. Monsma and D. Sibley, unpublished observations, 1992). Assuming that translation begins at the second Met, the open reading frame would yield 446 residues (Fig. 4), equivalent to the human D_1 dopamine receptor (10–12).

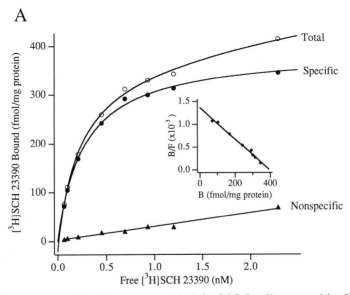

FIG. 5   Expression of the D_1 receptor cDNA in COS-7 cells assayed by [³H]SCH 23390 binding. (A) Saturation isotherms of the total, nonspecific, and specific binding of [³H]SCH 23390 to transfected COS-7 cell membranes. The inset shows a Scatchard transformation of the specific binding data. In this experiment, which was representative of three, the calculated K_D and B_{max} values were 0.3 nM and 400 fmol/mg protein, respectively. (B) Competition analysis of various dopaminergic ligands for [³H]SCH 23390 binding in COS-7 cell membranes. In this experiment, [³H]SCH 23390 (0.5 nM) was incubated with increasing concentrations of the indicated ligands. The K_i values, derived from graphically obtained IC$_{50}$ values [Y. C. Cheng and W. H. Prusoff, *Biochem. Pharmacol.* **22,** 3099 (1973)], are as follows: (+)-SCH 23390, 0.2 ± 0.01 nM; (+)-butaclamol, 2.8 ± 0.2 nM; (−)-SCH 23388, 41 ± 1.2 nM; spiperone, 290 ± 7 nM; dopamine + GppNHp, 0.64 ± 0.09 μM; and (−)-butaclamol, 31 ± 0.8 μM [F. J. Monsma, Jr., L. C. Mahan, L. D. McVittie, C. R. Gerfen, and D. R. Sibley, *Proc. Natl. Acad. Sci. U.S.A.* **87,** 6723 (1990)].

B

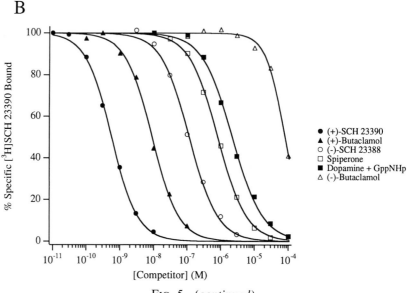

FIG. 5 *(continued).*

Hydropathy analysis of this protein revealed the presence of seven hydrophobic regions, which exhibited 44% homology with the rat D_2 dopamine receptor and 44, 43, and 40% homology with human β_1-, β_2-, and β_3-adrenergic receptors, respectively. Expression of this clone in a mammalian cell line resulted in the appearance of specific, saturable, high-affinity binding sites for [^{3}H]SCH 23390 that were not present in untransfected cells (Fig. 5A). The K_d for [^{3}H]SCH 23390 binding (0.3 nM) agrees well with that observed in rat striatal membranes as well as in NS20Y cell membranes. In addition, the ability of a variety of dopaminergic ligands to compete for [^{3}H]SCH 23390 binding sites in transfected cell membranes (Fig. 5B) was virtually identical to that observed in both striatal and NS20Y cell membranes. Finally, transfected cells also gained the ability to respond to dopamine by activation of adenylate cyclase and elevation of cAMP levels, a response which has classically been associated with D_1 receptor activation. Based on these results, we conclude that this clone represents the classic D_1 dopamine receptor of the rat striatum.

The application of the polymerase chain reaction with degenerate oligonucleotide primers has permitted us to obtain sequence information representing a dopamine receptor expressed by the NS20Y neuroblastoma cell line, allowing the subsequent cloning of a full-length clone from a rat cDNA library. Moreover, other fragments were obtained which represented previously cloned (e.g., the D$_2$ dopamine receptor) G-protein-coupled receptors as well as novel putative G-protein-coupled receptors. These results indicate that, with the use of judiciously designed primer pairs and the appropriate choice of starting template, this method should be generally applicable to the identification and cloning of novel members of virtually any receptor protein family.

References

1. P. H. Andersen, J. A. Gingrich, M. D. Bates, A. Dearry, P. Falardeau, S. E. Senogles, and M. G. Caron, *Trends Pharmacol. Sci.* **11,** 231 (1990).
2. H. G. Dohlman, J. Thorner, M. G. Caron, and R. J. Lefkowitz, *Annu. Rev. Biochem.* **60,** 653 (1991).
3. J. R. Bunzow, H. H. M. Van Tol, D. K. Grandy, P. Albert, J. Salon, M. Christie, C. A. Machida, K. A. Neve, and O. Civelli, *Nature (London)* **336,** 783 (1988).
4. F. J. Monsma, Jr., L. C. Mahan, L. D. McVittie, C. R. Gerfen, and D. R. Sibley, *Proc. Natl. Acad. Sci. U.S.A.* **87,** 6723 (1990).
5. F. J. Monsma, Jr., D. L. Brassard, and D. R. Sibley, *Brain Res.* **492,** 314 (1989).
6. A. C. Barton and D. R. Sibley, *Mol. Pharmacol.* **38,** 531 (1990).
7. H. Okayama, M. Kawaichi, M. J. Brownstein, F. Lee, T. Yokata, and K. Arai, *In* "Methods in Enzymology" (R. Wu and L. Grossman, eds.), Vol. 154, p. 3. Academic Press, San Diego, 1987.
8. J. Sambrook, E. F. Fritsch, and T. Maniatis, "Molecular Cloning: A Laboratory Manual," 2nd Ed. Cold Spring Harbor Laboratory, Cold Spring Harbor, New York, 1989.
9. Y. Takebe, M. Seike, J. Fujisawa, P. Hoy, K. Yokota, K. Arai, M. Yoshida, and N. Arai, *Mol. Cell. Biol.* **8,** 466 (1988).
10. A. Dearry, J. A. Gingrich, P. Falardeau, R. T. Fremeau, Jr., M. D. Bates, and M. G. Caron, *Nature (London)* **347,** 72 (1990).
11. Q. Y. Zhou, D. R. Grandy, L. Thambi, J. A. Kushner, H. H. M. Van Tol, R. Cone, D. Pribnow, J. Salon, J. R. Bunzow, and O. Civelli, *Nature (London)* **347,** 76 (1990).
12. R. K. Sunahara, H. B. Niznik, D. M. Weiner, T. M. Stormann, M. R. Brann, J. L. Kennedy, J. E. Gelernter, R. Rozmahel, Y. Yang, Y. Israel, P. Seeman, and B. F. O'Dowd, *Nature (London)* **347,** 80 (1990).

[4]　Molecular Cloning of a Third Dopamine Receptor (D3)

Pierre Sokoloff, Marie-Pascale Martres, Bruno Giros, Marie-Louise Bouthenet, and Jean-Charles Schwartz

Introduction

Dopamine (DA) is an important neurotransmitter in the brain involved in the control of motor functions, emotion, and cognition. Dysfunctions of DA systems have been implicated in various neurological and psychiatric disorders such as Parkinson's disease and schizophrenia. Until recently it was widely accepted that DA affects its target cells in brain and endocrine tissues via interaction with only two receptor subtypes, termed D_1 and D_2, differing from one another by their pharmacological specificity and their opposite effect on adenylate cyclase (1, 2). It was also generally agreed that the therapeutic efficacy of antipsychotics derives from their high-affinity binding to D_2 receptors. The idea that more than a single molecular entity, the D_2 receptor, was responsible for the various actions of antipsychotics had been put forward (3), but it remained controversial in spite of its substantial clinical relevance.

This situation has started to be modified with the advent of molecular biology in this field, which has confirmed the existence of additional DA receptors. Their existence throws new light onto the modes of action and side effects of many drugs used in neurology and psychiatry. This is particularly the case for the D_3 receptor that we recently identified in rat (4) and human brain (5), and which appears, from its pharmacological specificity and its selective expression in the limbic parts of the brain, to be a major biological substrate for antipsychotic action.

Molecular Cloning of Rat and Human D3 Receptors

Molecular cloning of various receptor cDNAs has been achieved in two different ways: either an efficient expression system allowed the selection of cDNA by characterizing the encoded product in binding or in functional studies, or sequence similarity was used for selection of homologous cDNAs, hybridizing with known probes at low stringency. The latter approach was

Methods in Neurosciences, Volume 12

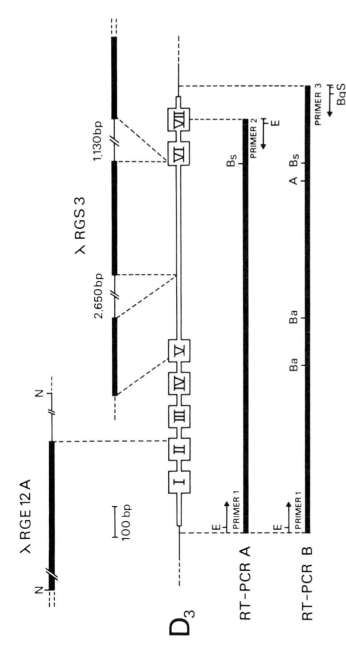

FIG. 1 Strategy for cloning the D_3 receptor. Schematic representation of the organization of the clones obtained after genomic DNA screening and PCR amplification. The D_3 receptor sequence is drawn as seven boxes representing transmembrane domains, open boxes representing coding sequences, and single lines representing untranslated sequences. λ RGE 12A and λ RGS 3 are genomic DNA clones; RT-PCR A and B represent cDNA clones obtained from reverse transcription and PCR amplification of poly(A)$^+$ mRNA. Solid boxes represent coding regions; thin lines, introns in genomic clones. Dashed lines indicate continuity of the sequence. Primers 1, 2, and 3 were used for RT-PCR amplification. Restriction sites are as follows: A, *Ava*I; Ba, *Bam*HI; Bg, *Bgl*II; Bs, *Bst*XI; E, *Eco*RI; N, *Nhe*I; S, *Sal*I.

successful for the cloning of the D_2 receptor cDNA (6,7), which, in turn, paved the way for the cloning of a series of DA receptor genes encoding for D_1 (8–11), D_3 (4,5), D_4 (12), and D_5 receptors (13).

The cloning strategy for the D_3 receptor involved a combination of screenings of cDNA and genomic libraries with reverse transcription–polymerase chain reaction (RT-PCR) (illustrated in Fig. 1). A clone isolated from a rat brain cDNA library (14) was initially used to screen a genomic library. The sequence of a positive clone (λ RGE 12A) was shown to contain the 5′ part of a gene homologous to the D_2 receptor gene. RT-PCR was performed with a specific primer in the sequence of this clone (primer 1) and a degenerate primer (primer 2) designed in the seventh transmembrane domain (TM7) of the D_2 receptor. The PCR product was subcloned, sequenced, and used in screening of a rat genomic library which provided a clone (λ RGS 3) containing the 3′ end of the coding region. The full-length cDNA was finally obtained by RT-PCR with specific primers flanking the coding region (primers 1 and 3) and RNA from olfactory tubercle as template (4). A similar approach was used for the cloning of the human D_3 receptor cDNA. Specific sequences flanking the coding region were obtained by screening a human genomic library with corresponding rat probes. The full coding sequence was then obtained in RT-PCR, using primers designed in these sequences and RNA from human mammillary bodies (5).

Structure of D_3 Receptor

Sequence analysis of the D_3 dopamine receptor predicts a structure consisting of seven transmembrane domains forming three intracytoplasmic loops (Fig. 2). These features, common to a large family of receptors coupled to a G protein (15,16) have been validated by structural studies for rhodopsin (17), with which they display sequence homology and, presumably, a common phylogenetic origin. The D_3 protein has 446 residues in rat but only 400 residues in human, the main difference residing at the level of the third putative intracytoplasmic loop (i_3), presumably involved in the coupling to the G protein. The sequence homology of the D_3 receptor is 52 and 39% with the D_2 and D_4 receptors, respectively, contrasting with a sequence homology of only 28 and 30% with D_1 and D_5, respectively.

The transmembrane helices constitute the ligand binding domain, particularly three amino acid residues thought to interact with catecholamines: an aspartic residue (Asp^{110} in the human D_3 receptor) in TM3, which forms an ion pair with the protonated amine group of DA, and two serine residues (Ser^{193} and Ser^{196}) in TM5, which presumably form a hydrogen-bonding interaction with the two phenol groups of DA. This last interaction, specific for DA

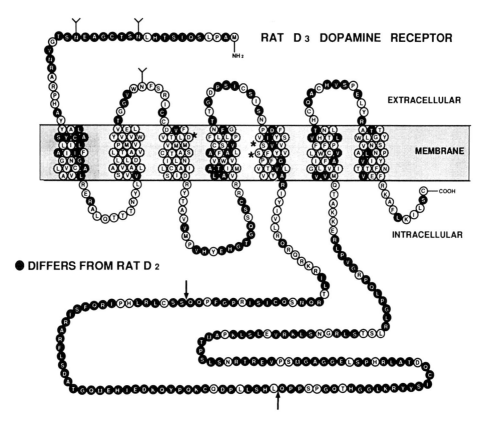

FIG. 2   Proposed membrane topography of the rat D_3 dopamine receptor and its relationship with the D_2 receptor. Darkened circles represent residues which differ between the rat D_2 and D_3 receptors. The portion of the third intracytoplasmic loop depicted by arrows is absent in the human D_3 receptor, but otherwise its structure is highly homologous to that of the rat D_3 receptor. Asterisks denote amino acid residues presumably involved in the binding of dopamine.

and agonists, could cause a conformational change in the helix, transmitted to the i_3 loop.

The various genes of the DA receptor family can be classified into two groups according to their organization: (a) intronless genes, that is, those of the D_1 and D_5 receptor, in which the coding nucleotide sequence is continuous, and (b) genes having their coding sequence contained in discontinuous DNA segments (exons) interspersed among sequences (introns) that do not form a part of the mature mRNA. This last organization, found in the rhodopsin gene as well as the D_2, D_3, and D_4 receptor genes, may potentially lead, via a mechanism of alternative splicing (in which a given exon in the pro-mRNA is either present or absent in the final mRNA), to the biosynthesis of several distinct proteins encoded by a unique gene (see below). Remarkably enough, the D_2 and D_3 receptor genes have four introns, out of six and five, respectively, located at strictly similar positions, suggesting relatively recent divergence from a common ancestral gene.

The human D_3 receptor was assigned to chromosome 3 (5) and localized to its long arm (3q13.3) by *in situ* hybridization (18), whereas D_1, D_2, D_4, and D_5 receptors are localized to chromosomes 5, 11, 11, and 4, respectively (7, 10, 12, 13). Several polymorphisms were recently identified on the human D_3 receptor gene, among which a *Bal*I polymorphism corresponds to a single amino acid mutation in the extracellular N-terminal part of the receptor protein (19). The exact role of this part is not clearly understood, but it is unlikely that binding of dopamine or antipsychotic or transduction signals would be affected by this amino acid substitution. Nevertheless, it cannot be excluded that integration of the receptor protein within the membrane may be disturbed. This *Bal*I polymorphism is conveniently studied by PCR and is inherited in a codominant way according to Mendelian laws (19). Hence, it can be used in genetic linkage and association studies to assess a possible relationship between the D_3 receptor gene and several mental diseases, such as schizophrenia, bipolar illness, or Tourett's syndrome, for which genetic factors are of major etiological importance.

Splice Variants of D_3 Receptor mRNA

Alternative splicing was shown to occur in the case of the D_2 receptor, potentially leading to two distinct receptors differing by a stretch of 29 amino acids at the level of the third intracytoplasmic loop, called $D_{2(444)}$ (or D_{2L} for D_2 long, or D_{2A}) and $D_{2(415)}$ (or D_{2S} for D_2 short, or D_{2B}). These two isoforms of the D_2 receptor display identical pharmacology but are differently expressed among cerebral areas; they may interact differently with various G proteins (14, 20, 21), and their relative abundance is affected by neuroleptic treatments (22).

In the case of the rat D_3 receptor, alternative splicing potentially give rise, in addition to the 446 amino acid receptor, to two truncated proteins of 109 and 428 amino acids, with sizable deletions of 113 bp in TM3 and 54 bp in the second extracellular domain (O2), respectively. The proteins potentially encoded by these two transcripts have been designated as D_3(TM3-del) and D_3(O2-del), respectively (23).

Two distinct alternative splicing mechanisms underlie the production of these two mRNAs. In the case of D_3(TM3-del), the process involves combinatorial exons, the "cassette" exon being the second exon. Because the latter does not comprise $n \times 3$ nucleotides, this introduces a frameshift in the sequence, and the splice product encodes a 109-amino acid protein. By contrast, in D_3(O2-del) mRNA, the in-frame 54-bp deletion does not correspond to a full exon: alternative splicing occurs within the fourth exon where an internal acceptor site can be used by the splicing machinery, thereby giving rise to an mRNA encoding a 428-amino acid protein.

Whereas the structure of D_3(TM3-del) makes it unlikely that the protein may function as a receptor, this is not so clear in the case of D_3(O2-del), whose structure may still be compatible with the occurrence of seven TM domains, as revealed by the hydropathy profile. However, CHO clones stably expressing D_3(O2-del) mRNA failed to show any dopaminergic binding activity, as assessed with various radioactive ligands. These truncated products of the D_3 receptor gene could be formed at random during biosynthesis of the functionally active D_3 receptor, or, alternatively, they might control the abundance of the active D_3 receptor. Finally, because multiple D_3 receptor gene transcripts are also found in human brain (5), it cannot be excluded that defects in the alternative splicing mechanisms, leading to the formation of inactive receptors, might occur during psychiatric diseases.

Anatomical Distribution of D_3 Receptor mRNA in Rat Brain

The distribution of D_3 receptor gene transcripts in rat brain areas, as established using Northern or PCR analysis or visualized by *in situ* hybridization histochemistry (4, 24) markedly differs from those of the D_1 (25) or D_2 receptor (24, 26) gene transcripts. For instance, only a weak D_3 receptor hybridization signal was detected in striatum, which contains the highest densities of DA axons and D_2 receptor mRNA (Fig. 3). By contrast, the D_3 receptor mRNA is highly expressed in the olfactory tubercle–islands of Calleja complex, the bed nucleus of stria terminalis, and nucleus accumbens. These areas constitute, with the ventral and ventromedial parts of the caudate putamen, the "ventral striatum,"a region receiving afferents from the prefrontal or allocortex and amygdala and its major DA inputs from the A10 cell group in the ventral tegmental area. It projects to ventral pallidum and to the mediodorsal thalamic nucleus which selectively innervates the prefrontal cortex (27). This connectivity has led to the designation of this region as the "limbic" part of the striatal complex, in which D_3 receptors may, therefore, mediate a large part of DA signals. The remainder of the striatal complex, which is mainly innervated by DA projections from the substantia nigra, receives its cortical inputs from the somatic neocortex and is highly enriched in D_1 or D_2 receptors. This suggests a major participation of D_3 receptors in dopaminergic transmissions in limbic areas known to be associated with cognitive, emotional, and endocrine functions, whereas D_2 receptors are associated with the motor control in the extrapyramidal system.

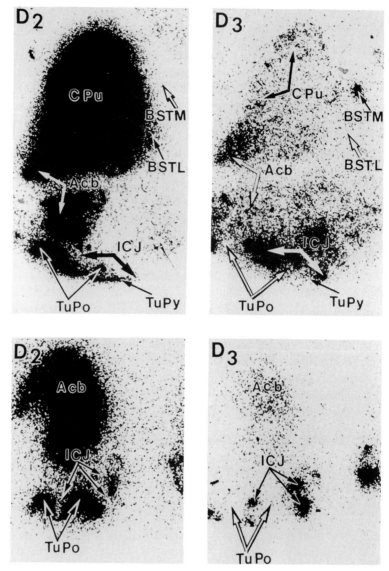

FIG. 3 Compared distributions of D_2 and D_3 receptor mRNAs established by *in situ* hybridization in sagittal (top) and frontal (bottom) sections performed in rat telencephalon. Note the nonoverlapping complementary distributions of the two transcripts in the ventral striatum, particularly at the level of olfactory tubercle–islands of Calleja complex and basal nucleus of the stria terminalis. Acb-Accumbens nucleus; BSTL and BSTM, bed nucleus of the stria terminalis, lateral or medial part; ICJ, islands of Calleja; CPu, caudate putamen; TuPo and TuPy, polymorph and pyramidal layers of the olfactory tubercle.

In several other brain regions, such as stria terminalis (Fig. 3) or mammillary bodies, D_2 and D_3 receptor messengers appear to be expressed in distinct parts of these structures in a mutually exclusive manner. The two receptor subtypes differ by the much higher affinity of dopamine for the D_3 receptor and, possibly, by their intracellular signaling systems (see below). Hence, it seems likely that different kinds of signal might be generated by DA in neighboring but topographically distinct cerebral structures.

High levels of D_3 receptor mRNA are also found in cerebellar lobules 9 and 10 in the Purkinje cell layer (24), whereas [^{125}I]iodosulpride binding is found in the molecular layer (28) which contains dendrites of these cells. This suggests that Purkinje cells in this area express the D_3 receptor. Interestingly, no specific D_3 receptor signal could be detected by Northern and PCR analyses in the pituitary, a prototype localization of D_2 receptors. This allows the prediction that selective D_3 receptor ligands, when available in therapeutics, will not affect, like the currently used neuroleptics, the activity of mammotrophs.

D_3 Receptor as Second Autoreceptor

The expression of D_3 receptor mRNA in substantia nigra shown by PCR and Northern analyses (4) as well as by *in situ* hybridization (24) suggested that D_3 receptors are expressed by DA neurons themselves. This was verified after degeneration of DA neurons with 6-hydroxydopamine, which induced a marked ipsilateral reduction of the D_3 receptor signal in both the substantia nigra ($-65 \pm 10\%$) and the ventral tegmental area ($-69 \pm 14\%$). In the same tissue extracts, the D_2 receptor mRNA levels were similarly affected namely, by -88% and -65%, respectively (4), indicating that both receptors act as autoreceptors. Such a role for the D_3 receptor is consistent with its pharmacological profile (see below). D_2 and D_3 autoreceptors might variously participate in the actions attributed to DA autoreceptors, such as inhibitions of impulse flow, DA synthesis and release at either nerve terminals or dendrites, and cotransmitter release.

Pharmacology of D_3 Receptor

The pharmacology of the rat (4) and human D_3 receptors (29) was studied in transfected CHO cells expressing a high level of sites labeled with high affinity by [^{125}I]iodosulpride, formerly considered as a D_2 receptor-selective ligand (28). This pharmacology was also recently assessed in rat brain membranes using [^{3}H]7-hydroxydipropylaminotetralin {[^{3}H]7OH-DPAT}, a D_3 receptor-selective ligand (28a). The values obtained with this ligand are essentially the same as those reported in Table I for the D_3 receptor. The D_3 receptor can be considered, like the D_4 receptor, as a "D_2-like" receptor; it poorly recognizes "D_1-specific" ligands such as SKF 38393 or SCH 23390, whereas it binds "D_2-specific" agonists such as quinpirole or antagonists such as sulpiride (Table I). However, several salient features of the D_3 receptor pharmacology should be emphasized.

D_3 receptor binds dopamine with high affinity, and this binding is only weakly modulated by guanine nucleotide (4, 29). The very high affinity of dopamine (Ki-8 nM) and low modulatory effect of guanine nucleotide is also found at D_3 receptor on brain membranes (28a), where it is presumably functionally coupled. This suggests that this feature is an intrinsic property of the D_3 receptor, possibly related to a peculiar mode of intracellular signaling (see the following). This high affinity and selectivity toward the D_3 receptor is shared by several agonists, such as pergolide, TL 99, or quinerolane, which were previously considered autoreceptor-selective agonists in some animal models, such as the butyrolactone-induced increase of DA synthesis

TABLE I   Pharmacology of Human D_2 and D_3 Receptor Subtypes[a]

	K_i value (nM)		
	D_2	D_3	K_iD_2/K_iD_3
Agonists			
Dopamine	540	23	24
Dopamine + Gpp(NH)p[b]	2000	30	60
Apomorphine	70	70	1
Bromocriptine	5	7	1
Pergolide	20	2	10
TL 99	66	2.3	30
Quinpirole	1400	40	40
Antagonists			
Clozapine	70	500	0.1
Haloperidol	0.6	3	0.2
Prochlorperazine	0.4	1.8	0.2
Chlorpromazine	2.3	5.9	0.4
(−)-Sulpiride	10	20	0.5
Amisulpride	1.3	2.4	0.5
Pipothiazine	0.2	0.3	0.7
Pimozide	10	11	0.9
UH 232	40	10	3

[a] Values for human D_2 and D_3 receptors transfected in CHO cells are from Ref. 29.
[b] Gpp(NH)p (5'-guanidylimidodiphosphate), a guanine nucleotide analog of 5'-guanosine triphosphate, was used at 0.1 mM.

(30, 31). This may account for the role of D_3 receptors as autoreceptors, which have been shown sensitive to DA in very low concentrations (32). This suggests that some functions attributed to autoreceptor stimulation actually involve the D_3 receptor. In agreement, AJ76 and UH232, the only antagonists exhibiting (limited) D_3 receptor selectivity, have behavioral stimulating properties in animals attributed to autoreceptor blockade (33). These pharmacological data suggest that the D_3 receptor plays a major role in the feedback inhibition of DA transmission.

Most antipsychotics display high affinities at the D_3 receptor, indicating that this receptor is probably blocked during the treatment of schizophrenia and related disorders. The degree of this blockade would depend, however, on the antipsychotics used because their recognition by the D_3 receptor relatively to that of the D_2 receptor is variable. The compounds for which the ratios between K_i values for D_2 and D_3 receptors (K_iD_2/K_iD_3 ratios) are the highest would exert a more complete blockade of DA transmission in limbic areas, where the D_3 receptor is selectively expressed. Conversely, those for which the ratios are the lowest would preferentially block the D_2 receptor present in other dopaminergic areas, including the extrapyramidal system, mainly implicated in the control of motor function. This could be one of the molecular bases for the distinction of "atypical" neuroleptics. Consistent with this hypothesis is the observation of a high K_iD_2/K_iD_3 ratio measured with atypical neuroleptics such as sulpiride or amisulpride. Nevertheless, the peculiar clinical properties of clozapine are more likely to derive from its higher affinity for D_4 than any other receptor subtype (12).

Interestingly, among antipsychotics having the highest K_iD_2/K_iD_3 ratios are amisulpride, carpipramine, pipothiazine, and pimozide, which all exhibit definite disinhibitory actions sought in the treatment of negative symptoms in schizophrenia. Conceivably, the more efficient blockade of D_3 autoreceptors by these compounds could facilitate DA transmission in some brain areas, which might lead to the alleviation of negative symptoms (34). To address these questions, further studies will be necessary, however, using more selective compounds the design of which should be facilitated by the use of clonal cell lines expressing a single receptor subtype.

Signaling Pathway of the D_3 Receptor

Via interaction with a Gi protein, the D_2 receptor is linked to numerous signaling pathways including inhibition of adenylate cyclase or phospholipase C and activation of K^+ channels (35). More recently, the D_2 receptor expressed in transfected CHO cells was also shown to mediate an enhancement of arachidonic acid release, provided that such release has been initiated by increasing intracellular Ca^{2+} (36, 37). The potential importance of this novel eicosanoid pathway lies in the fact that it may account for the synergistic interaction between D_1 and D_2 receptors (36).

In CHO cells transfected with D_3 receptors, no or inconsistent inhibition of either adenylate cyclase or phospholipase C was evident (4, 29), and the arachidonate response was weak (36). This suggests that the D_3 receptor couples to a G protein which is distinct from the Gi involved in the D_2 receptor signaling pathway and absent from the recipient CHO cell, a hypothesis supported by the observation that the modulation of agonist binding by guanine nucleotides is weak in this cell (4). However, in CHO cells transfected with the human D_3 receptor (which markedly differs from its rat counterpart at the level of the i_3 loop), a modest but reproducible effect of guanine nucleotides is observed (29). Furthermore, this modulatory effect is enhanced in CHO cells cotransfected with the α_0 subunit of a G protein, although adenylate cyclase inhibition still cannot be confirmed (P. Sokoloff, M.-P. Martres, B. Giros, M.-L. Bouthenet, and J.-C. Schwartz, unpublished observations). From these data it appears that D_3 receptor signalization may involve pathways different from those activated by stimulation of D_2 receptors, via interaction with a distinct G protein(s).

Conclusions

The recent and rapid enlargement of the DA receptor family illustrates how the complexity of receptor families was severely underestimated by most pharmacologists in the pre-molecular biology days and raises questions about the functional significance of such a heterogeneity. The heterogeneity probably reflects the diversity of the messages transmitted, which comes from the existence of subtypes differing in their affinity for the neurotransmitter dopamine, from their differential interactions with various G proteins (allowing distinct chemical signals to be triggered), and from the selective expression of the various subtypes in distinct neuron populations. Several features of the D_3 receptor suggest a peculiar mode of action for this receptor.

The very high affinity of D_3 for dopamine, unusual for an aminergic neurotransmitter, suggests that dopamine might act at some distance from the terminals which release it. This quasi-hormonal mode of action is supported by the dense expression of D_3 receptor mRNA within the core of islands of Calleja, whereas dopamine axons surround these islands, making only a few contacts with the granule cells (38). Because the D_3 receptor is an autoreceptor, it is conceivable that it may be not only present at the synaptic cleft but also distributed all along the axon. The low modulatory effects of guanine nucleotides, as well as the absence of clear effects related to activation of a Gi protein on stimulation of the D_3 receptor, suggests a peculiar mode of coupling, which remains to be identified.

The existence of three pharmacologically distinct "D_2-like" subtypes, namely, D_2, D_3, and D_4, instead of a single D_2 receptor, which was formerly recognized as *the* target for antipsychotic agents, raises important issues. Among the three, which is (are) responsible for the beneficial therapeutic effects, and which is (are) responsible for each unwanted side effect? At this early stage of our knowledge much caution is needed, but two clues point to the D_3 as a key receptor in schizophrenia: its selective expression in a phylogenetically old part of the brain known as the limbic system and its relatively preferential binding of several atypical antipsychotics. The last criterion, on the other hand, points to the D_4 receptor as a key target for clozapine, a compound with a spectrum characterized by its activity in a subpopulation of patients resistant to other antipsychotics.

The discovery of novel putative targets for antipsychotics opens opportunities for developing new therapeutical agents, which may be more effective and safer in the treatment of several major neurological and psychiatric diseases. Furthermore, the probable role of the D_3 receptor as a target for antipsychotics raises the possibility that its gene might be affected in various psychiatric diseases, a hypothesis being actively explored in several laboratories.

References

1. P. F. Spano, S. Govoni, and M. Trabucchi, *Adv. Biochem. Psychopharmacol* **19,** 155 (1978).
2. J. W. Kebabian and D. B. Calne, *Nature (London)* **277,** 93 (1979).
3. J. C. Schwartz, M. Delandre, M. P. Martres, P. Sokoloff, P. Protais, M. Vasse, J. Costentin, P. Laibe, C. G. Wermuth, C. Gulat, and A. Lafitte, *in* "Catecholamines: Neuropharmacology and Central Nervous System. Theoretical Aspects" (E. Usdin, A. Carlsson, A. Dahlstrom, and J. Engel, eds.), p. 59, Alan R. Liss, New York, 1984.
4. P. Sokoloff, B. Giros, M. P. Martres, M. L. Bourthenet, and J. C. Schwartz, *Nature (London)* **347,** 146 (1990).
5. B. Giros, M. P. Martres, P. Sokoloff, and J. C. Schwartz, *C. R. Acad. Sci. Ser. 3* **511,** 501 (1990).
6. J. R. Bunzow, H. H. M. Van Tol, D. K. Grandy, P. Albert, J. Salon, McD. Christie, C. A. Machida, K. A. Neve, and O. Civelli, *Nature (London)* **336,** 783 (1988).
7. D. K. Grandy, M. Litt, L. Allen, J. R. Bunzow, M. Marchionni, H. Makam, I. Reed, R. E. Magenis, and O. Civelli, *Am. J. Hum. Genet.* **45,** 778 (1989).
8. Q. K. Zhou, D. K. Grandy, L. Thambi, J. A. Kushner, H. H. M. Van Tol, R.

Cone, D. Pribnow, J. Salon, J. R. Bunzow, and O. Civelli, *Nature (London)* **347**, 76 (1990).

9. A. Dearry, J. A. Gingrich, P. Falardeau, R. T. Fremeau, M. D. Bates, and M. G. Caron, *Nature (London)* **347**, 72 (1990).

10. R. K. Sunahara, H. B. Niznik, D. M. Weiner, T. M. Stormann, M. R. Brann, J. L. Kennedy, J. E. Gelernter, R. Rozmahel, Y. Yang, Y. Israel, P. Seeman, and B. F. O'Dowd, *Nature (London)* **347**, 80 (1990).

11. F. J. Monsma, L. C. Mahan, L. D. McVittie, C. R. Gerfen, and D. R. Sibley, *Proc. Natl. Acad. Sci. U.S.A.* **87**, 6723 (1990).

12. H. H. M. Van Tol, J. R. Bunzow, H. C. Guan, R. K. Sunahara, P. Seeman, H. B. Niznik, and O. Civelli, *Nature (London)* **350**, 610 (1991).

13. R. K. Sunahara, H. C. Guan, B. F. O'Dowd, P. Seeman, L. G. Laurier, G. Ng, S. R. George, J. Torchia, H. H. M. Van Tol, and H. B. Niznik, *Nature (London)* **350**, 614 (1991).

14. B. Giros, P. Sokoloff, M. P. Martres, J. F. Riou, L. J. Emorine, and J. C. Schwartz, *Nature (London)* **342**, 923 (1989).

15. J. Findlay and E. Eliopoulous, *Trends Pharmacol. Sci.* **11**, 492 (1990).

16. D. C. Strader, S. I. Sigal, and A. F. R. Dixon, *Am. J. Respir. Cell. Mol. Biol.* **1**, 81 (1989).

17. H. Saibil, *Semin. Neurosci.* **2**, 15 (1990).

18. M. Leconiat, P. Sokoloff, J. Hillion, M. P. Martres, B. Giros, C. Pilon, J. C. Schwartz, and R. Berger, *Hum. Genet.* **87**, 618 (1991).

19. L. Lannfelt, P. Sokoloff, M. P. Martres, C. Pilon, B. Giros, and J. C. Schwartz, *Psychiatr. Genet.* **2**, 16 (1991).

20. R. Dal Toso, B. Sommer, M. Ewert, A. Herb, D. B. Pritchell, A. Bach, B. D. Shivers, and P. H. Seeburg, *EMBO J.* **8**, 4025 (1989).

21. F. J. Monsma, L. D. McVittie, C. R. Gerfen, L. C. Manhan, and D. R. Sibley, *Nature (London)* **342**, 926 (1989).

22. M. P. Martres, P. Sokoloff, B. Giros, and J. C. Schwartz, *J. Neurochem.* **58**, 673 (1992).

23. B. Giros, M. P. Martres, C. Pilon, P. Sokoloff, and J. C. Schwartz, *Biochem. Biophys. Res. Commun.* **176**, 1584 (1991).

24. M. L. Bouthenet, E. Souil, M. P. Martres, P. Sokoloff, B. Giros, and J. C. Schwartz, *Brain Res.* **564**, 203 (1991).

25. R. T. Fremeau, G. E. Duncan, M. G. Fornaretto, A. Dearry, J. A. Gingrich, G. R. Brees, and M. G. Caron, *Proc. Natl. Acad. Sci. U.S.A.* **88**, 3772 (1991).

26. J. H. Meador-Woodruff, A. Mansour, J. R. Bunzow, H. H. M. Van Tol, S. J. Watson, and O. Civelli, *Proc. Natl. Acad. Sci. U.S.A.* **86**, 7625 (1989).

27. A. Björklund and O. Lindvall, *in* ''Handbook of Chemical Neuroanatomy'' (A. Björklund and T. Hökfelt, eds.), p. 55. Elsevier, Amsterdam, 1984.

28. M. P. Martres, M. L. Bouthenet, N. Salès, P. Sokoloff, and J. C. Schwartz, *Science* **228**, 752 (1985).

28a. D. Levesque, J. Diaz, C. Pilon, M. P. Martres, B. Giros, E. Souil, D. Schott, J. L. Morgat, J. C. Schwartz, and P. Sokoloff, *Proc. Natl. Acad. Sci. U.S.A.* **89**, 8155 (1992).

29. P. Sokoloff, M. Andrieux, R. Besancon, C. Pilon, M. P. Martres, B. Giros, and J. C. Schwartz, *Eur. J. Pharmacol. Mol. Biol.* **225,** 331 (1992).
30. M. E. Wolf and R. H. Roth, *in* "Dopamine Receptors" (I. Creese and C. M. Fraser, eds.), p. 45. Alan R. Liss, New York, 1987.
31. G. E. Martin, M. Williams, and D. R. Haubrich, *J. Pharmacol. Exp. Ther.* **223,** 298 (1982).
32. L. R. Skirboll, A. A. Grace, and B. S. Bunney, *Science* **206,** 80 (1979).
33. K. Svensson, A. M. Johansson, T. Magnusson, and A. Carlsson, *Naunyn-Schmiedeberg's Arch. Pharmacol.* **334,** 234 (1986).
34. A. Carlsson, *Neuropsychopharmacology* **1,** 179 (1988).
35. L. Vallar and J. Meldolesi, *Trends Pharmacol. Sci.* **10,** 74 (1989).
36. D. Piomelli, C. Pilon, B. Giros, P. Sokoloff, M. P. Martres, and J. C. Schwartz, *Nature* (*London*) **353,** 164 (1991).
37. R. Y. Kanterman, L. C. Mahan, E. M. Briley, F. J. Monsma, D. R. Sibley, J. Axelrod, and C. C. Feldek, *Mol. Pharmacol.* **39,** 364 (1991).
38. J. H. Fallon, S. E. Loughlin, and C. E. Ribak, *J. Comp. Neurol.* **218,** 91 (1983).

[5]   Molecular Biology of the Human Tachykinin Receptors

Norma P. Gerard, Xiao-Ping He, Hideya Iijima,
Levi A. Garraway, Lu Bao, Jean-Luc Paquet,
and Craig Gerard

Introduction*

The tachykinins are a family of low molecular weight peptide neurotransmitters characterized by the canonical C-terminal amino acid sequence -Phe-X-Gly-Leu-Met-NH$_2$, (1, 2), which is required for biological activity. Release of the peptides and their subsequent action on target tissues are associated with such diverse processes as transmission of sensory information, smooth muscle contraction, nociception, inflammation, sexual behavior, and possibly wound healing and nerve regeneration (3–6). In man, the relevant tachykinin ligands include substance P, neurokinin A, and nuerokinin B, each of which has a corresponding receptor for which it has primary specificity (7, 8). The NK-1 receptor is selective for substance P, the NK-2 receptor is specific for neurokinin A, and the NK-3 receptor is selective for neurokinin B (9). Because of the C-terminal sequence similarity among the ligands, however, there is some cross-reactivity among the receptors and their respective ligands, particularly at elevated concentrations. As a result, determining the specific tachykinin responsible for individual biological responses has been somewhat elusive. We felt, therefore, that the molecular characterization of the human tachykinin receptors would provide the tools to begin to address the problem.

Cloning Strategies for Tachykinin Receptors

Studies of the binding properties of the tachykinins to several membrane preparations showed that addition of GTP to the ligand–receptor complex resulted in rapid and complete reversal of specific binding (10, 11). These data suggested that the tachykinin receptors belong to the family of membrane proteins typified by rhodopsin, which are characterized by a requirement

---

* The nucleic acid sequences for the human tachykinin receptor genes have been submitted to GenBank. The NK-1 receptor gene has Accession Number M76675; the NK-2 receptor gene has Accession Number J05680.

Methods in Neurosciences, Volume 12

for association with a regulatory GTP-binding protein to form the high-affinity ligand-binding complex, leading to signal transduction (12, 13). These proteins are additionally characterized by a common structural motif consisting of seven hydrophobic amino acid sequences that are presumed to span the cell membrane, with an extracellular amino-terminal sequence and an intracellular carboxy-terminal tail (13–15). This knowledge, in addition to the assumption that the genes for the tachykinin receptors would not contain introns (an assumption which has proved useful for the cloning of other members of this receptor class; see Ref. 16), supported the approach we chose to clone the human tachykinin receptor genes.

Human NK-2 Receptor

Masu *et al.* originally reported the cloning of the bovine NK-2 receptor cDNA by expression in *Xenopus* oocytes, and the deduced amino acid sequence confirmed the presence of seven hydrophobic sequences consistent with a GTP-binding protein-dependent receptor (17). Subsequently Sasai and Nakanishi cloned the analogous molecule from rat by low-stringency hybridization with a rat stomach cDNA library (18). We approached the cloning of the human NK-2 receptor using methodology based on the polymerase chain reaction (PCR) (19). We analyzed the deduced protein sequences from the bovine and rat molecules for regions predicted to be conserved among species (Fig. 1) based on the content of infrequently mutating amino acids, particularly tryptophan and cysteine. Oligodeoxynucleotides were synthesized corresponding to nucleotides 91–108 (sense) and 538–555 (antisense), with *Eco*RI restriction sites at their 5′ termini, using standard cyanoethylphosphoramidite chemistry (Applied Biosystems Foster City, CA, Model 318A DNA synthesizer). The 27-mers are cleaved from the solid support with concentrated ammonium hydroxide and deprotected by heating at 65°C for 5–16 hr. The oligodeoxynucleotides are dried under reduced pressure and used without further purification.

RNA is prepared from human tracheal tissue obtained at autopsy by extraction with 4 M guanidinium thiocyanate containing 0.1 M Tris-HCl, pH 7.5, and 0.1% 2-mercaptoethanol (20, 21). Soldium lauryl sarcosinate is added to 0.5%, and 3.6 ml of the solution is layered on a 1.3-ml cushion of 5.7 M cesium chloride in 10 mM EDTA, pH 8.0, and centrifuged for 16 hr at 100,000 g (22). Poly(A)$^+$ RNA is purified using oligo(dT)-Sepharose (23) and transcribed into cDNA by the method of Gubler and Hoffman (24). Approximately 100 ng of the cDNA is mixed with 1 μg of the sense and antisense primers, deoxynucleotides, and 2.5 units (U) Taq DNA polymerase as recommended by the supplier (Perkin Elmer/Cetus, Norwalk, CT), and

A

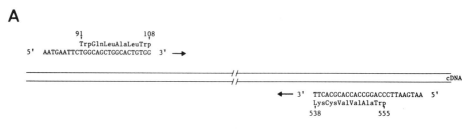

B

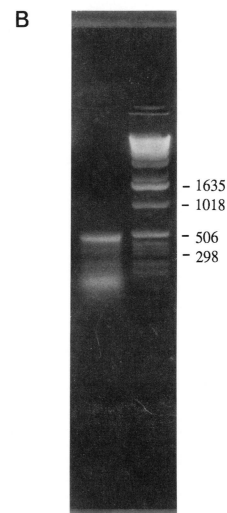

FIG. 1.   (A) Sequences of the oligodeoxynucleotide primers used in the PCR with human tracheal cDNA to generate the 465-bp partial-length human NK-2 receptor cDNA. Primer sequences are based on the bovine cDNA sequence (17). (B) Electro-

the PCR is carried out through 25 cycles consisting of 1 min at 95°C, 2 min at 37°C, and 3 min at 72°C, followed by a final extension of 7 min at 72°C. This reaction generates a 465-bp fragment of the human NK-2 receptor cDNA, encoding nucleotides 91–555, as shown in Fig. 1. The PCR product is purified by agarose gel electrophoresis and GeneClean (Bio 101, La Jolla, CA), digested overnight with *Eco*RI, purified again as before, and ligated to pBluescript (Stratagene, La Jolla, CA), for sequence analysis to confirm its identity. Our data indicates that the human cDNA is greater than 90% identical to the bovine molecule in this region, supporting the conclusion that this cDNA is, indeed, a part of the human NK-2 receptor (19). Attempts to generate a full-length human NK-2 receptor cDNA by PCR using oligonucleotides based on the bovine sequence were unsuccessful for reasons which are detailed below.

Recognition that a PCR-based approach was not likely to yield the full-length human NK-2 receptor cDNA, and that the message was sufficiently rare that screening a cDNA library would be difficult at best (the tachykinin receptor mRNA is estimated to make up only 0.0006 to 0.0003% of total RNA; Ref. 17), we opted to screen a human genomic library instead since the relative abundance of the gene should be higher. Also a determinant in this decision, as indicated above, was the expectation that the NK-2 receptor gene would not contain introns, as was the case for a number of other receptors in the family of G-protein-coupled reactions (16).

We screened a human genomic DNA library in EMBL-3 (generously provided by Dr. S. Orkin, Harvard Medical School) with the 465-bp partial-length cDNA labeled with ^{32}P by random priming and the Klenow fragment of DNA polymerase (25). Bacteriophage DNA is transferred to nitrocellulose filters in duplicate and hybridized with the probe in 5× SSC (0.75 M NaCl, 75 mM sodium citrate), containing 50% formamide, 20 mM Tris-HCl, pH 7.5, 1× Denhardt's solution, 10% dextran sulfate, and 0.1% sodium dodecyl sulfate (SDS) for 16 hr at 42°C. Filters are washed 3 times for 10 min in 2× SSC, 0.1% SDS followed by 30 min at 68°C in 0.2× SSC, 0.1% SDS, then exposed to X-ray film for 12–18 hr (Kodak, Rochester, NY, X-Omat) at −70°C with an intensifying screen.

Four positive clones were isolated from approximately 10^6 phage plaques (19) and the bacteriophage DNA purified from a liquid lysate as described (26). When the DNAs were digested with *Pst*I or *Eco*RI and analyzed on 1% agarose TAE gels, three of the clones had identical restriction patterns. One of these, clone NGNK-2, was subjected to more extensive characteriza-

---

phoresis of 10% of the PCR reaction mixture on a 1% agarose gel showing the predominant band at 465 bp corresponding to nucleotides 91–555 of the human NK-2 receptor cDNA. DNA size standards were electrophoresed simultaneously.

tion. The DNA was digested with *Pst*I, separated by agarose gel electrophoresis, blotted to nylon membranes (GeneScreen Plus, Du Pont, Wilmington, DE), and hybridized (using protocols described by Du Pont) with the partial cDNA, revealing two positive bands of 1.6 and 4.8 kb. These DNA fragments were purified using GeneClean following agarose gel electrophoresis, ligated to pBluescript, and subjected to sequence analysis using the dideoxy nucleotide chain termination sequencing methodology (27) and either the primers used for PCR or the vector primers, SK and KS.

The results are shown schematically in Fig. 2. The 1.6-kb fragment sequenced with the PCR primer at residues 91–108; both SK and KS primers yielded sequences in noncoding regions. The 4.8-kb fragment sequenced with the PCR primer at 538–555 and also generated noncoding sequence with the vector primers. These data indicated that the human NK-2 receptor gene contained at least two introns, and comparison with the bovine cDNA sequence indicated that the 1.6-kb *Pst*I fragment contained exon 1, approximately 550 bp of 5′ flanking sequence, and some 3′ intron sequence. The 4.8-kb *Pst*I fragment contained exon 2 with intron sequences at both 5′ and 3′ ends; neither *Pst*I site occurred within the coding sequence.

To identify DNA fragments containing the remaining exon(s), synthetic oligodeoxynucleotides were prepared corresponding to the bovine sequences at nucleotides 924–939 and at 972–987, corresponding to deduced amino acid sequences CCLNHR and CCPWVT, respectively, chosen because of their cysteine content. A third probe was selected corresponding to nucleotides 597–621, amino acid residues 199–205 (LIVIAL), which is contained in the fifth putative membrane-spanning sequence of the molecule. Exon 3 hybridized with the probe for nucleotides 597–621 and was cloned as a 1.6-kb *Eco*RI fragment since the corresponding *Pst*I fragment of the genomic clone was very small (<200 bp). Exons 4 and 5 were identified in *Pst*I fragments of 1.2 and 1.8 kb, respectively. All three DNA fragments were subcloned into pBluescript and sequenced with oligonucleotide primers based on the bovine sequence. Analysis of the sequences obtained and comparison with cDNAs from the rat and bovine molecules indicated that the entire coding region was contained in five exons, as shown diagrammatically in Fig. 2.

To prove that we had cloned the entire human NK-2 receptor gene, PCR primers were prepared corresponding to the 5′ and 3′ ends of the human NK-2 receptor coding sequence and used in the polymerase chain reaction with cDNA from human stomach or lung as described above. This yielded a 1.2-kb product with a single open reading frame and cDNA sequence identical to that determined from exon sequences of the gene (19). As indicated above, attempts to generate full-length human NK-2 receptor cDNA based on the 5′ and 3′ ends of the bovine cDNA coding sequence were

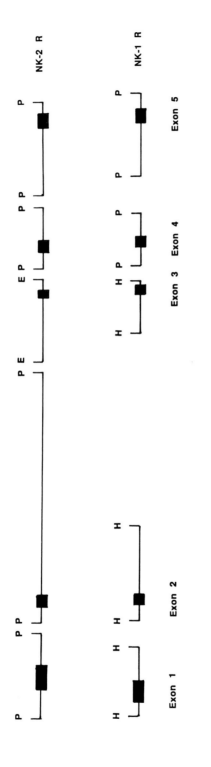

FIG. 2. Schematic diagram of the intron/exon structure of the human NK-2 (top) and NK-1 (bottom) receptor genes. Restriction fragments are drawn to scale with each of the exons (P, *PstI*; E, *EcoRI*; H, *HindIII*). The sizes and positions of the exons in each of the restriction fragments are accurate to within 100 bp; distances between each of the fragments are not necessarily accurate.

unsuccessful. In retrospect, this was not surprising because, as shown in Fig. 3, the sequences are most divergent in these regions.

Human Substance P (NK-1) Receptor

We approached the cloning of the human substance P receptor in a manner similar to that used for the NK-2 receptor (28). Synthetic oligonucleotides were prepared corresponding to regions of the rat NK-1 receptor cDNA (29, 30) that were not expected to mutate among species and simultaneously not to hybridize with the human NK-2 receptor. Sense and antisense primers were synthesized corresponding to nucleotides 85–105 (TWQIVLW) and 538–555 (CMIEWP), respectively, with $EcoRI$ restriction sites at their 5' ends. PCR was performed with these primers and cDNA from human lung tissue or from human IM9 lymphoblastoid cells [American Type Culture Collection, (ATCC), Rockville, MD, CCL 159], which also express high-affinity receptors for substance P (31). IM9 cells are maintained in RPMI containing 10% fetal calf serum, at 37°C in an atmosphere of 5% CO_2. The initial PCR conditions used are identical to those described above for the NK-2 receptor, and yielded a faint band at 470 bp. Ten percent of the initial PCR product was subjected to secondary PCR consisting of 25 cycles of amplification under the same conditions. The 470-bp PCR product obtained was subcloned into pBluescript for sequence confirmation and subsequently used to screen the genomic DNA library as before, yielding seven positive clones. These clones were digested with $Hind$III, electrophoresed on 1% agarose gels, blotted to GeneScreen Plus, and hybridized with the ^{32}P-labeled partial-length cDNA probe. Two positively hybridizing fragments were observed in these clones, one at 1.3 kb, contained in four of the clones, and one of 1.8 kb, contained in the other three. No clones were isolated which contained both of these fragments. Subcloning and sequence analysis showed that the 1.3-kb fragment contains exon 1 and 220 bp of 5' flanking sequence, and the 1.8-kb piece contains exon 2 (Fig. 2). Hybridization of Southern blots prepared from these clones with oligodeoxynucleotides based on the rat cDNA sequence, however, revealed no additional DNA fragments containing coding sequence.

To identify the remaining exons, we prepared a second partial-length cDNA by PCR using oligodeoxynucleotides based on the rat sequences at nucleotides 422–439 (sense) and 988–1005 (antisense), as described above, to produce a probe encompassing nucleotides 422–1005, and then we screened the human genomic DNA library as before. This screen yielded an additional four clones which were analyzed by Southern blots following $Hind$III digestion, hybridized with either the partial-length cDNA from nu-

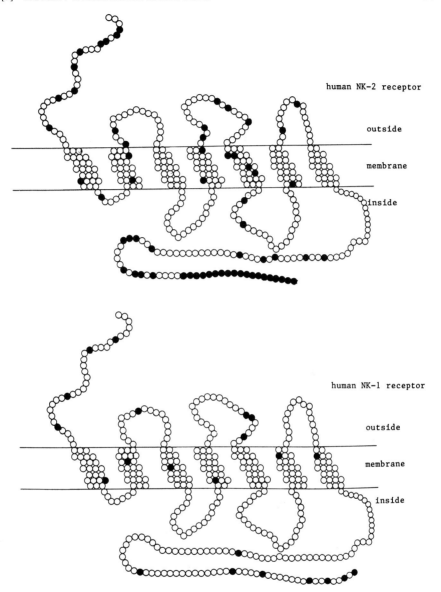

FIG. 3. Schematic representation of the human NK-1 and NK-2 receptor protein structures. Putative membrane-spanning sequences are shown between the double lines. Extracellular sequences are shown above the upper line and cytoplasmic sequences below the lower line. Differences in deduced amino acid sequences between human and rat are shown by filled circles. Sequence divergence for bovine (NK-2) or mouse (NK-1 and NK-2) occur in similar positions.

cleotides 422–1005 or with synthetic oligodeoxynucleotides specific for each of the exons. Two of the new clones contained only exon 3, one contained exons 3 and 4, and one contained exons 3, 4, and 5; none contained exons 1 or 2. Each of the positively hybridizing DNA fragments was subcloned into pBluescript and subjected to sequence analysis (28). Comparison of the organization of the NK-1 receptor gene with that determined for the NK-2 receptor gene shows identical intron splice junctions within the coding sequence. The NK-1 receptor gene appears to span 45–60 kb of DNA, while the NK-2 receptor gene is contained in approximately 12 kb. These findings are shown diagrammatically in Fig. 2.

To ensure that we had identified all of the exons comprising the human NK-1 receptor gene, we prepared oligonucleotide primers based on the 5′ and 3′ sequences determined from the human NK-1 receptor gene sequence and used them in the PCR with cDNA from human lung or IM9 cells. The cDNAs made in this manner yielded sequences with a single open reading frame of approximately 1200 bp. The sequences derived from lung and IM9 cells were identical to one another as well as to that determined from the individual exons. In contrast to the NK-2 receptor, which has 46 and 53 differences, respectively, in the deduced amino acid sequences between human and rat or human and bovine, the NK-1 receptor has only 21 amino acid substitutions between the human and rat molecules (Fig. 3). In addition, the human NK-2 receptor is 14 amino acids longer than the bovine and 9 amino acids longer than the rat molecules (19), whereas the human and rat NK-1 receptors are the same length (28).

Expression of Human Tachykinin Receptors in Cultured Cells

Transient Expression

To certify that each of the cloned cDNAs encodes human tachykinin receptors, we expressed them in mammalian cells which do not normally express these molecules and demonstrated their functional properties. We have employed several systems for this purpose. For transient expression, the cDNA molecules including only the NK-1 or NK-2 receptor coding sequences are prepared by PCR as described above using oligodeoxynucleotide primers containing either BamHI restriction sites at the 5′ ends of both primers or a HindIII site at the 5′ end of the sense primer and an XbaI site at the 5′ end of the antisense primer. Following purification and digestion with the appropriate restriction enzymes, the constructs are ligated to the expression vector pCDNA (In Vitrogen, San Diego, CA). This expression vector utilizes the cytomegalovirus (CMV) promoter and the simian virus 40 (SV40) origin of replication to provide high-level expression of cloned genes in mammalian cells (32).

Plasmids are transformed into competent *Escherichia coli* MC1061/p3, expanded as previously described (33), and purified by alkaline lysis and centrifugation in cesium chloride (34). Restriction and/or sequence analyses are used to confirm that the orientation of the insert is appropriate for translation and that no mutations are introduced by the PCR reaction, and the plasmids are transfected in COS-7 cells (ATCC CRL 1651) using DEAE-dextran (35). Cells are maintained in DMEM (Dulbecco's modified minimal essential medium, high glucose) supplemented with 6 mM glutamine, nonessential amino acids, penicillin, streptomycin, and 10% fetal calf serum in an atmosphere of 5% CO_2. Cells are trypsinized and replated at 750,000 per 10-cm dish 16–24 hr before transfection. Two micrograms of plasmid DNA is incubated with cells in 3 ml DMEM containing 10% NuSerum (Collaborative Research, Lexington, MA), 400 μg/ml DEAE-dextran (Sigma, St. Louis, MO), and 100 μM chloroquine (Sigma) for 2.5 hr at 37°C in an atmosphere of 5% CO_2. The mixture is aspirated and replaced with phosphate-buffered saline (PBS) containing 10% dimethyl sulfoxide (DMSO). After 5–10 min at room temperature, the solution is replaced with DMEM containing 10% fetal calf serum and returned to the incubator for 48–96 hr.

Control transfections are performed using plasmid constructs containing the cDNA for bacterial β-galactosidase. Detection of β-galactosidase is performed using a modification of the method described by Lim and Chae (36). Briefly, the cells are washed once with PBS, fixed for 5 min in freshly prepared 2% formaldehyde, 0.2% glutaraldehyde in PBS, washed 3 times with PBS, and stained with 0.1% X-Gal (5-bromo-4-chloro-3-indolyl-β-D-galactose), 1 mM $MgCl_2$, 3.3 mM potassium ferricyanide, 3.3 mM potassium ferrocyanide in PBS at 37°C for 1–24 hr. Cells transfected with bacterial β-galactosidase stain green with this treatment, whereas untransfected cells remain unstained, even after 24 hr at 37°C. Data obtained with COS-7 cells suggest that this method yields a transfection efficiency of 1–5%.

An alternative method for transient expression employs 293 cells (ATCC CRL 1573). These cells are maintained in the same medium as used for COS-7 cells, but subculturing is performed without trypsin, as 293 cells are not particularly adherent. Transfections are carried out as described by Rose *et al.* (37), using cationic liposomes consisting of phosphatidylethanolamine and dimethyldioctadecylammonium bromide (Sigma) at a ratio of 2.5:1 (w/w). Liposomes are prepared by drying the lipid mixture from chloroform and resuspending in sterile water by sonication on ice at 1.4 mg (total lipid weight)/ml. For each 10-cm dish of transfected cells, 45 μl of liposomes and 15 μg of plasmid DNA are mixed with 3 ml of DMEM in polystyrene tubes and allowed to stand for 5–10 min at room temperature to allow the DNA to adsorb to the liposomes. Growth medium is aspirated from the cells, which had been subcultured and plated 16–24 hr previously at a density of approximately 750,000 cells 10-cm dish, and replaced with the DNA-containing solution. Cells are returned to the incubator for 3 hr. At the end of this time, 3 ml DMEM containing 20% fetal calf serum is added to restore the serum concentration to 10%, and the cells are cultured for an additional 48–96 hr. This method routinely yields an apparent transfection efficiency of 70–80% using the β-galactosidase-containing plasmid.

Binding experiments are carried out on COS-7 cells by removing the culture medium and washing the cell layer with 50 mM Tris-HCl, pH 7.4, containing 0.15 M NaCl, 3 mM MnCl$_2$, and 0.02% bovine serum albumin (BSA). Cells are incubated at room temperature with gentle rocking in 2 ml of the same buffer containing the protease inhibitors chymostatin, leupeptin, and pepstatin at 4 μg/ml, bacitracin at 40 μg/ml, and phosphoramidon at 1 μM (10), for 45 min with 0.1 nM ^{125}I-labeled ligand [neurokinin A or Bolton–Hunter-labeled substance P (BHSP), both at a specific radioactivity of 2200 Ci/mmol; New England Nuclear (NEN), Boston, MA] and increasing concentrations of unlabeled peptide. The cell layers are washed 3 times with the same buffer without protease inhibitors, solubilized with 2 ml of 1 M NaOH, transferred to 12 $\times$ 75 mm tubes, and counted in a γ counter.

Transfected 293 cells are suspended in 50 mM Tris-HCl, pH 7.4, containing 0.15 M NaCl, 3 mM MnCl$_2$, 0.02% BSA, and the protease inhibitors indicated above, at 2 $\times$ 10^6 cells/ml, and incubated on ice with the same concentrations of labeled and unlabeled ligands for 45 min. Bound ligand is separated from free by rapid filtration of 0.5-ml aliquots of the cell suspension through glass fiber filters (Whatman, Clifton, NJ, GF/F) that had been soaked in 5% polyethyleneimine (Sigma). Filters are washed 3 times with the same buffer without protease inhibitors and counted in a γ counter.

Transient transfection of the NK-1 receptor coding sequence in pCDNA in COS-7 cells yielded cells which express binding sites for substance P with a Kd of 0.35 $\pm$ 0.07 nM. Untransfected cells display no detectable binding for the tachykinins. Neurokinin A and neurokinin B also interact with the cloned NK-1 receptor, but with 100- and 500-fold lower affinity, respectively. These findings are consistent with data published using membrane preparations expressing the naturally occurring receptor (7, 8, 10). Binding of [^{125}I]BHSP to the cloned human NK-1 receptor is also blocked by the substance P receptor antagonist CP-96,345 (38), with an IC$_{50}$ of approximately 2 nM.

Similarly, when the cDNA corresponding to the NK-2 receptor coding sequence in pCDNA was transiently transfected in the same cells and tested for binding as above, the apparent affinity for neurokinin A calculated from Scatchard analyses was 0.71 $\pm$ 0.17 nM. The human NK-2 receptor has approximately 600-fold higher affinity for neurokinin A than substance P or neurokinin B (7, 8).

Stable Transfections

COS-7 cells were initially used to generate stably transfected lines expressing either the human NK-1 or NK-2 receptors (28). The coding sequence for each of the receptor cDNAs is made by PCR using oligodeoxynucleotides containing *Bam*HI restriction sites at the 5' termini. The PCR products are purified as described above and ligated to pMam-neo-Blue (Clontech, Palo Alto, CA), which contains coding sequences for neomycin resistance and bacterial β-galactosidase. The plasmids are linearized with *Nde*I (Boehringer-Mannheim, Indianapolis, IN), and 20 μg is transfected per dish in COS-7 cells at approximately 80% confluence using 10 μg/ml Lipofectin (GIBCO BRL, Gaithersburg, MD) in 5 ml serum-free DMEM at 37°C in 5% CO_2 (39). After 12–18 hr, the medium is restored to 10% fetal calf serum, and transfected cells are selected for neomycin resistance with 400 μg/ml geneticin (GIBCO BRL). Clones are expanded from single surviving cells and tested for specific binding as described above. Several of the neomycin-resistant clones expressed tachykinin receptors for several passages, but the lines were relatively unstable and eventually reverted to nonreceptor-bearing cells (N. P. Gerard, unpublished observations, 1991).

Because of this apparent instability of COS-7 cells, we also generated stable cell lines in NIH 3T3 cells (ATCC CRL 1658). These cells are transfected with the same linearized plasmid constructs as above using the calcium phosphate transfection methodology (40). Transfected cells are selected for neomycin resistance with 400 μg/ml geneticin. Twenty-four of the surviving cells were expanded from single cells and tested for insertion of cDNA in the β-galactosidase coding sequence (absence of blue staining; see above) as well as for ligand binding activity. Of the five clones expressing the highest receptor number, only one, clone 13, had relatively normal growth characterisitics. These cells express approximately 30,000 sites/cell, with a K_d of 1.5 n*M*, and appear to be stable for 5 to 10 passages before receptor expression is lost (N. P. Gerard, unpublished observations, 1991).

Receptor Subtypes

Data based on pharmacologic studies have suggested the possibility of multiple subtypes for both the NK-1 and NK-2 receptors (41). We questioned whether they might arise from different genes for these molecules or whether the genes we cloned give rise to receptor subtypes by alternative splicing of the mRNAs. Southern blots of human genomic DNA hybridized under high-stringency conditions, as described above, with the cDNA from either receptor provides evidence for only one gene since the only bands observed can be accounted for by exons 1–5. The sequence identity among receptor subtypes should be high enough to expect that, if they existed, they should also hybridize under high-stringency conditions. If receptor subtypes exist, they must therefore arise by alternative splicing. Indeed, a recent study reported the existence of a truncated form of the human NK-1 receptor cloned from glioblastoma cells (42). This cDNA encodes exons 1–4 but eliminates the coding sequence included in exon 5, and it appears to use the intron following exon 4 as the 3′ untranslated region. Interestingly, the genes for both NK-1 and NK-2 receptors have a stop codon (TGA) in frame just after the arginine codon (AGG) that ends the coding sequence of exon 4 (19, 28), which could explain the origin of the truncated molecule. We have not performed RNase protection experiments to evaluate the possible existence of such products. Nonetheless, these findings suggest the possibility for tachykinin receptor subtypes that arise from alternative splicing events in mRNA formation, and the cDNAs will need to be transfected and tested for binding and signal transduction as we have described for the full-length molecules.

Acknowledgments

This work was supported in part by National Institutes of Health Grants HL41587, HL36162, and HL41277. L.A.G. is a Predoctoral Fellow of the Howard Hughes Medical Institute. J.L.P. is the recipient of a postdoctoral fellowship from INSERM. C.G. is the recipient of a Faculty Scholars Award from R.J.R. Nabisco.

References

1. M. M. Chang, S. E. Leeman, and H. D. Nial, *Nature* (*London*) New Biol. **232**, 86 (1971).
2. V. Erspamer, *Trends Neurosci.* **4**, 267 (1981).
3. B. Pernow, *Pharmacol. Rev.* **35**, 85 (1983).
4. P. W. Manthy, C. R. Mantyh, T. Gates, S. R. Vigna, and J. E. Maggio, *Neuroscience* (*Oxford*) **25**, 817 (1988).
5. P. W. Mantyh, D. J. Johnson, C. G. Boehmer, M. D. Catton, H. V. Vinters, J. E. Maggio, H. P. Too, and S. R. Vigna, *Proc Natl. Acad. Sci. U.S.A.* **86**, 5193 (1989).
6. P. J. Skerrett, *Science* **249**, 625 (1990).

7. S. H. Buck, E. Burcher, C. W. Shults, W. Lovenberg, and T. L. O'Donohue, *Science* **226,** 987 (1984).

8. E. Burcher, S. H. Buck, W. Lovenberg, and T. L. O'Donohue, *J. Pharmacol. Exp. Ther.* **236,** 819 (1986).

9. J. L. Henry *in* "Substance P and Neurokinins" (J. L. Henry, *et al.,* eds.), p. xvii. Springer-Verlag, New York, 1987.

10. S. R. Coats and N. P. Gerard, *Am. J. Respir. Cell Mol. Biol.* **1,** 269 (1989).

11. M. A. Cascieri and T. Liang, *J. Biol. Chem.* **258,** 5158 (1983).

12. R. A. F. Dixon, I. S. Sigal, M. R. Candelore, R. B. Register, W. Scattergood, E. Rands, and C. D. Strader, *EMBO J.* **6,** 3269 (1987).

13. H. G. Dohlman, J. Thorner, M. G. Caron, and R. J. Lefkowitz, *Annu. Rev. Biochem.* **60,** 653 (1991).

14. H. Wang, L. Lipfert, C. C. Malbon, and S. Bahouth, *J. Biol. Chem.* **264,** 4424 (1989).

15. D. Kunz, N. P. Gerard, and C. Gerard, *J. Biol. Chem.* **267,** 9101 (1992).

16. P. C. Ross, R. A. Figler, M. H. Corjay, C. M. Barber, N. Adam, D. R. Harcus, and K. R. Lynch, *Proc. Natl. Acad. Sci. U.S.A.* **87,** 3052 (1990).

17. Y. Masu, K. Nakayama, H. Tamaki, Y. Harada, M. Kuno, and S. Nakanishi, *Nature (London)* **329,** 836 (1987).

18. Y. Sasai and S. Nakanishi, *Biochem. Biophys. Res. Commun.* **165,** 695 (1989).

19. N. P. Gerard, R. L. Eddy, T. B. Shows, and C. Gerard, *J. Biol. Chem.* **265,** 20455 (1990).

20. A. Ullrich, J. Shine, J. Chirgwin, R. Pictet, E. Tischer, W. J. Rutter, and H. M. Goodman, *Science* **196,** 1313 (1977).

21. J. M. Chirgwyn, A. E. Przybyla, R. J. MacDonald, and W. J. Rutter, *Biochemistry* **18,** 5294 (1979).

22. V. Glisin, R. Crkvenjakov, and C. Byus, *Biochemistry* **13,** 2633 (1972).

23. H. Aviv and P. Leder, *Proc. Natl. Acad. Sci. U.S.A.* **69,** 1408 (1972).

24. U. Gubler and B. Hoffman, *Gene* **25,** 263 (1983).

25. A. P. Feinberg and B. Vogelstein, *Anal. Biochem.* **132,** 6 (1983).

26. B. A. White and S. Rosenzweig *BioTechniques* **7,** 694 (1989).

27. E. Y. Chen and P. H. Seeburg, *DNA* **4,** 165 (1985).

28. N. P. Gerard, L. A. Garraway, R. L. Eddy, T. B. Shows, H. Iijima, J.-L. Paquet, and C. Gerard, *Biochemistry* **30,** 10640 (1991).

29. Y. Yokota, Y. Sasai, K. Tanaka, T. Fujiwara, K. Tsuchida, R. Shigemoto, A. Kakizuke, H. Ohkubo, and S. Nakanishi, *J. Biol. Chem.* **264,** 17649 (1989).

30. A. D. Hershey and J. E. Krause, *Science* **247,** 958 (1990).

31. D. G. Payan, J. P. McGillis, and M. L. Organist, *J. Biol. Chem.* **261,** 14321 (1986).

32. B. Seed, *Nature (London)* **329,** 840 (1987).

33. D. Hanahan, *J. Mol. Biol.* **166,** 557 (1983).

34. H. C. Birnboim and J. Doly, *Nucleic Acids Res.* **7,** 1513 (1979).

35. N. P. Gerard and C. Gerard, *Nature (London)* **349,** 614 (1991).

36. K. Lim and C. B. Chae, *BioTechniques* **7,** 576 (1989).

37. J. K. Rose, L. Buonocore, and M. A. Whitt, *BioTechniques* **10,** 520 (1991).

38. R. M. Snider, J. W. Constantine, J. A. Lowe III, K. P. Longo, W. S. Lebel, H. A. Woody, S. E. Drozda, M. C. Desai, F. J. Vinick, R. W. Spencer, and H.-J. Hess, *Science* **25,** 435 (1991).

39. P. L. Felgner, T. R. Gadek, M. Holm, R. Roman, H. W. Chan, M. Wenz, J. P. Northrop, G. M. Ringold, and M. Danielson, *Proc. Natl. Acad. Sci. U.S.A.* **84,** 7413 (1987).

40. F. L. Graham and A. J. van der Eb, *Virology* **52,** 456 (1973).

41. L. L. Iverson, K. J. Watling, A. T. McKnight, B. J. Williams, and C. M. Lee, *Top. Med. Chem.* **65,** 1 (1987).

42. T. M. Fong, S. A. Anderson, H. Yu, R.-R. C. Huang, and C. D. Strader, *Mol. Pharmacol.* **41,** 24 (1991).

[6] Purification and Molecular Cloning of Bombesin/Gastrin-Releasing Peptide Receptor

James F. Battey, Richard Harkins, and Richard I. Feldman

Introduction

Bombesin is a tetradecapeptide first isolated from the skin of the frog *Bombina bombina* (1). Over ten peptides have subsequently been characterized with structural homology to bombesin. Two of these, gastrin-releasing peptide (GRP) (2) and neuromedin B (NMB) (3), have been found in mammals. In mammalian tissues, two distinct receptors with binding preference for either GRP (the GRP-R) or NMB (the NMB-R) have been pharmacologically distinguished (4), and complementary DNA (cDNA) clones encoding both receptors have been isolated (5–8).

Bombesin receptors have a role in regulating numerous biological processes including gastrointestinal hormone release, smooth muscle contraction, and modulation of neuronal firing rates [reviewed in Lebacq-Verheyden *et al.* (9)]. Furthermore, bombesin-like peptides are potent mitogens for Swiss 3T3 fibroblasts (10–12) and human bronchial epithelial cells (13), and they have been implicated as autocrine growth factors for some human small cell lung carcinomas (14–16). Swiss 3T3 fibroblasts are the best characterized source of GRP receptors, exhibiting higher numbers of receptors per cell than other bombesin-responsive cells (11, 17). The Swiss 3T3 GRP-R is a glycoprotein appearing as a diffuse band after sodium dodecyl sulfate–polyacrylamide gel electrophoresis (SDS-PAGE) with an apparent molecular weight of 75,000–100,000 (17, 18).

To further characterize the biochemistry, structure, and function of the Swiss 3T3 GRP-R and related subtypes, we have purified a murine GRP-R from Swiss 3T3 fibroblast membranes (19). The purified protein was used to obtain a partial amino acid sequence, which was used to isolate cDNA clones for the GRP-R (5). Here we describe methods used for the solubilization, purification, and partial protein sequencing of the Swiss 3T3 GRP-R. In addition, we describe methods to generate a cDNA library enriched for GRP-R clones which was used to obtain the initial partial GRP-R cDNA clones by screening with oligonucleotide probes. Finally, procedures for isolating longer GRP-R cDNA clones encoding the entire protein coding region are discussed.

Solubilization of Swiss 3T3 Gastrin-Releasing Peptide Receptor

Preparation of Crude Swiss 3T3 Cell Membranes

Typically, membranes are prepared from 100 roller bottles of cells, and four such preparations are combined in a single receptor purification to yielded about 0.5 μg protein (19). However, the protocols described below could easily be scaled down.

Swiss 3T3 fibroblasts (20) are cultured at 37°C in Dulbecco's modified Eagle's medium (DMEM) containing 4.5 g/liter glucose and supplemented with 1 mM sodium pyruvate, 2 mM glutamate, and 10% (v/v) fetal bovine serum in an environment of 10% (v/v) CO_2. When growing cells for GRP-R purification, cells are cultured on the surface of roller bottles (1300 cm^2, Falcon, Oxnard, CA). When cells reach confluence, they are washed twice with calcium/magnesium-free phosphate buffered saline (PBS), incubated with 50 ml PBS plus 0.04% EDTA and 1% glucose for about 15 min at room temperature, and dislodged from the surface of the bottles by vigorous agitation. Cells are then collected by centrifugation (800 g for 10 min), washed twice with PBS, and suspended in 3–4 volumes (~500 ml) of lysis buffer [50 mM HEPES, pH 7.5, 2 mM MgCl$_2$, 1 mM EGTA, 50 μg/ml leupeptin, 2.5 μg/ml pepstatin, 10 μg/ml aprotinin, and 0.5 mM phenylmethylsulfonyl fluoride (PMSF)].

The cells are efficiently broken using a cell disruption bomb (920 ml size, Parr Instruments, Moline IL) that allows cells to be released from a chamber pressurized to 900 psi of nitrogen, causing rapid decompression and lysis. The cells are kept on ice while several batches of cells are broken and subsequently pooled. The lysate is centrifuged at 39,000 g for 30 min at 4°C, and the membrane pellet is suspended in lysis buffer with the aid of a Potter–Elvehjem homogenizer powered by a variable-speed electric drill. An aliquot is taken for determination of protein concentration using the bicinchoninic acid protein assay (BCA) (Pierce, Rockford, IL), with bovine serum albumin serving as the protein standard. Membranes are pelleted and finally suspended at a concentration of 15 mg/ml protein in 50 mM HEPES, pH 7.5, 1 mM EGTA, 0.25 M sucrose, 50 μg/ml leupeptin, 2.5 μg/ml pepstatin, 10 μg/ml aprotinin, and 0.5 mM PMSF. Membranes could be frozen in liquid nitrogen and stored for at least 1 year at −80°C without significant loss of ^{125}I-labeled GRP binding activity. Yields of membrane protein ranged from 500 to 800 mg per 100 roller bottles.

Assay of ^{125}I-Labeled GRP Binding to Swiss 3T3 Membranes

The following protocol uses a binding medium with a low salt concentration since physiological salt levels lowered the K_d of ^{125}I-labeled GRP binding to the receptor by about 10-fold (19). To determine binding parameters under physiological conditions, roughly 10 times more membranes should be used.

Crude membranes (20–30 μg protein) are suspended in siliconized tubes in 500 μl of binding medium (50 mM HEPES, pH 7.5, 2 mM EDTA, 10 mg/ml bovine serum albumin, and 30 μg/ml bacitracin). The binding reaction is initiated by addition of ^{125}I-labeled GRP (20 pM, Amersham, Arlington Heights, IL). After incubating for 60 min at 37°C, the reactions are placed on ice for several minutes and filtered through Whatman (Clifton, NJ) GF/B glass filter fibers which had been soaked (for at least 5 min) in a solution of 0.3% polyethyleneimine. The filters are rapidly washed 4 times with ice-cold 50 mM Tris, pH 7.5. A large number of assays could be performed using a device which filters many samples at once and automatically dispenses wash solution (e.g., the receptor binding harvester made by Brandel, Gaithersburg, MD). The radioactivity on filters is determined using a γ counter.

Nonspecific binding is determined by inclusion of 100 nM unlabeled GRP and is typically 1.5–2% of the total ^{125}I-labeled GRP added. For this assay, a receptor concentration is determined experimentally that is well below saturating levels (<20% of the total ^{125}I-labeled GRP should be bound and >80% should remain free). The amount of receptor present is calculated from the following equation:

Receptor (pmol) =
$$K_d \times [CPM_b/(CPM_t - CPM_b)] + L \times [CPM_b/CPM_t] \times 0.5$$

where K_d (nM) is the dissociation constant for ^{125}I-labeled GRP determined from previous binding data by the method of Scatchard (21), CPM_b and CPM_t are counts per minute of ^{125}I-labeled GRP bound and the total added, respectively, and L is the concentration of ^{125}I-labeled GRP added in nanomolar (typically 0.02 nM).

Extraction of GRP Receptor from Swiss 3T3 Membranes

Crude Swiss 3T3 membranes are washed twice in a high salt buffer (50 mM HEPES, pH 7.5, 1.0 M NaCl, 2 mM EDTA, 50 μg/ml leupeptin, 2.5 μg/ml pepstatin, 10 μg/ml aprotinin, and 0.5 mM PMSF) in order to remove extrinsic membrane proteins, washed once in the same buffer without NaCl, and finally suspended at 7 mg/ml protein in 50 mM HEPES, pH 7.5, 2 mM EDTA, 1 mM EGTA, 100 mM NaCl, 25 μg/ml leupeptin, 30 μg/ml bacitracin, 10 μg/ml aprotinin, 2.5 μg/ml pepstatin, and 0.5 mM PMSF. This salt washing procedure removes about half of the crude membrane protein without loss of GRP receptor binding activity.

The membranes are then solubilized by slowly mixing in a solution of 20% 3-[(3-cholamidopropyl) dimethylammonio]-1-/propanesulfonate (CHAPS) (Boehringer-Mannheim, Indianapolis, IN), and 2% cholesteryl hemisuccinate (CHS) (Sigma, St. Louis, MO) to yield a final detergent concentration of 0.75% CHAPS/0.075% CHS. After incubation for 30 min at room temperature, the mixture is cooled to 0°C and centrifuged at 100,000 g for 60 min at 4°C to remove insoluble material. The supernatant containing the soluble receptor could be frozen in liquid nitrogen and stored for many months without loss of activity.

Assay of ^{125}I-Labeled GRP Binding to the Soluble GRP Receptor

The following protocol is similar to the assay of the membrane-bound form of the GRP-R described previously. The soluble receptor extract is assayed in a final volume of 500 μl of binding medium (50 mM HEPES, pH 7.5, 2mM EDTA, 10 mg/ml bovine serum albumin, 30 μg/ml bacitracin, 0.075% CHAPS, 0.0075% CHS, and 20 pM ^{125}I-labeled GRP tracer). The amount of detergent added to the binding medium with the soluble extract is taken into account when determining how much additional detergent to add to the final binding medium. Samples are incubated at 15°C for 30 min, then placed on ice for several minutes. Receptor–ligand complexes are recovered by rapidly filtering the samples through Whatman GF/B filters soaked in 0.3% polyethyleneimine. After filtration, GF/B filters are washed 4 times with 4 ml ice-cold 50 mM Tris, pH 7.5. Nonspecific background is determined by inclusion of 100 nM unlabeled GRP. The background is typically 1.5–2% of the initial radioactive tracer. Filter-bound radioactivity is determined using a γ counter.

Comments

The ligand binding activity of the GRP-R is very sensitive to detergents, a property shared with many other seven-transmembrane domain receptors. Many detergents could extract the GRP-R from Swiss 3T3 membranes, but only the zwitterionic detergent CHAPS in the presence of CHS solubilized detectable levels of the GRP-R that could still bind ^{125}I-labeled GRP with high affinity (19). Solubilization of the receptor with CHAPS alone yielded an inactive form of the receptor that could then be efficiently activated by the addition of CHS, showing that the effect of CHS is to promote a more native conformation of the ligand binding site of the receptor in solution (19). ^{125}I-Labeled GRP binding to Swiss 3T3 membranes was inhibited by about 80% with the addition of 5'-guanyl yl imidodiphosphate [Gpp (NH) p] (19, 22), which is a hallmark for the coupling of a receptor to guanine nucleotide regulatory proteins (G proteins) (23). However, ligand binding to the receptor after extraction with CHAPS and CHS was not inhibited by Gpp(NH)p or other guanyl nucleotides, indicating that the soluble receptor is no longer coupled to G proteins.

Several parameters in the protocol had significant effects on the yield of soluble receptor, including temperature, NaCl, and Mg^{2+}. Although 0.75% CHAPS was the optimal detergent concentration to release the GRP-R from Swiss 3T3 membranes, ligand binding activity was severely inhibited unless the detergent was diluted. Receptor binding activity as a function of detergent concentration showed a relatively sharp maximum at 0.075% CHAPS/ 0.0075% CHS. Membrane extractions using CHAPS concentrations above 0.75% resulted in significant irreversible loss of binding activity. The salt washing of membranes can be omitted if the soluble receptor is to be used without purification. The ligand binding affinity and specificity of the soluble GRP-R was similar to those of the membrane-bound form (Table I). Rela-

TABLE I   Relative Affinities of Membrane-Bound and Soluble Forms of GRP
            Receptor for Various Peptides

	IC$_{50}$ (nM)	
Peptide	Soluble receptor	Membrane receptor
GRP1–27	0.13	0.13
[Nle$^{14, 27}$]GRP13–27	0.10	ND^a
[Nle$^{14, 27}$,Leu^{17}]GRP14–27	10	ND
[Nle$^{14, 27}$]GRP13–27-OH^b	8.4	18
GRP18–27	10	4.9
NAcGRP20–27	20	12
[Lys^3]Bombesin	3.8	2.1
Bombesin	1.3	ND
GRP1–16	>1000	>1000
Substance P	>1000	>1000
[D-Arg^1,D-Pro^2,D-Trp^{7,9},Leu^{11}] substance P	>1000	>1000
Physalemin	>1000	>1000

^{a} ND, Not determined.

^{b} This peptide contains a free C terminus, in contrast to the other GRP peptides shown, which are amidated.

tively low concentrations of NaCl reversibly inhibited the binding of ^{125}I-labeled GRP to both the soluble GRP-R (EC_{50}150 mM) and Swiss 3T3 membranes (EC_{50}40 mM). The yield of receptor solubilized was about 30%, as determined by Scatchard analysis.

Several proteins besides the GRP-R with moderate binding affinities for ^{125}I-labeled GRP could be solubilized from Swiss 3T3 membranes by extraction with Triton X-100 or Nonidet P-40 (R. I. Feldman and J. Wu, unpublished data, 1990). One such protein was characterized by cross-linking to ^{125}I-labeled GRP, displaying an apparent molecular weight on SDS-PAGE of about 110,000 and a K_d for ^{125}I-labeled GRP in 1% Nonidet P-40 of about 50 nM. Another protein with a apparent molecular weight similar to that of the GRP-R on SDS-PAGE (~75,000) was isolated from Triton X-100 membrane extracts by eluting a GRP affinity column with a pH 5 buffer. Both proteins were distinct from the GRP-R, since they did not contain N-linked carbohydrates. In addition both proteins were also found in membranes from cells (Balb 3T3 fibroblasts) that did not display high-affinity GRP binding sites with the pharmacologic properties predicted for the GRP-R.

Purification of Gastrin-Releasing Peptide Receptor

The following protocol uses a combination of lectin and ligand affinity chromatography to purify the GRP-R (19). All steps are performed at 4°C over the course of 4 days. The receptor is not stable during purification, so prolonged storage between steps should be avoided. The purified receptor is a glycoprotein that migrates as a diffuse band on SDS-PAGE with an M_r of 75,000–100,000 (Fig. 1) and binds ^{125}I-labeled GRP with an affinity similar to that of the receptor in crude membrane extracts or in Swiss 3T3 membranes (Table I). Table II summarizes the steps used in purification.

Fractionation by Polyethylene Glycol Precipitation

Precipitation by polyethylene glycol (PEG) is a convenient method of concentrating the crude membrane extract containing soluble GRP-R, while lowering the detergent concentration without significant loss of ^{125}I-labeled GRP binding activity. The receptor is more stable at lower detergent concentrations, especially at later stages of purification. This procedure also resulted in a two fold enrichment of the GRP-R.

Ice-cold PEG 8000 (50%, w/v, in water) is added to the soluble extract prepared from salt-washed membranes, until a final concentration of 20% (w/v) PEG 8000 is achieved. The suspension is thoroughly mixed, and the

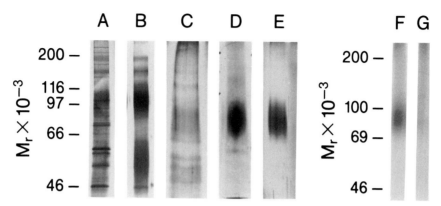

FIG. 1 Electrophoretic analysis of various steps of the GRP-R purification, and products of ^{125}I-labeled GRP cross-linking to the soluble GRP-R. Lanes A–E show a silver-stained SDS–polyacrylamide gel (7.5% acrylamide) of the GRP-R at various stages of purification. Lane A, soluble extract after PEG fractionation; lane B, WGA-agarose eluate; lane C, first [Nle^{14,27}]GRP13–27-agarose column eluate; land D, second [Nle^{14,27}]GRP13–27-agarose column eluate; lane E, receptor after Superose 6 chromatography. The GRP-R in the crude soluble extract was affinity labeled by cross-linking to ^{125}I-labeled GRP as described previously (19), subjected to SDS-PAGE on a 7.5% gel, and visualized by fluorography. Lane F shows cross-linked products. Lane G shows cross-linked products in the presence of 10 nM unlabeled GRP.

TABLE II Purification of GRP Receptor from Swiss 3T3 Cell Extracts

Step	Total activity[a] (pmol)	Total protein[b] (mg)	Specific activity (pmol/mg)	Yield/ step (%)	Yield overall (%)	Purification (-fold)
Membranes	1026	3270	0.31	100	100	1
PEG precipitation	280	281	1.0	27	27	3.2
WGA-agarose	183	42	4.4	65	18	14
[Nle^{14,27}]GRP13–27-agarose	79	0.06[c]	1316	43	7.7	4200
Desalt	50	ND[d]	ND	63	4.8	ND
[Nle^{14,27}]GRP13–27-agarose	36	ND	ND	72	3.5	ND
Ultrafiltration	19	ND	ND	52	1.9	ND
Superose 6	20	0.00073[e]	2.7×10^4	105	1.9	87,000

[a] Calculated from ^{125}I-labeled GRP1–27 binding data obtained as described in the text, using a k_d value of 0.038 nM.

[b] Protein was determined as described in the text, except where indicated.

[c] Determined from the ratio of A_{280} values of the column flow-through and eluate.

[d] ND, not determined.

[e] Determined from the A_{280} of the protein, assuming that the ε_{280} of the receptor was the same as bovine serum albumin or chicken ovalbumin standards.

precipitate that forms within several minutes (containing GRP-R) is collected by centrifugation at 100,000 g for 10 min. The pellet is suspended in 25 mM HEPES, 25 mM Tris, pH 7.5, 2 mM EDTA, 0.075% CHAPS, 0.0075% CHS, 5μg/ml leupeptin, and 10 μg/ml bacitracin in 1/4 the original volume of extract, using a Potter–Elvehjem homogenizer.

Lectin Affinity Chromatography

A column (1.6 × 10 cm) is prepared containing 20 ml wheat germ agglutinin (WGA)-agarose resin (3–5 mg of lectin/ml wet gel; E-Y Laboratories, San Mateo, CA) and is equilibrated at 4°C with lectin column buffer (25 mM HEPES, 25 mM Tris, pH 7.5, 2 mM EDTA, 0.25% CHAPS, 0.025% CHS, 5 μg/ml leupeptin, and 10 μg/ml bacitracin). Immediately prior to loading the column, the soluble extract is diluted with one volume of lectin column buffer and the detergent concentration adjusted to 0.25% CHAPS/0.025% CHS. The detergent concentration is increased to reduce aggregation of proteins in the extract. (An increased concentration of detergent is tolerated by the receptor for the brief period needed to load the column). The extract is applied to the column at a flow rate of 1.4 ml/min. The column is then washed with about 10 column volumes of affinity column buffer (25 mM HEPES, 25 mM Tris, pH 7.5, 2 mM EDTA, 0.075% CHAPS, 0.0075% CHS, 5 μg/ml leupeptin, and 10 μg/ml bacitracin), then eluted with affinity column buffer augmented with 5 mM N,N',N''-triacetylchitotriose (attempts to elute GRP-R with N-acetyl-D-glucosamine resulted in poor recoveries). ^{125}I-Labeled GRP binding activity is assayed as described above and closely parallels the peak of total protein eluted.

Preparation of [Nle^{14,27}]GRP13–27-Agarose Resin for Ligand Affinity Chromatography

A beaded agarose matrix (30–100 μm) containing aldehyde moieties at the end of the five-atom hydrophilic spacers (Actigel Superflow; Stereogene, San Gabriel, CA) is coupled to the GRP analog [Nle^{14,27}]GRP13–27 using a reduced Schiff base linkage to either the ε-amino group of lysine 13 or the amino terminus of the peptide. Ten milliliters of resin is washed with 5 volumes of 100 mM sodium phosphate, pH 7.0, and then suspended in 10 ml of 100 mM sodium phosphate, pH 7.0, 100 mM NaCNBH$_3$, and 2 mg/ml [Nle^{14,27}]GRP13–27 for 2 hr with gentle mixing to promote coupling. The resin is washed sequentially several times with a low pH buffer (100 mM

sodium acetate, pH 4.0, and 0.5 M NaCl) and a high pH buffer (100 mM Tris, pH 8.0, and 0.5 M NaCl). The efficiency of coupling is usually over 90%. The resin could be stored in 100 sodium phosphate, pH 7.0, and 0.04% sodium azide at 4°C for months. Affinity columns may be recycled by washing in affinity column buffer supplemented with 1 M NaCl and used many times without apparent loss of efficiency.

Sites on the resin that do not react with peptide should not be blocked with compounds such as ethanolamine since this chemistry converts unreacted aldehyde groups to charged secondary amines, greatly increasing the nonspecific binding of proteins to the resin. The irreversible formation of Schiff base linkages with proteins passing through the column is prevented by including a primary amine such as Tris in the affinity column buffer.

Preliminary experiments indicated that GRP1–27 was a suitable affinity chromatography ligand, but it contained methionine residues that may oxidize and limit the stability of the column. When resins activated by N-hydroxysuccinimide were tried instead of Actigel, ligand matrices that bound the GRP-R efficiently were generated, but these resins did not release GRP-R under nondenaturing conditions.

Ligand Affinity Chromatography

[Nle^{13,27}]GRP13–27-agarose is poured into a column (1.6 × 5 cm) and equilibrated at 4°C with affinity column buffer (25 mM HEPES, 25 mM Tris, pH 7.5, 2 mM EDTA, 0.075% CHAPS, 0.0075% CHS, 5 μg/ml leupeptin, and 10 μg/ml bacitracin). The column is loaded with WGA-purified GRP-R at a flow rate of 0.1 ml/min and washed with about 20 column volumes of affinity column buffer at a flow rate of 0.5 ml/min. The bound receptor is eluted with the affinity column buffer plus 0.5 M NaCl at a flow rate of 0.2 ml/min. The receptor coelutes with the main peak of 280 nm absorbance. Fractions containing ^{125}I-labeled GRP binding activity are pooled (~10–15 ml volume).

Although NaCl elutes the GRP-R with its binding activity intact, proteins nonspecifically bound to the column coelute with the receptor, thereby limiting the amount of purification achieved in this step. Further purification is achieved using chromatography on a second, smaller affinity column after desalting the fractions containing receptor. Desalting is effected by concentrating the receptor to 1 ml using a Centriprep-10 device (Amicon, Danvers, MA), followed by addition of 14 ml affinity column buffer, concentration to a 1 ml a second time, and addition of 4 ml affinity column buffer to a final volume of 5 ml.

The second [Nle^{13,27}]GRP13–27-agarose column (1.0 × 3 cm) is equilibrated with affinity column buffer. The sample is loaded onto the column at a flow rate of 0.03 ml/min. The column is washed with 20 column volumes of affinity column buffer at a flow rate of 0.5 ml/min, then eluted with affinity column buffer plus 0.5 M NaCl at a flow rate of 0.1 ml/min. Fractions containing GRP-R binding activity are pooled (~6 ml). SDS-PAGE analysis showed a diffuse band of GRP-R (M_r 75,000–100,000) with a slight amount of contamination (Fig. 1, lane D).

Gel Filtration Chromatography

The remaining contaminants can be significantly reduced by gel filtration. The receptor is concentrated to about 1 ml on a Centriprep-10 ultrafiltration device and is further concentrated to 0.3 ml using a Centricon-10 device. The sample is chromatographed on a Superose 6 HR 10/30 column (Pharmacia LKB Biotechnologies, Inc., Piscataway, NJ), run in 20 mM HEPES, pH 7.5, 2 mM EDTA, 0.075% CHAPS, 0.0075% CHS, and 100 mM NaCl at a flow rate of 0.4 ml/min. The receptor exhibits an apparent molecular weight of about 200,000 on the column, and eluted in a volume of 2–3 ml. The plot of absorbance at 280 nm showed a shoulder consisting of GRP-R protein followed by a much larger peak, consisiting of detergent aggregates. The binding activity of the receptor can be preserved at this step by freezing in liquid nitrogen.

Partial Amino Acid Sequencing of Purified Gastrin-Releasing Peptide Receptor

Amino-Terminal Sequence of Purified GRP Receptor

Purified by GRP-R is concentrated and separated from residual detergent before attempting sequential Edman degradation. The sample is concentrated to about 100 μl by ultrafiltration using a Centricon-10 device as recommended by the manufacturer (Amicon). Residual detergent and other contaminants that might interfere with sequencing are removed by washing the receptor in the same Centricon-10 device by two rounds of dilution with 2 ml water and reconcentration.

To perform NH$_2$-terminal sequence analysis, the sample (95 μl) is loaded onto a Biobrene [Applied Biosystems, Inc. (ABI), Foster City, CA] precycled glass filter and subjected to 30 cycles of automated Edman degradation on an ABI 475A gas-phase sequencer equipped with an ABI Model 120A on-line high-performance liquid chromatography detection (HPLC) system for identification of phenylthiohydantoin (PTH)-amino acids as described previously (24). Quantitation of PTH-amino acids is performed by an ABI Model 900 data system using 60 pmol of a set of known PTH-amino acid standards (ABI).

Following two separate NH$_2$-terminal sequence runs on two independent purification preparations of GRP-R, the following consensus NH$_2$-terminal amino acid sequence was obtained for 17 residues (X denotes a residue where an accurate assignment could not be made):

$$^1A \ P \ N \ X^5 \ S \ X \ L \ N \ L^{10} \ D \ V \ D \ P \ F^{15} \ L \ S$$

The N-X-S sequence, located at the position of the indeterminate residues, is known to be a consensus sequence for carbohydrate addition. The presence of an oligosaccharide chain in the GRP-R glycoprotein probably obscured the signal for three of the sequencing cycles in the vicinity of this sequence.

Sequence Analysis of Internal Tryptic Peptide Fragments of GRP Receptor

Additional amino acid sequence data are obtained after trypsin digestion of the purified GRP-R and isolation of tryptic fragments by reversed-phase chromatography. For this analysis, about 40 mol of purified receptor is used (1.6 μg GRP-R protein). The purified receptor sample (in 3 ml) is reduced to about 100 μl by ultrafiltration using a Centricon-10 device. The sample is washed with 2 ml water, concentrated to 100 μl, diluted to 1 ml with water, and concentrated again to 138 μl. The receptor is digested into fragments by addition of trypsin (0.1 μg), followed by incubation at 37°C for 2 hr. An additional 0.1 μg trypsin is added, followed by incubation for 5 hr. Finally, a third 0.2-μg aliquot of trypsin is added and the sample incubated at 37°C for 22 hr. The digested GRP-R is rapidly frozen in liquid nitrogen and stored at −80°C. Analytical SDS-PAGE analysis of a portion of the digested GRP-R revealed complete loss of the intact GRP-R protein (75,000–100,000) as expected. However, the digestion was not complete under the conditions employed, since a new, high molecular weight band (40,000) was observed after digestion.

The entire trypsin-digested GRP-R sample is thawed to room temperature and reduced by incubation in 10 mM dithiothreitol (DTT) at 37°C for 30 min. Digested peptides are fractionated by reversed-phase HPLC using a 2.1 mm × 3 cm C$_4$ column (Brownlee, Santa Clara, CA, Aquapore Butyl, 300 Å pore size), and a linear gradient of 0.05% trifluoroacetic acid (TFA) in water (solvent A) to 0.05% TFA in 100% acetonitrile (solvent B). The conditions for the HPLC gradient are 0% solvent B to 100% solvent B in 60 min at a flow rate of 0.2 ml/min. Effluent fractions are detected at 215 nm, collected at 1-min intervals, and stored at 4°C.

Numerous distinct absorbance peaks were identified (Fig. 2). Fifteen separate fractions (or pools of fractions) corresponding to these peaks were subjected to automated amino acid sequence analysis. For peptide sequence analysis, HPLC fractions are pooled and concentrated on a Speed-Vac vacuum centrifuge (Savant, Farmingdale, NY) to a final volume of about 50 μl. The entire sample is loaded onto a glass fiber filter which had been treated and precycled with Biobrene (ABI). Automated amino acid sequence analysis is performed as before. Three of the fifteen samples gave unambiguous sequences. Combined HPLC fractions 56 through 59 gave the sequence MASFLVFYVIPLAII (T56/59), fraction 44 yielded the sequence XXPFIQLTSVGVSVFT (T44), and the sequence derived from fraction 50 was XVPNLFISXLALGXLLXXVT (T50), where X denotes a residue that could not be unambiguously identified. The remaining twelve samples yielded sequence signals that were too low to be reliable, some of which might have been nonprotein contaminants such as residual detergent.

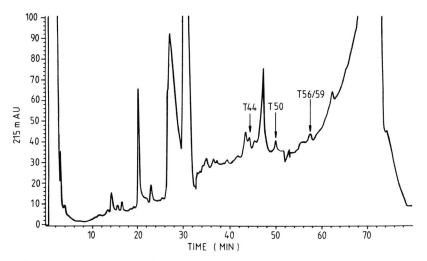

Fig. 2   Separation of tryptic fragments of the GRP-R by C_4 reversed-phase HPLC, monitored by absorbance at 215 nm. Fractions with peaks that yielded a readable amino acid sequence are marked by arrows.

Construction of cDNA Library Enriched Gastrin-Releasing Peptide Receptor

We predicted that the GRP-R mRNA would be a relatively rare transcript in Swiss 3T3 poly(A)$^+$ mRNA (~1 part per 50,000), given the estimates of the number of high-affinity ^{125}I-labeled GRP binding sites (50,000 per cell). Thus, the frequency of GRP-R cDNA clones expected in a representative Swiss 3T3 cDNA library would be low. Labeled oligonucleotide probes are relatively inefficient when used in cDNA library plaque hybridization screening, identifying only a small fraction of the positive clones present. Moreover, using standard conditions for hybridization and washing, oligonucleotide probes generally hybridize to many other clones in representative libraries besides the intended cDNAs. To increase the efficiency of oligonucleotide screening for GRP-R cDNAs, a cDNA library enriched about 30- to 100-fold for GRP-R clones is constructed using subtractive hybridization followed by polymerase chain reaction amplification.

In contrast to Swiss 3T3 fibroblasts, Balb/3T3 fibroblasts show undetect-

1. Synthesize random hexamer primed first strand Swiss 3T3 cDNA

2. Hybridize Swiss 3T3 cDNA with an excess of Balb 3T3 mRNA

3. Purify unhybridized Swiss 3T3 cDNA by HAP chromatography

4. Repeat steps 2 and 3 with unhybridized Swiss 3T3 cDNA (step 3) and fresh Balb 3T3 mRNA.

5. Re-anneal unhybridized Swiss 3T3 cDNA from step 4 to Swiss 3T3 mRNA, forming mRNA:cDNA heteroduplexes

6. Purify Swiss 3T3 mRNA:cDNA heteroduplexes by HAP chromatography

7. Convert heteroduplex to DNA duplex by nick translation

8. Polish and phosphorylate termini of double stranded cDNA

9. Ligate an adaptor to the cDNA termini for PCR amplification

10. Amplify cDNA using PCR and a primer complementary to the ligated adaptor

11. Clone amplified cDNA into lambda gt10 to create a subtracted cDNA library for screening with oligonucleotide probes

FIG. 3   Steps to generate a subtracted cDNA library enriched for GRP-R cDNA clones. The sequence of procedures used to enrich Swiss 3T3 cDNA for species not expressed by Balb/3T3 fibroblasts is shown, along with steps to form a library from these differentially expressed cDNAs.

able levels of high-affinity GRP binding sites. In other respects, the two murine embryonic fibroblast cell lines are very similar in their gene expression properties. Thus, an enrichment for GRP-R cDNA clones could be obtained by eliminating cDNAs expressed at equal levels in both Swiss 3T3 and Balb/ 3T3 cells and constructing a cDNA library from cDNAs uniquely expressed by Swiss 3T3 cells. A flowchart for the steps used to construct this library is shown in Fig. 3. A detailed analysis of the results obtained using this particular method for differential cDNA library construction is presented elsewhere (25).

First-Strand cDNA Synthesis and Purification

Swiss 3T3 cells are grown to confluence in DMEM supplemented with 10% fetal bovine serum. The cells are harvested by lysis in 4 M guanidine thiocyanate and poly(A)$^+$ mRNA prepared using standard methods (26). Thirty micrograms of Swiss 3T3 poly(A)$^+$ mRNA dissolved in 120 μl water is heated to 70°C for 2 min, then placed on ice. Hexamer-primed first-strand cDNA synthesis using Moloney reverse transcriptase is performed by adding reagents in the following order: 60 μl of 5× RT buffer [provided with Super-Script reverse transcriptase; (Bethesda Research Laboratories (BRL), Gaithersburg, MD], 30 μl of 0.1 M DTT, 5 μl random hexamer primer (1 μg/μl), 60 μl dGTP, dATP, dCTP, and dTTP (2.5 mM in each dNTP), 5 μl RNasin [30 units (U)/μl; Promega, Madison, WI], 5 μl [α-^{32}P]dCTP (3000 Ci/mmol), and 15 μl SuperScript reverse transcriptase (200 U/μl). The reaction is mixed gently and incubated at 42°C for 60 min.

The RNA template for cDNA synthesis is eliminated by alkaline hydrolysis. Thus, 18 μl of 0.5 M EDTA and 12 μl of 10 N NaOH are added to the reverse transcription reaction, and the sample is incubated at 70°C for 30 min. After RNA hydrolysis, 100 μl of 2 M Tris, pH 7.4, is added and mixed, followed by addition of 60 μl of 2 N HCl and mixing. The remaining cDNA is purified by phenol–chloroform extraction and subsequent chloroform extraction. High molecular weight cDNA is purified from smaller species and nucleotides using push column chromatography (Primerase column, Stratagene, La Jolla, CA). After the push column, cDNA is precipitated by adding 0.1 volume of 5 M NaCl and 2.5 volumes ethanol, incubating on ice for 30 min, followed by centrifugation in a microcentrifuge for 15 min. The pellet is rinsed in 80% ethanol, microfuged for 2 min, dried in a vacuum centrifuge, and resuspended in 10 μl of water. About 10 μg of labeled first strand cDNA is recovered.

Subtractive Hybridization, and Purification of cDNA from cDNA–mRNA Hybrids by Hydroxyapatite Chromatography

For subtractive hybridization, a large excess of Balb/3T3 mRNA (subtractive driver) is annealed to the Swiss 3T3 cDNA at high concentration for many hours, to allow RNA–cDNA hybrids to form for all genes expressed by both fibroblast cell lines. After hybridization, only Swiss 3T3 cDNAs from genes not expressed by Balb 3T3 cells would remain single-stranded. Balb/3T3 poly(A)$^+$ mRNA is prepared from confluent Balb/3T3 fibroblasts (about 50 150-cm^2 flasks) using standard methods (26).

To initiate subtractive hybridization, the following solutions are added to the 10 μl of Swiss 3T3 cDNA: 65 μl Balb/3T3 poly(A)$^+$ RNA (4 μg/μl), 20 μl of 5× hybridization buffer (3 M NaCl, 100 mM Tris, pH 7.4, 10 mM EDTA), and 5 μl of 10% SDS. The hybridization reaction is overlayered with mineral oil, heated to 95°C for 5 min, and incubated at 70°C for 20 hr. Swiss 3T3 cDNA not hybridized to mRNA is then purified by hydroxyapatite (HAP) chromatography.

About 10 g HAP powder (HTP, Bio-Rad, Richmond, CA) is mixed with 50 ml of 50 mM phosphate, pH 7.0, 150 mM NaCl. The powder is allowed to settle, and the fines are removed by decanting the overlaying buffer. Fifty milliliters of 50 mM phosphate, pH 7.0, 150 mM NaCl is added again, and the mixture is incubated in boiling water for 15 min. The fines are removed as before, the slurry suspended in 50 ml of 50 mM phosphate, pH 7.0, 150 mM NaCl, and the mixture stored at 4°C for as long as several months. For optimal separation, HAP chromatography must be performed at 60°C, either in a water-jacketed column or in a homemade system consisting of 3-ml plastic syringes immersed in a 60°C water bath (25).

After hybridization, mineral oil is removed and the aqueous phase mixed with 2 ml HAP mixture prepared as described. The sample–HAP mixture is incubated at 60°C for 10 min with occasional mixing, then pipetted into the column equilibrated at 60°C. Solutions containing no more than 0.1 M phosphate, pH 7.0, elute single-stranded Swiss 3T3 cDNA but are not sufficient to elute cDNA–RNA hybrids from the HAP resin. Thus, the HAP matrix is packed by positive pressure, and the flow-through containing unhybridized cDNA is collected and saved. Additional unhybridized Swiss 3T3 cDNA is recovered by washing the packed HAP column 6 times at 60°C with 1 ml of 0.1 M phosphate, pH 7.0, 150 mM NaCl forced through the matrix using gentle positive pressure. The cDNA–RNA hybrids are then eluted by washing the column as before, but using a buffer containing higher concentrations of phosphate (0.45 M phosphate, pH 7.0, 150 mM NaCl) at

60°C. Less than 2% of the input counts remained bound to the HAP after the high phosphate washes.

Fractions containing single-stranded cDNA are applied directly to a fresh HAP column as before to remove any residual cDNA–RNA hybrids. Single-stranded cDNA is eluted a second time with six washes using 1 ml of 0.1 M phosphate, pH 7.0, 150 mM NaCl. About 3% of the original cDNA (300 ng) is eluted in the low phosphate washes (unhybridized cDNA) after the second HAP column. The remaining cDNA (about 9.7 μg) is found in the high phosphate washes, forming cDNA–RNA hybrids. Ten micrograms of Balb/3T3 carrier RNA is added to the 300 ng unhybridized cDNA. In preparations for the second round of subtractive hybridization, the unhybridized cDNA sample (present in the low phosphate washes) is desalted and concentrated using a Nensorb 20 column as recommended by the manufacturer (Du Pont–NEN, Boston, MA). After binding to the Nensorb 20 column, cDNA is eluted from the column using 50% ethanol and dried by vacuum centrifugation.

After drying, an additional 250 μg of Balb/3T3 poly(A)$^+$ mRNA is added to the unhybridized cDNA. The subtractive hybridization and HAP purification of unhybridized cDNA are repeated a second time. At the end of the second subtractive hybridization, 0.5% (50 ng) of the original cDNA is present as unhybridized cDNA. After the addition of 10 μg of Balb/3T3 carrier RNA, the twice subtracted cDNA is purified and desalted using Nensorb chromatography as before, then dried by vacuum centrifugation.

Hybridization of Subtracted cDNA to Swiss 3T3 mRNA to form RNA–cDNA Hybrids for Second-Strand cDNA Synthesis

Fifty micrograms of Swiss 3T3 poly(A)$^+$ mRNA is added to the 50 ng subtracted cDNA, and the sample is dried in a vacuum centrifuge. RNA and cDNA are dissolved in 15 μl water. Four microliters of 5× hybridization buffer and 1 μl of 10% SDS are added, forming the hybridization reaction. The reaction is overlayered with 50 μl mineral oil, heated to 95°C, and incubated at 70°C for 20 hr to form cDNA–RNA duplexes. HAP chromatography (performed as described previously) is used to purify the hybrids, which elute selectively with the *high* phosphate wash (0.45 M phosphate, pH 7.0, 150 mM NaCl). After HAP elution, the hybrids are desalted and concentrated using Nensorb chromatography as before, then dried by vacuum centrifugation.

Synthesis of Second-Strand cDNA by Nick Translation Using DNA Polymerase I and RNase H

The following solutions are added to the dried pellet containing Swiss 3T3 RNA–cDNA duplexes: 25 μl of 2× second-strand buffer (40 mM Tris, pH 7.4, 10 mM MgCl$_2$, 20 mM ammonium sulfate, 200 mM KCl, 100 μg/ml nuclease-free bovine serum albumin, and dGTP, dATP, dCTP, and dTTP, 1 mM in each dNTP), 5 μl [α-^{32}P]dCTP (3000 Ci/mmol), 15 μl water, 1μl RNase H (2 U/μl, BRL), 4 μl DNA polymerase I (5 U/μl, Boehringer-Mannheim). The reaction is incubated at 14°C for 2 hr. The reaction is terminated by adding 2 μl of 0.5 M EDTA, 5 μl of 5 M NaCl, and 10 μg of Balb/3T3 RNA carrier. After terminating the reaction, nucleic acids are purified by phenol–chloroform extraction, chloroform extraction, and ethanol precipitation. The pellet is rinsed with 80% ethanol, then dried by vacuum centrifugation.

Polishing and Phosphorylation of cDNA Termini, Followed by Adapter Ligation

The following solutions were added to the dried pellet: 5 μl of 10× T4 polymerase buffer (700 mM Tris, pH 7.7, 100 mM MgCl$_2$, 50 mM DTT), 10 μl dGTP, dATP, dTTP, and dCTP, each dNTP at 2.5 mM, 5 μl of 10 mM rATP, 26 μl water, 2 μl T4 DNA polymerase (5 U/μl, BRL), and 2 μl T4 polynucleotide kinase (5 U/μl, Pharmacia). The reaction is incubated at 37°C for 30 min. The reaction is terminated by adding 2 μl of 0.5 M EDTA, 5 μl of 5 M NaCl. Nucleic acids are purified by phenol–chloroform extraction, chloroform extraction, and ethanol precipitation. The pellet is rinsed with 80% ethanol, then dried by vacuum centrifugation.

The amplification adapter is prepared by mixing together 5 μg each of two synthetic oligonucleotides (5'-OH AGCTAGAATTCGGTACCGTCGACC 3'-OH, 5'-p GGTCGACGGTACCGAATTCT 3'-OH) in 18 μl water and 2 μl of 10× T4 polymerase buffer. The mixture is incubated at 95°C for 5 min, then cooled gradually to room temperature. The adapter is stored at −20°C until needed.

The polished cDNA pellet is resuspended in 6 μl water, after which 1 μl of amplification adapter, 2 μl of 5× T4 ligase buffer (BRL), and 1 μl T4 ligase (BRL) are added. The reaction is incubated at 4°C for 16 hr. The reaction is terminated by heat-inactivating the T4 ligase at 72°C for 10 min.

Adaptor-ligated cDNA fragments are separated from adaptor monomers and dimers by Sepharose CL-4B chromatography. A 4-ml column is packed and equilibrated with TE (10 mM Tris pH 7.4, 0.1 mM EDTA). For calibration and sample chromatography, 200-μl aliquots are collected. The column is calibrated with blue dextran (which runs with the void volume) and [α-^{32}P]dCTP (which runs in the included volume). The first third of the fractions between the void fraction and the included fraction contains adaptor-ligated cDNA fragments, but not adapters or ligated adaptor dimers. These fractions are pooled, after which 10 μg Balb/3T3 RNA carrier, 1/10 volume of 5 M NaCl, and 2.5 volumes ethanol are added. The adaptor-ligated cDNA is precipitated by incubation on ice for 30 min, followed by centrifugation in a microfuge for 15 min at 4°C. The adaptor-ligated cDNA pellet is washed with 80% ethanol and resuspended in 20 μl water. The sample is stored at −20°C until needed for polymerase chain reaction (PCR) amplification.

PCR Amplification of Adapter-Ligated Subtracted cDNA

Adaptor-ligated cDNA is used as template for PCR amplification. In a 450-μl tube combine 5 μl adaptor-ligated subtracted cDNA, 10 μl of 10× PCR buffer [100 mM Tris-HCl, pH 8.3, 500 mM KCl, 15 mM MgCl$_2$, 0.01% (w/v) gelatin], 16 μl of 1.25 mM dGTP, dATP, dCTP, and dTTP, 1 μl amplification oligonucleotide (5′-OH AGCTAGAATTCGGTACCGTC-GACC 3′-OH, 0.5 μg/μl in water), 67 μl water, 1 μl Taq DNA polymerase (Ampli-Taq, United States Biochemical, Cleveland, OH). The reaction is overlayered with 50 μl mineral oil and cycled in a Perkin-Elmer (Norwalk, CT) thermal cycler for 30 cycles (94°C, 1 min; 50°C, 1 min; 72°C, 2 min). Ten microliters of the amplified subtracted cDNA product is analyzed by agarose gel electrophoresis. Several micrograms of amplified cDNA is observed in the gel, forming a heterogeneous smear of fragments ranging from 300 to 800 bases in length.

Extensive amplification of a heterogeneous sample (such as subtracted cDNA) can result in some terminal cycles where some component of the amplification reaction has been exhausted (amplification oligonucleotide, Taq polymerase activity, etc.) When this occurs, denaturation is not followed by quantitative annealing of amplification oligonucleotide and subsequent copying of the sample to form duplex product molecules. A PCR chase reaction is performed to ensure that the subtracted cDNA PCR product is predominantly duplex DNA before EcoRI digestion and cloning. For this reaction, 30 μl amplified subtracted cDNA product, 7 μl of 10× PCR buffer, 16 μl of 1.25 mM dNTPs, 2 μl amplification oligonucleotide (0.5 μg/μl), 43 μl water, and 2 μl Taq DNA polymerase are combined. The reaction is incubated in the thermal cycler at 94°C for 2 min, 50°C for 2 min, 72°C for 5 min, 50°C for 2 min, and finally 72°C for 10 min. The chased DNA is purified by phenol–chloroform extraction, followed by chloroform extraction. Ten microliters of 5 M NaCl and 250 μl ethanol are added, and the product is concentrated by ethanol precipitation. After rinsing the pellet with 80% ethanol, the pellet is dried by vacuum centrifugation and resuspended in 20 μl water.

Cloning Amplified Subtracted cDNA into λ-gt10 to Generate Subtracted Library

The chased cDNA is digested with 200 U of *Eco*RI at 37°C for 1 hr as recommended by the supplier (BRL). The reaction is terminated by phenol–chloroform extraction followed by ethanol precipitation. The pellet is resuspended in 20 μl water, then electrophoresed on a 1% agarose gel (FMC Bioproducts, Rockland, ME Seakem GTG grade agarose). Fragments larger than 300 bases are either electroeluted from conventional agarose gels using standard methods (26) or purified from low melting point agarose gels using the Magic PCR Preps DNA purification system (Promega). After purification, 1 μl yeast tRNA carrier (10 μg/μl) and 1/10 volume of 5 M NaCl are added to the eluate, and cDNA is purified and concentrated by phenol–chloroform extraction, chloroform extraction, and ethanol precipitation. The pellet is washed with 80% ethanol, dried by vacuum centrifugation, and resuspended in 50 μl water. The *Eco*RI-digested cDNA is ligated into *Eco*RI-digested, calf intestinal alkaline phosphatase (CIAP)-treated, λ-gt10 vector (Stratagene), packaged *in vitro*, and plated on c600 hfl host bacteria as recommended by the supplier (Stratagene) to form the subtracted cDNA library.

Screening cDNA Library with Long "Best Guess" Oligonucleotide Probe

An oligonucleotide probe is used to select GRP-R clones from the subtracted cDNA library. The strategy for probe design involves the synthesis of a relatively long oligonucleotide (44 bases) where a best guess of the DNA sequence is based on codon usage frequency data (27). A 15-amino acid sequence from the GRP-R tryptic peptide T56/59 is used to predict the sequence of the sense and antisense strand of cDNA encoding the protein (Fig. 4). The antisense strand oligonucleotide is synthesized for use as the probe. For hybridization, the probe is labeled with [α-^{32}P]ATP and T4 polynucleotide kinase (probe specific activity ~10^8 cpm/μg), and purified from unincorporated [γ^{32}P]ATP by phenol–chloroform extraction and ethanol precipitation. The probe pellet is resuspended at 10^6 cpm/μl and stored at $-20°C$ until needed.

The subtracted cDNA library is screened essentially as described (28). For screening, the library is plated on 10 dishes at 10^4 plaques per 150-mm dish. After plaque formation, the bacteriophage are transferred and immobilized onto nitrocellulose filters (Schleicher and Schuell, Keene, NH) using established methods (26). Filters are prehybridized for several hours at 42°C in filter hybridization buffer [20% formamide, 6× SSC (1× SSC is 0.15 M NaCl, 15 mM sodium citrate, pH 7.0), 50 mM sodium phosphate, pH 6.8, 0.1% (w/v) bovine serum albumin (Sigma), 0.1% (w/v) Ficoll (Sigma), 0.1% (w/v) polyvinylpyrrolidone (Sigma), 100 μg/ml (w/v) boiled sonicated herring sperm DNA, and 10% (w/v) dextran sulfate]. After prehybridization, labeled oligonucleotide (10^6 cpm/ml hybridization buffer) is added, then filters are hybridized at 42°C for 12 hr. Filters are washed twice in 2× SSC, 0.1% SDS at room temperature for 15 min, followed by two washes in 0.2× SSC, 0.1% SDS at 37°C for 15 min. The filters are air-dried and exposed to Kodak (Rochester, NY) X-OMAT film for 48 hr using Cronex intensifying screens (Du Pont, Wilmington, DE).

Positive plaques are identified and purified by successive rounds of dilution, plating, and rescreening with the probe until the plaques are pure. Bacteriophage DNA is prepared, digested with *Eco*RI, and subcloned into plasmid vectors using established methods (26). Plasmid miniprep DNA is prepared (26), and the cDNA insert sequences are determined using a commercially available kit (Sequenase, United States Biochemical).

Two of the five positive clones we obtained contained overlapping DNA sequence from the same gene, which encoded the 15-amino acid sequence used for probe design. These clones are partial GRP-R cDNAs, with an open reading frame that matches the GRP-R amino acid sequence perfectly (15/15 amino acids) and the best guess probe sequence in 37 of 44 positions (85% identity) (Fig. 4). The other three clones came from three different mRNAs,

```
               MetAlaSerPheLeuValPheTyrValIleProLeuAlaIleIle
GRP-R:  5'ATGGCTTCCTTTCTGGTTTTCTACGTTATCCCACTGGCGATCAT 3' sense


        5'ATGGCcTCCTTcCTGGTcTTCTAtGTgATCCCcCTGGCcATCAT 3' sense
Guess:
        3'TACCGgAGGAAgGACCAgAAGATaCAcTAGGGgGACCGgTAGTA 5' a-sense
```

FIG. 4 Design of a long best guess oligonucleotide probe based on the amino acid sequence of a purified GRP-R tryptic fragment (T56/59). The amino acid sequence is shown, with the cDNA sequence found in GRP-R cDNA clones immediately below. Sense and antisense best guess oligonucleotide probes are below the sequence found in isolated clones, with differences between the clone and guess sequences indicated by lowercase letters.

none of which encodes an open reading frame matching the GRP-R peptide sequence. However, all three unrelated clones contain a nucleotide sequence showing greater than 75% identity with the oligonucleotide probe. These results underscore the advantages of screening libraries enriched for rare cDNAs (i.e., Swiss+/Balb− subtracted cDNA library) when isolating cDNA clones with best guess oligonucleotide probes.

Isolation of cDNA Clones Encoding Complete Gastrin-Releasing Peptide Receptor

The partial GRP-R cDNA clones isolated from the subtracted library encode a long open reading frame extending beyond several termination codons in the 3′ direction but not extending to the initiator methionine codon in the 5′ direction. We use gene-specific primer-directed cDNA library construction to enrich for cDNA clones that contain the complete open reading frame of the Swiss 3T3 GRP-R. An antisense synthetic oligonucleotide primer complementary to mRNA sequences located about 170 bases 3′ to the termination codon is synthesized. This primer, instead of oligo(dT), is used to prime reverse transcription of Swiss 3T3 mRNA as the first step in cDNA library construction.

The gene-specific primer offers two advantages over the more conventional oligo(dT) primer generally used for first-strand cDNA synthesis. RNA blot analysis using partial GRP-R cDNA probes indicated that one of the GRP-R mRNA was rather large (7.2 kb). Very long first-strand cDNA reverse transcripts would be needed to include the open reading frame of the protein if an oligo(dT) primer initiating at the 3′ end of the mRNA in the poly A tail were used. The frequency of clones of this length in libraries is very low, particularly for a relatively rare mRNA such as that for the GRP-R. In contrast, the gene-specific oligonucleotide should initiate first-strand cDNA synthesis immediately 3′ to the termination codon, so that a minimal length reverse transcript (1.4 kb) would be needed to capture the full length of the protein open reading frame. In addition, the gene-specific primer should only anneal to a small subset of sequences in Swiss 3T3 mRNA including the GRP-R, providing further enrichment for GRP-R clones in the resulting cDNA library.

First-Strand cDNA Synthesis Using GRP Receptor Gene-Specific Primer

To initiate first-strand synthesis, 5 μg Swiss 3T3 poly(A)$^+$ and 200 ng gene-specific primer (5′ TACTTTGAGATACAATGG 3′) are combined in 11 μl water, incubated at 70°C for 2 min, and placed on ice. To this primer–template mix are added the following solutions: 10 μl of 5× RT buffer [provided with BRL SuperScript enzyme], 5 μl of 0.1 M DTT, 20 μl dGTP, dATP, dTTP, and dCTP, 1.25 mM in each of the four dNTPs, 1 μl RNasin (Promega), and 3 μl SuperScript reverse transcriptase (200 U/μl, BRL). The reaction is incubated at 42°C for 1 hr. Two microliters of 0.5 M EDTA and 5 μl of 5 M NaCl are added to the sample to terminate the reaction. First-strand cDNA is purified by phenol–chloroform extraction, chloroform extraction, and ethanol precipitation. The pellet is rinsed with 80% ethanol, dried in a vacuum centrifuge, and resuspended in 25 μl water.

Second-Strand Synthesis

The following solutions are added to the 25 μl first-strand cDNA sample: 25 μl of 2× second-strand buffer (40 mM Tris, pH 7.4, 20 mM ammonium sulfate, 200 mM KCl, 10 mM MgCl$_2$, 100 μg/ml nuclease-free bovine serum albumin, and 1 mM in each of the four dNTPs), 2 μl [α-^{32}P]dCTP (3000 Ci/mmol), 1 μl RNase H (2 U/μl, BRL), and 4 μl DNA polymerase I (5 U/μl, Boehringer-Mannheim). The reaction is incubated for 2 hr at 14°C, then terminated by the addition of 2 μl of 0.5 M EDTA followed by phenol–chloroform extraction. The sample is desalted using a push column as recommended by the manufacturer (Primerase, Stratagene), and ethanol precipitated. The pellet is rinsed with 80% ethanol, and dried by vacuum centrifugation.

Polishing and Phosphorylation of cDNA Termini, Ligation of EcoRI Adaptor, and Forming λ-gt10 Library

The cDNA termini are polished and phosphorylated using T4 DNA polymerase and T4 polynucleotide kinase in the same reaction, as described previously after second-strand synthesis while preparing the subtracted cDNA library. Following this reaction, the cDNA is resuspended in 6 μl water. EcoRI-XmnI adaptors are prepared as recommended by the manufacturer (New England Biolabs, Beverly, MA). Alternatively, ready-to-use EcoRI–NotI adaptors are purchased (Invitrogen, San Diego, CA), and resuspended at 0.5 μg/μl. Adaptor ligation to cDNA is performed as follows: 6 μl polished cDNA, 1 μl EcoRI adaptors (0.5 μg/μl), 1 μl of 10× ligase buffer (500 mM Tris, pH 7.4, 70 mM MgCl$_2$, 10 mM DTT), 1 μl of 10 mM ATP, pH 7.0, and 1 μl T4 DNA ligase (5 U/μl, BRL) are combined and incubated at 4°C for 16 hr. The ligase is heat-inactivated by incubating the sample at

72°C for 10 min. The adaptor-ligated cDNA termini are phosphorylated by adding the following solutions to the heat-inactivated reaction: 2 μl of 10 mM ATP, pH 7.0, 1 μl of 10× T4 ligase buffer (Stratagene), 4 μl water, and 3 μl T4 polynucleotide kinase (10 U/μl, Pharmacia). The reaction is incubated at 37°C for 30 min, then heat-inactivated by incubation at 72°C for 10 min.

Adaptor-ligated cDNA fragments are separated from adaptor monomers and dimers by Sepharose CL-4B chromatography, using a 4-ml column as described previously for subtracted cDNA library preparation. Fractions containing relatively large adaptor-ligated, ^{32}P-labeled cDNA fragments (fractions immediately following the void volume) are pooled. Ten micrograms of yeast tRNA carrier is added to the pooled fractions, and cDNA is purified by sequential phenol–chloroform extraction and chloroform extraction. After extraction, 1/10 volume of 5 M NaCl and 2.5 volumes ethanol are added. Nucleic acids are precipitated for 30 min on ice, followed by microcentrifugation for 15 min at 4°C. The cDNA pellet is washed with 80% ethanol and resuspended in 10 μl water. EcoRI-adapted cDNA insert is ligated into

```
                                                            I
  1   MAPNNCSHLN LDVDPFLSCN DTFNQSLSPP KMDNWFHPGF IYVIPAVYGL
                                            II
 51   IIVIGLIGNI TLIKIFCTVK SMRNVPNLFI SSLALGDLLL LVTCAPVDAS
                        III
101   KYLADRWLFG RIGCKLIPFI QLTSVGVSVF TLTALSADRY KAIVRPMDIQ
                  IV
151   ASHALMKICL KAALIWIVSM LLAIPEAVFS DLHPFHVKDT NQTFISCAPY
                                        V
201   PHSNELHPKI HSMASFLVFY VIPLAIISVY YYFIARNLIQ SAYNLPVEGN
                                            VI
251   IHVKKQIESR KRLAKTVLVF VGLFAFCWLP NHVIYLYRSY HYSEVDTSML
                VII
301   HFVTSICAHL LAFTNSCVNP FALYLLSKSF RKQFNTQLLC CQPGLMNRSH

351   STGRSTTCMT SFKSTNPSAT FSLINRNICH EGYV*
```

FIG. 5 Predicted amino acid sequence of the Swiss 3T3 GRP-R deduced from cDNA clones. The 384-amino acid sequence is shown, with putative transmembrane domains shaded and labeled I–VII. The tryptic peptide (T56/59) generating the probe used for screening is enclosed by a rectangle (amino acids 213–227). Other sequences determined by amino acid sequencing of purified tryptic fragments are also found in the deduced amino acid sequence, including T44 (XXPFIQLTSVGVSVFT, amino acids 118–131) and T50 (XVPNLFISXLALGXLLXXVT, amino acids 74–93). The amino-terminal sequence determined by Edman degradation (APNXXSXLNLDVDPFLS) is consistent with that deduced from cDNA sequence analysis (amino acids 2–18).

EcoRI–CIAP λ-gt10 vector, packaged *in vitro*, and plated on c600 hfl host bacteria as recommended by the supplier (Stratagene) to generate the cDNA library.

Screening cDNA Library

A ^{32}P-labeled GRP-R cDNA fragment probe is prepared from one of the initial GRP-R clones isolated from the subtracted library. The probe is used to screen 5×10^5 plaques from the gene-specific cDNA library, using well-established high-stringency wash and hybridization conditions (26). Positive clones are plaque purified and subcloned into the EcoRI site of the pGEM 4 plasmid vector (Promega). Of ten positive clones we obtained, the two containing the longest inserts (1.7 kb long) were selected for nucleotide sequence analysis and functional characterization.

DNA sequence analysis of both clones reveals a single long open reading frame predicted to encode a 384-amino acid polypeptide with a predicted molecular weight of 43,000 (5), containing seven hydrophobic transmembrane domains typically found in G-protein-coupled receptors. The predicted amino acid sequence is shown in Fig. 5, indicating the location of hydrophobic transmembrane segments as well as the location of the peptide sequence (in transmembrane domain V) used to design the original oligonucleotide probe.

References

1. A. Anastasi, V. Erspamer, and M. Bucci, *Experientia* **27,** 166 (1971).
2. T. J. McDonald, H. Jornvall, G. Nilsson, M. Vagne, M. Ghatei, S. R. Bloom, and V. Mutt, *Biochem. Biophys. Res. Commun.* **90,** 227 (1979).
3. N. Minamino, K. Kangawa, and H. Matsuo, *Biochem. Biophys. Res. Commun.* **144,** 541 (1983).
4. T. von Schrenck, P. Heinz-Erian, T. Moran, S. A. Mantey, J. D. Gardner, and R. T. Jensen, *Am. J. Physiol.* **256,** G747 (1989).
5. J. F. Battey, J. M. Way, M. H. Corjay, H. Shapira, K. Kusano, R. Harkins, J. Wu, T. Slattery, E. Mann, and R. Feldman, *Proc. Natl. Acad. Sci. U.S.A.* **88,** 395 (1991).
6. E. R. Spindel, E. Giladi, P. Brehm, R. H. Goodman, and T. P. Segerson, *Mol. Endocrinol.* **4,** 1956 (1990).
7. E. Wada, J. M. Way, H. Shapira, K. Kusano, A.-M. Lebacq-Verheyden, D. Coy, R. T. Jensen, and J. F. Battey, *Neuron* **6,** 421 (1991).
8. M. H. Corjay, D. J. Dobrzanski, J. M. Way, J. Viallet, H. Shapira, P. Worland, E. A. Sausville, and J. F. Battey, *J. Biol. Chem.* **266,** 18771 (1991).
9. A.-M. Lebacq-Verheyden, J. Trepel, E. A. Sausville, and J. F. Battey, *in* "Pep-

tide Growth Factors and Their Receptors II,'' Handbook of Experimental Pharmacology, Vol. 95/II (M. Sporn and A. B. Roberts, eds.), p. 71. Springer-Verlag, Berlin and Heidelberg, 1990.

10. E. Rozengurt and J. Sinnett-Smith, *Proc. Natl. Acad. Sci. U.S.A.* **80,** 2936 (1983).

11. I. Zachary and E. Rozengurt, *Proc. Natl. Acad. Sci. U.S.A.* **82,** 7616 (1985).

12. I. Zachary, P. J. Woll, and E. Rozengurt, *Dev. Biol.* **124,** 295 (1987).

13. J. C. Willey, J. F. Lechner, and C. C. Harris, *Exp. Cell Res.* **153,** 245 (1984).

14. D. N. Carney, F. Cuttitta, T. W. Moody, and J. D. Minna, *Cancer Res.* **47,** 821 (1987).

15. S. Weber, J. E. Zukerman, D. G. Bostwick, K. G. Bensch, B. I. Sikic, and T. A. Raffin, *J. Clin. Invest.* **75,** 306 (1985).

16. F. Cuttitta, D. N. Carney, J. Mulshine, T. W. Moody, J. Fedorko, A. Fischler, and J. D. Minna, *Nature (London)* **316,** 823 (1985).

17. R. M. Kris, R. Hazan, J. Villines, T. W. Moody, and J. Schlessinger, *J. Biol. Chem.* **262,** 11215 (1987).

18. I. Zachary and E. Rozengurt, *J. Biol. Chem.* **262,** 3947 (1987).

19. R. I. Feldman, J. M. Wu, J. C. Jenson, and E. Mann, *J. Biol. Chem.* **265,** 17364 (1990).

20. G. J. Todaro and H. Green, *J. Cell Biol.* **17,** 299 (1963).

21. G. Scatchard, *Ann N.Y. Acad. Sci.* **51,** 660 (1949).

22. J. B. Fischer and A. Schonbrunn, *J. Biol. Chem.* **263,** 2808 (1988).

23. A. G. Gilman, *Annu. Rev. Biochem.* **56,** 615 (1987).

24. R. M. Hewick, M. W. Hunkapiller, L. E. Hood, and W. J. Dreyer, *J. Biol. Chem.* **256,** 7990 (1981).

25. C. Timblin, J. F. Battey, and W. M. Kuehl, *Nucleic Acids Res.* **18,** 1587 (1990).

26. L. Davis, M. Dibner, and J. F. Battey, ''Basic Methods in Molecular Biology.'' Elsevier, New York, 1986.

27. R. J. Lathe, *J. Mol. Biol.* **183,** 1 (1985).

28. W. I. Wood, *in* ''Methods in Enzymology'' (S. L. Berger and A. R. Kimmel, eds.), Vol. 152, p. 443. Academic Press, San Diego, 1987.

[7]   Expression and Characterization of
       Recombinant Nerve Growth Factor
       Receptors: Selection of Recombinant
       Baculoviruses and Membrane
       Binding Assays

Barbara L. Hempstead, Dorota K. Poluha,
Nancy E. Crosbie, and Alonzo H. Ross

Introduction

Two major nerve growth factor receptors (NGFR) have been identified (for review, see Ref. 1). The first is known as the low-affinity NGFR or gp75^{NGFR}. It is a 75,000-Da protein containing a signal peptide, an extracellular domain with a single N-linked glycosylation site, four homologous cysteine-rich internal repeats, a single transmembrane segment, and a 155-amino acid cytoplasmic domain (2). The second NGFR is a 140,000-Da tyrosine kinase (gp140^{trk}) encoded by the *trk* protooncogene (3–5). Despite some controversy (5), it appears that both gp75^{NGFR} and gp140^{trk} are required for high-affinity binding of NGF (6) and for regulation of Trk tyrosine kinase activity (7).

Two current questions in this field are (1) What are the structures of these receptors, and how do they bind NGF? (2) How do gp75^{NGFR} and gp140^{trk} interact to form a functional receptor? To facilitate structural studies of the NGFR, we expressed the extracellular domain of the gp75^{NGFR} in the baculovirus–insect cell system (8). In this chapter, we describe a simple method for the selection of recombinant baculoviruses. To answer the second question, recombinant NGFRs have been expressed in a variety of cell types (5, 6, 9–11). We describe an NGF binding assay to determine whether the recombinant proteins reconstitute high-affinity NGF binding.

Expression of Soluble Extracellular Domains of Nerve Growth Factor Receptors

The Baculovirus–Insect Cell Expression System

The baculovirus–insect cell system is popular because of the excellent yields of properly folded recombinant proteins (12, 13). This system takes advantage

Methods in Neurosciences, Volume 12

of the dual life cycle of the baculovirus. Early in the infection, virus buds from the infected cells. In the later stages of infection, occlusions are produced consisting of viral cores surrounded by a coat protein, polyhedrin. In the wild, they persist after the death of the infected insect, are ingested by other insects, and thereby propagate the virus. In the laboratory, these occlusions are not required. Variant viruses which express a recombinant protein instead of polyhedrin are generated by homologous recombination.

Preparation and Selection of Recombinant Viruses

To prepare recombinant viruses, the cDNA for the desired protein is inserted into a shuttle vector between the polyhedrin promoter and the 3' region of the polyhedrin gene (for review of the available vectors, see Ref. (12). To optimize the levels of expression, the cDNA should include as little noncoding sequence as possible. Insect cells are cotransfected with the shuttle vector and wild-type baculovirus DNA. By homologous recombination, a few of the resulting viruses lack the polyhedrin gene and instead include the cDNA encoding the recombinant protein.

The critical step for this procedure is the selection of the recombinant virus. The classic method is a plaque assay in which viruses which do not express polyhedrin and, hence, do not form occlusions are visually selected (13). A monolayer of Sf9 cells is infected with virus and then overlayed with agarose. Since occlusions are visible by light microscopy, the resulting recombinant plaques are clearly distinguishable by microscopy from wild-type occlusion-containing plaques. Although this procedure becomes easier with experience, it sometimes is difficult to distinguish recombinant plaques from clumps of uninfected cells. Some investigators have found staining the plaques with neutral red (S. Meyer, personal communication, 1991) or 3-(4,5-dimethylthiazol-2-yl)-2,5-diphenyltetrazolium bromide (14) helps to select recombinant plaques.

We prefer to use the limiting dilution procedure of Fung *et al.* (15).

Sf9 cells seeded in a 96-well plate are infected with a series of dilutions of the virus-containing cotransfection culture supernatant. DNA samples from the infected cells are spotted on a membrane and analyzed by hybridization with the cDNA insert (see Fig. 1). This method very clearly distinguishes wells with and without recombinant virus. At greater dilutions of the viral inoculum, fewer positive wells are detected, but the hybridization frequently is more intense because there are fewer wild-type viruses competing with recombinant viruses to infect the insect cells. The positive wells which were infected with the most dilute inoculae and which display the smallest number of occlusions are selected. The resulting viral inoculae are used for a new

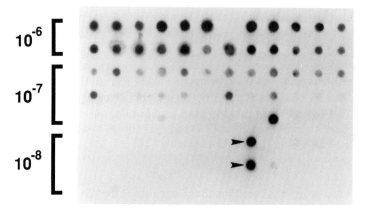

FIG. 1   Detection of recombinant baculoviruses by hybridization. In the third round of purification of a recombinant baculovirus encoding the intracellular domain of gp75^{NGFR}, Sf9 cells in a 96-well plate were infected with viral inoculum diluted 10^{-6}, 10^{-7}, or 10^{-8}. DNA from these wells was spotted on Duralose and hybridized with a gp75^{NGFR} probe. Two wells (arrowheads) from the 10^{-8} dilution were strongly positive and did not contain any wild-type occlusions. The viruses from these wells were pure.

round of infections, and, usually, a pure recombinant virus is obtained after three rounds of infections. To avoid isolating nonexpresser viruses, a secondary screen for protein expression is very helpful, if not essential. We have had some cDNAs which consistently resulted in nonexpresser viruses. To screen for deletions and rearrangements, polymerase chain reaction (PCR) amplification of the resulting viral DNA may be useful (16). Because the limiting dilution method of virus selection is much less laborious than the plaque procedure, it is possible to purify simultaneously several viruses and then choose the virus with the greatest production of recombinant protein.

The detailed protocol for the limiting dilution selection of baculovirus is as follows.

1. In each well of a 96-well cell culture plate, seed 1.5×10^4 cells in 100 μl of medium. After 1 hr, confirm the attachment of the cells with an inverted microscope. Cells should not be confluent, since they will grow during the incubation period.

2. Make 10-fold serial dilutions of the cotransfection supernatant using complete insect cell medium. For the first round of purification, the optimum dilutions generally are between 10^{-4} and 10^{-7}.

3. Gently remove the medium from the wells, add the dilutions of the

cotransfection supernatant, and return the plate to the 27°C incubator. The first occlusions are visible after 3 days, and by 5–8 days infections resulting from a single virus are evident. Longer incubations are not useful because uninfected cells overgrow the wells and die following depletion of the medium. Mark the wells with infected cells and relatively few occlusions.

4. Remove the virus-containing culture supernatants to a duplicate 96-well plate and store at 4°C. Lyse the cells by adding 50 μl of 0.5 N NaOH to each well and mixing. Neutralize the resulting lysates with 5 μl per well of 10 M ammonium acetate. Soak a Duralose membrane (Stratagene, La Jolla, CA) in 1 M ammonium acetate, 0.02 N NaOH. With suction, apply the lysates to the membrane in a slot–blot apparatus. Wash the membrane briefly with 4× SSC (1 × SSC: 150 mM NaCl, 15 mM sodium citrate, pH 7.0) and irradiate with UV light to irreversibly bind the DNA to the membrane.

5. Probe the membrane with a cDNA probe using standard conditions. The resulting positive DNA samples are easily detected by exposing the autoradiogram for 0.5 to 2 hr (Fig. 1).

6. Choose the virus-containing supernatants from wells with the strongest hybridization signal and the fewest occlusions. Prepare dilutions from 10^{-6} to 10^{-8} for the next round of infection. Generally, three rounds of selection are sufficient to obtain pure recombinant virus. After the second round of selection, it is usually possible to detect recombinant protein by Western blotting.

Production and Isolation of NGF Receptor Extracellular Domain

Once the recombinant virus is available, the production of large amounts of recombinant protein is relatively straightforward. The insect cells grow well in suspension or on plastic substratum. Although the suspension cultures reduce the cost of disposable plastics, we find that the increased chance of contamination and the difficulty in keeping the culture growing optimally negate this advantage. We have more reproducible yields of recombinant protein by infecting adherent Sf9 cells in 150-cm^2 tissue culture flasks.

The details of our procedure for the large-scale infection of Sf9 cells with the recombinant virus encoding the extracellular domain of gp75^{NGFR} are as follows.

1. To a confluent 150-cm^2 flask of Sf9 cells (~2.5 × 10^7 cells), add 4 ml of viral inoculum (for preparation, see Ref. 13) and gently rock the flasks at room temperature for 3–5 hr. Remove the inoculum, replace with 25 ml of fresh medium, and return the flask to the 27°C incubator. It is not necessary to have a separate chamber for infected cultures. We never have any problems with viruses "jumping" from one flask to another.

2. The infection can be followed by morphology. An early effect of infection is cessation of cell proliferation. Next, the nuclei become disorganized lacking clear nucleoli, the nuclear membrane becomes clearly visible, and the cells become larger. Late in the infection, some debris from lysed cells appears, but it is best to harvest the extracellular domain prior to large-scale cell lysis. We harvest the culture supernatant when most of the Sf9 cells lose their adherence to the plastic. Usually, this is on the fifth day of the infection. The supernatant is centrifuged for 15 min at 2000 g to eliminate cell debris. We then precipitate the culture supernatant protein with 60% ammonium sulfate. This is a convenient step, since the precipitated extracellular domain is quite stable at 4°C.

3. The extracellular domain is redissolved and purified by immunoaffinity chromatography and ion-exchange chromatography, as described by Vissavajjhala and Ross (8). Purification of recombinant proteins with an immunoaffinity column is extremely efficient. For proteins for which no antibody is available, it is now common practice to include a defined epitope at the N or C terminus (17).

An appealing alternative approach is to infect *Spodoptera frugiperda* larvae, which have been described as "low-cost protein factories" (18). Many investigators would not consider this possibility because of the necessity of setting up a breeding colony for moths. However, it is not that much work, if there is an experienced advisor available. The cost of such a breeding colony is remarkably modest compared to cell culture. We have infected larvae with the gp75[NGFR] extracellular domain virus and obtained recombinant protein; however, we have had problems with highly variable yields and have not been able to identify the source of the problem. Despite this difficulty, the low cost of the larvae makes this an important system for future development.

NGF Receptor Assays

Detection and quantitation of NGF receptors on ligand-responsive cells can be achieved by a variety of methods, including (a) immunological assays, using antisera specific for each NGF receptor subunit, (b) *in situ* analysis, using antisense RNA derived from each receptor cDNA, (c) binding assays, using radiolabeled ligand, and (d) functional assays, by assessing responses to NGF administration. Given the numerous recent reviews and papers on immunological assays (19), *in situ* hybridization (20), and functional assays (3, 5), this chapter focuses on binding assays.

The identification of specific cell surface receptors for NGF is an important advance in understanding the mechanism of action of NGF. Equilibrium binding analysis of radiolabeled NGF to the surface of intact, NGF-responsive cells has identified two receptor classes. The minority of binding sites (~6–10%) display high-affinity binding with a K_D of 10^{-11} M, whereas the majority display low-affinity binding with a K_D of 10^{-9} M. Initial binding studies (21) utilized intact cells dissociated from superior cervical and dorsal root ganglia. Comparable results have been obtained using the pheochromocytoma cell line PC12 (22, 23), and the majority of binding studies have used this cell line because of its relative ease of culture and cell homogeneity.

Several methods have been used to quantitate high- and low-affinity binding sites. Early studies utilized intact cells and allowed binding to proceed at physiological temperatures (37°C), conditions which favored the detection of low numbers of high-affinity sites (21). However, further analysis of NGF–receptor interactions indicated that NGF was rapidly internalized following binding to cell surface receptors at 37°C. With continued incubation, PC12 cells could accumulate high levels of NGF which would be inaccessible to competition with unlabeled ligand (24). As shown in Fig. 2, the amount

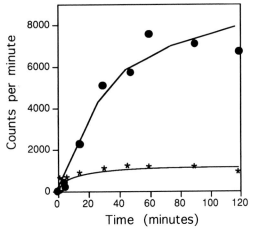

FIG. 2   Time course of surface binding and internalization of ^{125}I-labeled NGF. PC12 cells were incubated with 20 nM ^{125}I-labeled NGF in the presence or absence of 1 μM unlabeled NGF. Circles represent specific surface binding while stars represent specific internalized ligand, using the acid dissociation method of Haigler *et al.* [H. T. Haigler, F. R. Maxfield, M. C. Willingham, and I. Pastan, *J. Biol. Chem.* **255,** 1239 (1980)]. Each point represents the average of triplicate samples. Nonspecific binding (<20% of total binding) was subtracted from total binding to obtain the specific binding.

of internalized NGF exceeds that on the cell surface by 4-fold after 1 hr of incubation at 37°C. Given these results, binding studies performed at physiological temperatures may substantially overestimate the number of high-affinity binding sites.

One possible interpretation of these results is that high-affinity binding may simply represent inaccessible, internalized NGF. This would suggest that only one class of low-affinity receptors is represented on the surface of responsive cells (25). However, several studies convincingly argue against this interpretation. First, binding studies using intact cells, but performed at temperatures at which internalization is essentially blocked (2°C), disclose two binding classes, of high (K_D 10^{-11} M) and low (K_D 10^{-9} M) affinity (Table I). As predicted, the number of high-affinity sites is lower when binding studies are performed at low temperature (21). Second, studies using broken cell systems, where internalization would not occur, again reveal both high- and low-affinity sites (Fig. 3 and Table I). The greater number of both high- and low-affinity sites per cell is likely due to the presence of intracellular membranes (endoplasmic reticulum, Golgi) in addition to plasma membranes in these crude membrane preparations. Third, binding studies performed on detergent extracts of PC12 cells disclose both high- and low-affinity sites (26). These studies suggest that high-affinity sites are present on the cell surface and do not require ligand-mediated internalization or energy utilization.

The following binding methods utilize cell membrane preparations which eliminate the possibility of receptor internalization, yet allow the NGF–receptor reaction to proceed to steady-state conditions relatively rapidly by incubation at 30°C. In addition, this binding assay affords several technical advantages as compared to assays which utilize intact cells. First, membrane preparations are stable for months with appropriate storage, and therefore can be prepared in bulk. Second, these techniques can be readily modified and used in membrane fusion studies (27) to address questions of high-affinity

TABLE I   Equilibrium Binding Analysis of NGF Receptor Classes

Source of NGFR	Binding site 1[b]	Binding site 2[b]
Intact chick sensory neurons at 37°C[a]	K_D 2.3 × 10^{-11}	K_D 1.7 × 10^{-9}
	B_{max} 3000	B_{max} 45,000
Intact chick sensory neurons at 2°C[a]	K_D 9.8 × 10^{-12}	K_D 1.4 × 10^{-9}
	B_{max} 1000	B_{max} 45,000
Crude membranes from rat superior cervical ganglia	K_D 4 × 10^{-11}	K_D 5 × 10^{-9}
	B_{mas} 6000	B_{max} 65,000

[a] Data derived from Ref. 21.
[b] B_{max} represents the number of binding sites per cell.

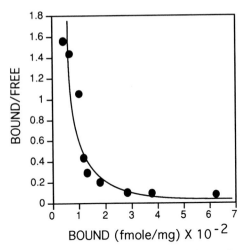

FIG. 3 Equilibrium binding analysis of NGF receptors in crude membrane preparations from rat superior cervical ganglia. Binding was performed as described in the text. Each binding reaction was carried out in triplicate in the presence or absence of unlabeled NGF, and specific binding represented greater than 70% of total binding.

binding site reconstitution using membranes expressing each receptor subunit independently.

Preparation of Radiolabeled NGF

Because NGF is a relatively unstable protein which binds to numerous surfaces, radiolabeling with iodine must be undertaken with care. The following method for radiolabeling of NGF using lactoperoxidase is adapted from Green and Greene (28) and results in biologically active ^{125}I-labeled NGF with a specific activity of 2800 to 3800 cpm/fmol (140–180 μCi/μg).

1. Add the following components to a final volume of 50 μl:
    10 μl of 1 mg/ml renin-free mouse 2.5 S NGF (Bioproducts for
    Science, Indianapolis, IN) in water
    1 mCi carrier-free Na^{125}I in 10 μl (>2000 Ci/mmol, Amersham,
    Arlington Heights, IL)
    6 μl of a 50 μg/ml lactoperoxidase solution (ICN, Costa Mesa,
    CA)
    25 μl of 0.017% (w/w) H$_2$O$_2$ (Sigma, St. Louis, MO) in 0.2 M
    sodium phosphate, pH 6.0

2. Incubate the mixture for 5 min at 22°C.

3. The reaction is halted by the addition of 50 μl of stop solution, consisting of 0.1 M NaI, 0.02% NaN$_3$, 0.1% bovine serum albumin (BSA) (Sigma 7888), 0.1% cytochrome c, 0.05% phenol red, and 0.3% acetic acid.

4. The incorporation of ^{125}I to NGF is determined by trichloroacetic acid (TCA) precipitation (2 μl of reaction mixture in 1 ml of 10% TCA); incorporation routinely ranges from 65 to 85%.

5. The ^{125}I-labeled NGF is separated from unincorporated ^{125}I by gel filtration using BioGel P-100 (Bio-Rad, Richmond, CA) equilibrated in 50 mM sodium phosphate, pH 7.5, 0.1 M NaCl, protamine sulfate (0.5 mg/ml, Sigma 4020), and BSA (1 mg/ml, Sigma 7888). A 0.7 × 25 cm column allows separation of radiolabeled NGF from unincorporated iodine, with 0.2-ml fractions being collected. The labeled NGF migrates at roughly half the volume of the phenol red indicator dye. A small shoulder migrating ahead of the major peak of NGF represents a small amount of iodinated lactoperoxidase, and these fractions are discarded.

6. Radioiodinated NGF is stored at 4°C and is used within 2 weeks.

Preparation of Cell Membranes

Crude membranes can be prepared from cultured cell lines as described by Hempstead *et al.* (9). This technique also is successful when using small (<1 g) quantities of fresh tissue.

1. Cells are washed in phosphate-buffered saline, 2 mM EDTA and suspended in 1 mM Tris, pH 8.0, 1 mM EDTA. Cells are broken by Polytron (Brinkmann, Westbury, NY) homogenization, setting 4, for 30 sec at 4°C.

2. Nuclei are removed by centrifugation at 2000 g for 5 min, and the crude membranes are pelleted by centrifugation at 40,000 g for 30 min.

3. By passage through a 25-gauge needle, membranes are resuspended to a final concentration of 5–10 mg/ml in 1 mM Tris, pH 8.0, 1 mM EDTA.

4. Membrane preparations can be used immediately or quick-frozen and stored at −70°C. Frozen preparations are stable for several months.

Equilibrium Binding of ^{125}I-Labeled NGF to Membrane Preparations

The following equilibrium binding technique is adapted from Hempstead *et al.* (9). The binding reaction proceeds to equilibrium within 60 min at 30°C. Radioimmunoassay grade polypropylene tubes (Falcon, Oxnard, CA, 2002) are used to minimize nonspecific binding of radiolabeled NGF.

1. Add the following to a final reaction volume of 100 μl:

> BSA (Sigma 7888), 1 mg/ml final concentration, in HEPES, pH
> 8.0, final concentration 50 mM
>
> ^{125}I-Labeled NGF, 0.02 to 10 ng
>
> 30 μg of cell membranes (diluted membranes are passed repetitively
> through a 25-gauge needle to assure a homogeneous suspension)

Unlabeled NGF (2 μg) is included in duplicate samples to assess nonspecific binding. Each reaction condition is determined in triplicate.

2. Reaction mixtures are incubated at 30°C with continual agitation for 60 min.

3. Binding is concluded by dilution of the reaction mixture with 1.5 ml of wash buffer (20 mM NaHPO$_4$, pH 7.4, 50 mM NaCl, 1 mg/ml BSA, and 1 mg/ml protamine sulfate) at 37°C. The inclusion of BSA and protamine sulfate in the wash solution minimizes nonspecific binding. Using a vacuum manifold, each reaction is filtered through Millipore (Bedford, MA) HVPL filters (0.45 μm) preequilibrated in wash buffer and washed three times with 10 ml of wash buffer in less than 60 sec. The large volume of wash buffer, as well as its temperature of 37°C, reduces nonspecific binding to less than 30% of total binding. In addition, the rapidity of washing is crucial, as NGF rapidly dissociates from the low-affinity receptor ($t_{1/2}$ <60 sec with membrane preparations).

Triplicate determinations, obtained in the presence or absence of unlabeled NGF, should reveal 60–90% specific binding. The data give rise to a saturation binding curve in which the receptor concentration is held constant and the concentration of ligand-occupied receptor can be determined at any ligand concentration. Therefore, an initial analysis of the data should be performed by plotting the NGF bound (y axis) versus the free NGF (total NGF added minus NGF bound, as the x axis). This binding curve should saturate at high ligand concentrations. A continual upward slope indicates that nonsaturating conditions exist; this usually is the result of an overabundance of receptors and can be corrected by using fewer cell membranes. Several techniques can be used to determine the four constants of interest (K_D and B_{max} of binding site 1, and K_D and B_{max} of binding site 2). Simply drawing a line through the top and bottom portions of the graph should not be used, as an overestimation of the number of high-affinity sites will occur. A more accurate method is to determine the limiting slopes for each end of the curve, and the four constants can then be determined mathematically (29). The LIGAND program (30) can be used to analyze the data by Scatchard analysis, using an iterative procedure to fit the data directly using a nonlinear least-squares program. For an accurate Scatchard analysis of a two-site curve,

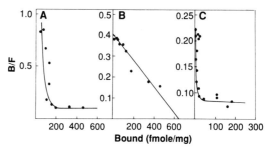

FIG. 4   Equilibrium binding analysis of NGF receptors from cell membrane preparations. Binding was performed as described in the text. Crude membranes were prepared from (A) PC12 cells, (B) 2_1 fibroblast cells, expressing the human low-affinity NGF receptor, and (3) NR1-1 cells, namely, NR18 cells transfected with the human low-affinity receptor. [Reprinted from Hempstead *et al.* (9).]

twelve or more experimentally derived points are preferred, particularly if there are only a small proportion of high-affinity sites.

A typical Scatchard plot of ^{125}I-labeled NGF binding to PC12 membranes is shown in Fig. 4A. This can be compared to a single site binding curve (Fig. 4B), generated using cells which express only one receptor subunit (i.e., 2_1 cells, fibroblasts expressing human low-affinity NGF receptors). Less than 2000 high-affinity sites can be reliably detected by this method, as shown in Fig. 4C using a PC12 mutant subclone, NR18 transfected with human low-affinity NGF receptors (NR1-1). These Scatchard analyses have been plotted using axes which demonstrate the high-affinity site. Additional data at higher ligand concentrations were replotted on an expanded plot to carefully quantitate low-affinity sites.

Acknowledgments

The authors would like to acknowledge the significant contributions of M. V. Chao and L. S. Schleifer in the development of the equilibrium binding technique described above. B.L.H. is supported by grants from the American Cancer Society, the March of Dimes, and the American Health Assistance Foundation. N.E.C. and A.H.R. are supported by grants from the National Institutes of Health, Cephalon Inc., and the Alzheimer's Association David Finkle Pilot Research Grant. D.K.P. is a postdoctoral fellow of the Muscular Dystrophy Association.

References

1. A. H. Ross, *Cell Regul.* **2,** 685 (1991).
2. D. Johnson, A. Lanahan, C. R. Buck, A. Sehgal, C. Morgan, E. Mercer, M. Bothwell, and M. Chao, *Cell (Cambridge, Mass.)* **47,** 545 (1986).

3. D. R. Kaplan, D. Martin-Zanca, and L. F. Parada, *Nature (London)* **350,** 158 (1991).
4. D. R. Kaplan, B. L. Hempstead, D. Martin-Zanca, M. V. Chao, and L. F. Parada, *Science* **252,** 554 (1991).
5. R. Klein, S. Jing, V. Nanduri, E. O'Rourke, and M. Barbacid, *Cell (Cambridge, Mass.)* **65,** 189 (1991).
6. B. L. Hempstead, D. Martin-Zanca, D. R. Kaplan, L. F. Parada, and M. V. Chao, *Nature (London)* **350,** 678 (1991).
7. M. M. Berg, D. W. Sternberg, B. L. Hempstead, and M. V. Chao, *Proc. Natl. Acad. Sci. U.S.A.* **88,** 7106 (1991).
8. P. Vissavajjhala and A. H. Ross, *J. Biol. Chem.* **265,** 4746 (1990).
9. B. L. Hempstead, L. S. Schleifer, and M. V. Chao, *Science* **243,** 373 (1989).
10. S. J. Pleasure, U. R. Reddy, G. Venkatakrishnan, A. K. Roy, J. Chen, A. H. Ross, J. Q. Trojanowski, D. E. Pleasure, and V. M. Y. Lee, *Proc. Natl. Acad. Sci. U.S.A.* **87,** 8496 (1990).
11. D. M. Loeb, J. Maragos, D. Martin-Zanca, M. V. Chao, L. F. Parada, and L. A. Greene, *Cell (Cambridge, Mass.)* **66,** 961 (1991).
12. V. A. Luckow, *in* "Recombinant DNA Technology and Applications" (C. Ho, A. Prokop, and R. Bajpai, eds.), p. 97. McGraw-Hill, New York, 1990.
13. M. D. Summers and G. E. Smith, "A Manual of Methods for Baculovirus Vectors and Insect Cell Culture Procedures." Texas Agricultural Experiment Station, College Station, Texas, 1987.
14. A. B. Shanafelt, *BioTechniques* **11,** 330 (1991).
15. M. C. Fung, K. Y. M. Chiu, T. Weber, T. W. Chang, and N. T. Chang, *J. Virol. Methods* **19,** 33 (1988).
16. A. C. Webb, M. K. Bradley, S. A. Phelan, J. Q. Wu, and L. Gehrke, *BioTechniques* **11,** 512 (1991).
17. J. Field, J. Nikawa, D. Broek, B. MacDonald, L. Rodgers, I. A. Wilson, R. A. Lerner, and M. Wigler, *Mol. Cell. Biol.* **8,** 2159 (1988).
18. J. A. Medin, L. Hunt, K. Gathy, R. K. Evans, and M. S. Coleman, *Proc. Natl. Acad. Sci. U.S.A.* **87,** 2760 (1990).
19. J. G. Heuer, S. Fatemie-Nainie, E. F. Wheeler, and M. Bothwell, *Dev. Biol.* **137,** 287 (1990).
20. D. Martin-Zanca, M. Barbacid, and L. F. Parada, *Genes Dev.* **4,** 683 (1990).
21. A. Sutter, R. J. Riopelle, R. M. Harris-Warrick, and E. M. Shooter, *J. Biol. Chem.* **254,** 5972 (1979).
22. A. L. Schechter and M. A. Bothwell, *Cell (Cambridge, Mass.)* **24,** 867 (1981).
23. G. E. Landreth and E. M. Shooter, *Proc. Natl. Acad. Sci. U.S.A.* **77,** 4751 (1981).
24. P. Bernd and L. A. Greene, *J. Biol. Chem.* **259,** 15509 (1984).
25. D. D. Eveleth and R. A. Bradshaw, *Neuron* **1,** 929 (1988).
26. N. V. Costrini and R. A. Bradshaw, *Proc. Natl. Acad. Sci. U.S.A.* **76,** 3242 (1979).
27. R. J. Robb, *Proc. Natl. Acad. Sci. U.S.A.* **83,** 3992 (1986).
28. S. H. Green and L. A. Greene, *J. Biol. Chem.* **261,** 15316 (1986).
29. D. L. Hunston, *Anal. Biochem.* **63,** 99 (1975).
30. P. J. Munson and D. Rodbard, *Anal. Biochem.* **107,** 220 (1980).

[8] Cloning and Analysis of Nicotinic Acetylcholine Receptor Gene Expression in Goldfish Retina during Optic Nerve Regeneration

Virginia Hieber and Daniel Goldman

Introduction

Among two groups of animals, boney fishes and amphibians, the eye continues to grow throughout life, and these animals also display an ability to regenerate the optic nerve following injury (1). This regenerative ability is not exhibited by higher animals. The regenerating goldfish optic nerve is a valuable experimental system for investigations of regeneration of adult central nervous system axons. The cellular events which accompany axonal outgrowth, synaptogenesis, and resumption of normal function can be followed in an easily accessible tissue, the retina.

In the goldfish the optic nerve tracts cross before reaching the optic tectum (Fig. 1), with axons from the right retina innervating the left tectum and axons from the left retina innervating the right tectum (1). Following optic nerve crush, retinal ganglion cell bodies undergo changes in RNA (2, 3) and protein synthesis (4, 5), leading in 3 to 4 days to outgrowth of new axons from the cell bodies (6). These extend along the optic nerve tract past the site of the lesion and continue on to reach the contralateral tectum within 1 to 2 weeks, depending on the temperature at which the goldfish are maintained (7). The initial pattern of connections made by retinal axons in the tectum is modified by an activity-dependent period of refinement over 2 to 3 months until a correct retinotopic map is reestablished (1, 8). Because of the complete crossing of the optic nerve tracts one may compare events in regenerating and control retinas and tecta in the same animal at different stages during the process by crushing one optic nerve and leaving the second intact.

Other surgical manipulations of the goldfish are possible to affect the usual course of regeneration. The optic nerve can be crushed more than once to sustain regenerative events in the ganglion cell, or the targeted tectum can be removed in order to delay synaptogenesis. In the latter situation the regenerating axons will eventually innervate the remaining or ipsilateral tectum and will form functional synapses on this tectum while driving a rearrangement of axons from the contralateral retina (9, 10).

Explants of retinal pieces can be placed in culture dishes to study neurite outgrowth from ganglion cells under a variety of culture conditions (11).

Methods in Neurosciences, Volume 12

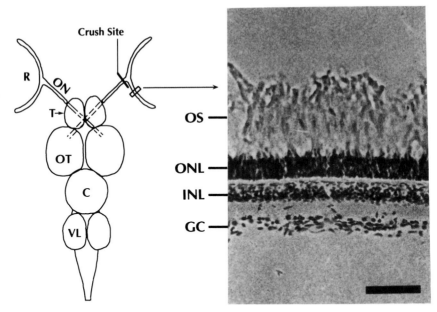

FIG. 1 Dorsal view of the goldfish brain showing the relation of the retina to the optic tectum. Optic nerves cross beneath the telencephalon at the front of the goldfish brain to innervate the contralateral tecta. R, Retina; ON, optic nerve; T, telencephalon; OT, optic tectum; C, cerebellum; VL, vagal lobe. The phase-contrast photomicrograph at upper right shows a section across the retina stained with hematoxylin and eosin. OS, Outer segment; ONL, outer nuclear layer; INL, inner nuclear layer, GC, ganglion cell layer. Bar, 50 μm.

Explants can also be sectioned and stained or carried through *in situ* hybridization with appropriate modifications of procedures used for the intact retina. Finally retinal ganglion cells can be dissociated from the intact retina and maintained in culture so that neurite extension can easily be observed microscopically (12).

We have chosen to use the regenerating optic nerve system in the goldfish because of its plasticity and ease of manipulation. This system is being used to investigate the molecular mechanisms regulating expression of synapse-specific proteins during synaptogenesis. Our initial studies (13) focused on characterizing retinal nicotinic acetylcholine receptor (nAChR) RNA expression during optic nerve regeneration. This expression was compared to that of α-tubulin RNA, which encodes a cytoskeletal protein. Following optic nerve crush, tubulin and nAChR RNA levels change in the crush relative to the control retina in a reproducible and time-dependent manner. An elevation of the α-tubulin RNA level is seen in retinal ganglion cells within 3 days after optic nerve crush. This is a time when axons are growing out from the site of the crush. The level continues to increase as axon elongation proceeds and reaches a maximum of approximately 5 times the control level at 15 days, by which time the optic tectum is covered by new axons from the regenerating optic nerve (Fig. 2). The RNA level then drops rapidly for 5 days and much more gradually for the following 2 to 3 months until a normal level is again reached.

The initial time course of nAChR RNA expression after optic nerve crush differs from that of α-tubulin. An increase in RNA levels for all subunits does not appear until 8 to 10 days postcrush. This increase occurs in the retinal ganglion cells and coincides with the time at which the regenerating axons reach the optic tectum and synapse formation begins. The nAChR RNA levels continue to rise to a maximum at approximately 15 days (i.e., 2 to 3 times control levels, Fig. 2). The levels then drop rapidly for 5 days, as for tubulin, and more gradually for 2 to 3 months to again reach control levels. Surprisingly, if one prevents optic nerve axons from reaching their targets in the optic tectum, by either repeated crushes of the regenerating nerve or by removal of the tectum, tubulin RNAs remain elevated and nAChR RNAs do not become induced during the 2 weeks following the initial optic nerve crush. This indicates that the optic tectum plays an important role in regulating presynaptic RNA expression. These differences between tubulin and nAChR RNA expression may be characteristic of two different classes of molecules responding to nerve injury and regeneration, with α-tubulin representing a cytoskeleton protein which participates in axonal elongation and the nAChRs being receptor proteins involved with synapse formation and function (13).

The rest of this chapter describes methods for optic nerve crush, retina isolation, and manipulation as well as procedures for preparing and characterizing goldfish retinal cDNA clones and studying their expression in the regenerating system.

Goldfish Procedures

Goldfish Selection and Maintenance

Goldfish are available from a number of breeders. We currently purchase fish from Grassyfork Fisheries (Martinsville, IN). No attempt is made by the breeders to control the genetics of the common goldfish (*Carassius auratus*). The fish are usually outbred for reasons of health and vigor. Fish are available in a range of sizes and ages. We generally use fish of 5–7 cm in body length (1–2 years of age). The regenerating optic axons grow at a

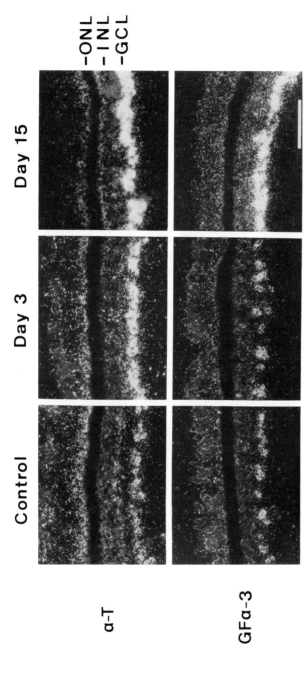

Fig. 2 *In situ* hybridization of α-tubulin and nAChR α-3 subunit RNAs in goldfish retina during optic nerve regeneration. The right optic nerve was crushed on day 0. Right and left retinas were removed at the indicated times postcrush and processed for *in situ* hybridization to ^{35}S-labeled RNA probes. The column labeled control corresponds to the left retina of experimental fish. Left retinas showed no detectable changes in tubulin or nAChR RNA levels over the time course of the experiment. All nAChR subunit RNAs showed maximal levels of expression in the right retina at day 15. Photomicrographs were taken using dark-field optics. α-T, α-Tubulin; ONL, outer nuclear layer, INL, inner nuclear layer, GCL, ganglion cell layer. Bar, 100 μm.

constant rate, and the time to reinnervation following crush varies with the length of the nerve and location of the crush. We use fish of approximately the same size and crush the nerve immediately adjacent to the eyeball (Fig. 1). Several environmental factors are controlled in maintenance of the fish. Fish are kept in aerated, continuously filtered water at 25°C, one fish or fewer per gallon. Because fish are extremely sensitive to chlorine, they are kept in a mixture of sodium-conditioned well water and pure water. The fish are fed liberally daily and exposed to a 14 hr light/10 hr dark cycle.

Optic Nerve Crush

Surgery and sacrifice of the fish are carried out after anesthetizing the fish in 0.05% tricaine methanesulfonate (3-aminobenzoic acid ethyl ester, Sigma, St. Louis, MO). Procedures are performed under a dissecting microscope for better visualization of the structures involved. The body of the fish is wrapped in wet paper toweling during the surgery to facilitate handling and prevent drying. The conjunctiva above the eye is torn with a pair of fine curved forceps. The eye is gently rolled down in the socket with the forceps to expose the optic nerve, which runs from the center back of the retina toward the contralateral tectum. Excess fluid or fat behind the eye is removed by suction through an 18-gauge needle (filed to a blunt tip) in order to make the optic nerve more visible. While the eye is held in place by suction through the needle, the crush is carried out by gripping the optic nerve with the forceps for 3 sec. After releasing the forceps the site of the crush should be obvious as an indentation in the nerve. The eye is rolled back into place in the eye socket and the fish returned to the home aquarium to recover. During the surgery care is taken to avoid tearing any blood vessels unnecessarily. Excessive bleeding or disruption of the blood supply to the retina by damage to the blood vessel which runs beside the optic nerve may result in cell death within the retina. Any fish which show signs of excessive bleeding or which later develop infection or abnormality of the eye, or other illness, are discarded.

Retina Removal

The goldfish retina is easily separated from the pigment epithelium when the fish is dark adapted for 30 min to 1 hr prior to euthanasia in tricaine. To remove the retina, the eye is briefly illuminated with a small dissecting light. The eye is pierced at the limbus with one point of a curved forceps with the other point extending across the front of the eye. The tips of the forceps are pressed together to hold the eye while a cut is made around the eye at the limbus. The front of the eye is then removed, and the lens is lifted out. A stream of saline from a wash bottle is used to separate the retina from the pigment epithelium. A pair of scissors is used to cut the connection of the retina to the back of the eye, and the retina is rinsed or lifted into a petri dish. A healthy retina is light pink and translucent. At this point the retina can be frozen in liquid nitrogen for later RNA isolation (see below) or immediately fixed for *in situ* hybridization as described later in this section.

Tectal Ablation

If desired, synaptogenesis can be delayed while allowing continued growth of the optic nerve. This can be accomplished by removal of the target tissue, the optic tectum. We use smaller fish for this type of surgery because the skull is thinner and easier to penetrate in the younger fish. The body of the anesthetized fish is wrapped in a paper towel wetted with aquarium water. Surgery is carried out under a dissecting microscope. The cranium roof is opened by cutting a three-sided flap of bone over the optic tectum with the tip of a No. 11 scapel blade. The flap is gently lifted to expose the brain without breaking the hinge of bone. The tectum is removed by suction through a 21-gauge needle filed to a flat end. This method results in removal of the torus longitudinalis which is attached to the medial edge of the tectum. Excess blood around the brain is removed by suction. Following surgery, the bone flap is pressed into place, tucking the edges slightly under the surrounding intact bone, and the fish is returned to the aquarium to recover. Healing of the edges of the flap to the skull occurs within 1 to 2 weeks. The completeness of the ablation can be checked at the end of the experiment by observation of the brain after removal from the skull.

Retinal Explanting

Retinal explants can be made from eyes either with or without prior optic nerve crush; however, neurite outgrowth is more rapid and reliable from retinas of fish which receive a conditioning optic nerve crush 7 to 10 days prior to euthanasia. In conditioned explants neurite outgrowth appears within 24 hr. Fish are dark adapted at least 30 min before surgery to facilitate removal of the retina. Since explantation requires additional time for processing the eye before removal of the retina, these procedures are carried out under red light to maintain dark adaptation. Sterile disposable tissue culture tubes and dishes are used, and instruments and solutions are sterilized.

Retinas for explant culture are removed from 4 to 5 fish at one time. After euthanasia of the fish, the eye is removed to a 35-mm culture dish. The surface of the eye is sterilized by immersion in 70% ethanol for no more than 30 sec. The eye is next passed through two changes of Dulbecco's phosphate-buffered saline, pH 7.2 (DPBS). Further dissection of the eye is carried out in a sterile hood. While grasping the eye with a pair of forceps, as described above, the front is cut off and the retina removed. The retina is rinsed well in sterile DPBS and cut halfway across with scissors, so that it can be placed flat on the stage of a McIlwain tissue chopper. The tissue chopper is set to make cuts every 500 μm. After cuts are made across the retina in one direction, the stage is turned 90° and a second set of cuts made at right angles to the first. This should produce square pieces of tissue 500 μm on a side. These are rinsed into a sterile container with DPBS. The tissue is allowed to settle to the bottom of the container and the saline above the tissue removed by aspiration through a Pasteur pipette. Ten milliliters of sterile DPBS is added to the container, the tissue pieces rinsed by gentle swirling, and the saline again removed when the tissue has settled. The tissue is rinsed a third time as described above. Finally, the tissue is resuspended in medium [Leibovitz medium (L-15, GIBCO, Grand Island, NY) containing 1% fetal calf serum, 20 mM HEPES buffer, pH 7.2, and 0.1% gentamycin sulfate; the medium can also be supplemented with 0.2 mM uridine and 0.1 mM 5-fluorodeoxyuridine in order to prevent overgrowth by fibroblasts] and transferred to a 35-mm culture dish.

The retina pieces are visually inspected with the aid of a dissecting scope. Regularly shaped pieces without adhering pigment epithelium are used for explanting. The pieces are placed individually on a culture dish which has been previously coated with poly-l-lysine (Sigma) and rinsed thoroughly with sterile water, then with DPBS, and finally with medium using a Pasteur pipette. The pieces are placed approximately 0.5 cm apart for optimal observation of neurite outgrowth. A grid placed beneath the dish facilitates uniform placement of the explants. After the dish has been filled with tissue pieces, they are kept undisturbed for 1 to 2 hr. Enough medium, about 0.5 ml, is added slowly to surround the tissue without floating the pieces off the plate. The dishes are placed in a humidified chamber in an incubator to maintain the explants at the desired temperature, which should be the same as that at which the fish are maintained.

RNA Probe Synthesis and in Situ Hybridization

Methods for RNA probe synthesis and *in situ* hybridization have been detailed in an earlier volume of this series (e.g., see Ref. 14) and so are described only briefly here. The methods for selection and characterization of α-tubulin and nAChR cDNA clones from the retinal cDNA library are described in following sections of this chapter. Restriction fragments from these clones were subcloned into pGEM or pSP73 vectors (Promega, Madison, WI). The subclones are linearized using restriction enzymes which generate a blunt or a 5' overhang since linearized vector templates with 3' overhangs have been reported to generate incorrect products. Sense and antisense RNA probes are prepared by transcription from the appropriate promoter of the linearized vectors in the presence of ^{35}S-labeled UTP and unlabeled ATP, CTP, and GTP according to the Promega protocol. Briefly, incubation mixtures are prepared containing 50 mM Tris-HCl, pH 7.5, 6 mM MgCl$_2$, 2 mM spermidine, 10 mM NaCl, 10 mM dithiothreitol (DTT), 20 units RNasin, 0.5 mM each ATP, CTP, and GTP, 200 μCi uridine 5'-α-thiotriphosphate (800 Ci/mmol), 0.5 μg linearized template DNA, and 1.0 μl SP6 or T7 polymerase in a final volume of 20 μl. Mixtures are incubated for 60 min at 37°C followed by treatment for 15 min with RNase-free DNase to remove the template DNA. After DNase treatment, phenol–chloroform extraction and ethanol precipitation, the probes are alkaline hydrolyzed to an average length of 100 to 200 bp using 0.4 M NaHCO$_3$, 0.6 M Na$_2$CO$_3$, pH 10.2. A trial experiment must be done initially for each RNA to determine the time required to produce a probe of the desired length. The size of hydrolyzed probes is determined on an agarose or acrylamide gel according to standard procedures (15).

Following removal of the retina (see above), the tissue is placed in ice-cold 4% paraformaldehyde–phosphate-buffered saline (PBS), pH 7.4, for 2 hr. The tissue is cryoprotected by overnight immersion in ice-cold 30% sucrose–PBS. The tissue is briefly rinsed in PBS and embedded and frozen in O.C.T. (Lab Tek Products) for sectioning. Fifteen-micron sections are cut in a cryostat and mounted on poly-L-lysine-coated slides. The section may be stored at -80°C or used immediately for *in situ* hybridization. Prior to *in situ* hybridization, the sections are digested with proteinase K (10 μg/ml in 0.1 M Tris-HCl, pH 7.5, 50 mM EDTA) for 5 min at 37°C. The sections are then treated with acetic anhydride (0.25 ml/100 ml) in 0.1 M triethanolamine, pH 8.0, for 10 min at room temperature, rinsed in 2× SSC (3 M NaCl, 0.3 M sodium citrate), dehydrated through alcohols, and air-dried.

The sections are covered with hybridization solution (10 mM Tris-HCl, pH 7.5, 50% formamide, 0.3 M NaCl, 1 mM EDTA, 10% dextran sulfate, 0.1 M DTT, 100 μg/ml *Escherichia coli* tRNA) containing 5 × 10^4 cpm/μl of radiolabeled probe, sealed with a coverslip, and incubated in a humidified chamber at 55°C for 21 hr. After the hybridization, the slides are rinsed with 2× SSC at room temperature for 10 min, incubated in 2× SSC, 50% formamide at 55°C for 30 min, rinsed well with 2× SSC to remove residual formamide, and then treated for 30 min at 37°C with RNase A (50 μg/ml in 10 mM Tris-HCl, pH 8.0, 1 mM EDTA, 0.5 M NaCl). The slides are washed in 0.5× SSC at 55°C for 1 hr. Following dehydration the slides are dipped in Kodak (Rochester, NY) NTB-2 emulsion (diluted 1:1 with water) and exposed at 4°C in the dark for 4 to 5 days or until sufficient silver grains have been produced. After development the sections are stained in hematoxylin and eosin, dehydrated, coverslipped, and examined microscopically using dark-field optics.

Explants are treated for *in situ* hybridization in a similar manner with the following modifications. The explants are fixed and cryoprotected in the petri dish in which they are maintained. When embedding them in OCT an effort is made to retain the original orientation with the ganglion cell layer parallel to the dish surface so that sectioning can be done across the layers of the explant to produce a section similar to that seen from the intact retina.

Library Construction, Clone Selection, and Analysis

Isolation of Retinal RNA

We have used two different procedures for purifying total RNA from goldfish retina and tectum (16, 17). The first method described below involves centrifuging the RNA through a cushion of CsCl and is used to generate high-quality RNA for cDNA libraries. The second method is less cumbersome and is used when working with small tissue samples (we have used it to isolate total RNA from a single retina) or when multiple samples are analyzed in S1 or RNase protection assays.

In the first procedure, if poly(A)$^+$ RNA is to be subsequently isolated, we generally start with 50–60 retinas. Frozen tissue is homogenized in 10 ml of GTC (5 M guanidine isothiocyanate, 50 mM HEPES, 0.17 mM N-lauroyl sarcosine, 50 μM DTT, pH 7.0). A Polytron (Brinkmann Instruments, Westbury, NY) is used for rapid homogenization. The homogenate is passed through 23-gauge and 25-gauge needles to shear the DNA and spun at 10,000 rpm for 15 min at 10°C in a Sorval SS34 rotor. The supernatant is layered over a 3-ml cushion of 5.7 M CsCl, 50 mM EDTA (pH 7.8) [the CsCl/EDTA solution is treated with 0.1% (v/v) diethyl pyrocarbonate (DEPC) overnight and autoclaved the next morning to inactivate ribonucleases] and centrifuged at 35,000 rpm for 15 hr at 20°C in a Beckman (Fullerton, CA) SW41 rotor. The RNA will sediment through the CsCl and form a clear pellet on the bottom of the tube, while the DNA will form an opaque band in the CsCl. After centrifugation the GTC solution above the CsCl is removed by aspiration. The sides of the tube are rinsed with 2 ml of fresh GTC, and this, along with the upper third of CsCl, is removed by aspiration. This last step is repeated an additional 2 times, the last of which removes all remaining CsCl.

The tube is then inverted and the top two-thirds of the tube cut off with a razor blade. The bottom third, containing the RNA pellet, is placed in a test tube rack and the pellet resuspended in 0.5 ml of water–DEPC [prepared by adding 0.1% (v/v) DEPC to water, shaking vigorously, and letting the solution sit overnight, followed by autoclaving the next morning]. The RNA pellet does not have to be completely in solution at this point. The suspended RNA is transferred to a tube containing 3.5 ml of 7.5 M guanidine-HCl, 50 mM HEPES, 50 μM DTT (pH 7.0) and vortexed vigorously until the RNA is in solution. Ninety microliters of 1 M acetic acid is added and the solution vortexed. The RNA is precipitated by adding 2 ml of ethanol and placing the solution at $-20°C$ for 30 min, then collected by centrifugation at 8000 rpm for 15 min at 4°C in a Sorval SS34 rotor. The RNA precipitate is washed with 95% ethanol and then dried under reduced pressure. The RNA is resuspended in 1 ml of water–DEPC. If the RNA does not go into solution readily, the pH is checked by spotting 1-μl aliquots on pH paper, and the pH is adjusted to about 7–8, if necessary, with 0.2 M NaOH. Care must be taken during this step since RNA is hyrolyzed in basic solutions. One-tenth volume of 2 M potassium acetate, pH 5.2, and 2.5 volumes of ethanol are added to precipitate the RNA. The tube is placed on ice for 20 min and then spun at 10,000 rpm in a Sorval SS34 rotor at 4°C for 10 min. The RNA precipitate can then be dried under vacuum and resuspended in water–DEPC. The RNA yield is determined by measuring the absorbance at 260 nm of a diluted aliquot. Generally one obtains 2–5 mg of RNA per gram of tissue. RNA at 40 μg/ml has an absorbance of 1 at 260 nm. The RNA can be stored in solution at $-80°C$.

In the second method, 1 g of tissue is homogenized in 10 ml of GTC [4 M guanidine isothiocyanate, 25 mM sodium citrate, pH 7.0, 0.5% (w/v) N-lauroyl sarcosine] using a Poltron (scale this and all subsequent volumes down if starting with less tissue). To the homogenate is added, sequentially, 1 ml of 2 M sodium acetate (pH 4.0), 10 ml of water-saturated phenol, and 2 ml of chloroform–isoamyl alcohol (24:1, v/v). The solution is mixed well after each addition and incubated on ice for 15 min. The homogenate is transferred to a Corex tube (1.5-ml microcentrifuge tube for small volumes) and spun at 10,000 g for 20 min at 4°C. The RNA will partition into the aqueous layer, and the DNA and proteins are in the organic and interface layers. The aqueous phase is transferred to a polypropylene tube and 1 volume of 2-propanol added. The tube is placed at $-20°C$ for 60 min and then spun at 10,000 g as above. The pellet is dissolved in 3 ml of GTC and the RNA precipitated with 3 ml of 2-propanol as above. The precipitate is collected by centrifugation at 10,000 g and washed with 75% ethanol. The pellet is dried under vacuum and resuspended in 0.5 ml of 0.5% sodium dodecyl sulfate (SDS, prepared in water–DEPC). Heating the RNA to 65°C may facilitate its solubilization. The yield of RNA is determined by measuring the absorbance of a diluted aliquot at 260 nm.

Poly(A)$^+$ RNA for library construction can be selected by passing the total RNA over an oligo(dT)-cellulose column. These columns can be prepared in a 5- or 10-ml disposable syringe. A 1-ml volume of packed oligo(dT)-cellulose is sufficient for the isolation of poly(A)$^+$ RNA from 8–10 mg of total RNA. Sterile tubing is attached to the end of the syringe and clamped shut. The syringe is filled with water–DEPC, and oligo(dT)-cellulose is added and allowed to hydrate while settling. The syringe is filled to 1 ml with the resin. The resin is washed with 2 ml of 0.1 N NaOH and equilibrated with 1× binding buffer (~20 ml); 2× binding buffer is 20 mM Tris-HCl (pH 7.5), 1 M NaCl, 2 mM EDTA, and 1% SDS (prepare by mixing the above minus the Tris and adding DEPC to 0.1%; shake vigorously and let sit overnight; autoclave the next morning and add Tris from a 1 M sterile stock). An equal volume of 2× binding buffer is added to the RNA in water–DEPC. The buffer is drained from the oligo(dT)-column until it is just above the cellulose.

The RNA solution is heated to 65°C for 5 min, quick-cooled on ice, and applied to the oligo(dT)-cellulose column. The RNA solution is allowed to flow into the column cellulose at about 1 ml/5 min using a screw clamp on the tubing to adjust the flow rate. The flow-through is collected, heated to 65°C, and applied to the column as above. This cycling is repeated 2 times. The flow-through contains the poly(A)$^-$ RNA which can be collected by ethanol precipitation. The column is washed with 1× binding buffer until the absorbance of the flow-through at 260 nm is below 0.01. The buffer in the column is drained just to the surface of the cellulose. The column tubing is clamped with a hemostat and 0.5 ml of elution buffer added (10 mM Tris-HCl, pH 7.5, 1 mM EDTA, 0.05% SDS; prepared by making the above solution lacking the Tris, treating with 0.1% DEPC overnight, autoclaving the next morning, and then adding Tris from a 1 M sterile stock). The hemostat is released and the eluate collected into a 1.5-ml Eppendorf tube. The tubing is clamped again, and this sequence is repeated a total of 6 times, collecting the eluate into a fresh tube each time. The absorbance of a 1:40 dilution from each tube is read at 260 nm. Peak fractions are pooled and the RNA precipitated by adding 0.1 volume of 2 M potassium acetate (pH 5.2), 2.5 volumes of ethanol, and cooling to −20°C for 60 min. Generally, 2–4% of the total RNA can be recovered from the oligo(dT)-cellulose column as poly(A)$^+$ RNA.

The integrity of the RNA should be checked on a 1.2% denaturing agarose gel. The gel is prepared by boiling agarose in 10 mM sodium phosphate buffer (pH 7.4) to dissolve it, cooling the solution to 65°C, and adjusting to 2.2 M formaldehyde (with 37% stock formaldehyde) and 0.5 mM EDTA (work in fume hood). The gel is poured and allowed to solidify in a fume hood. A running buffer containing 100 mM sodium phosphate (pH 7.4), 2.2 M formaldehyde, and 0.5 mM EDTA is used. RNA samples (~2–20 μg) are resuspended in 10 μl of formamide and then adjusted to 2.2 M formaldehyde, 10 mM sodium phosphate (pH 7.4), and 0.5 mM EDTA. Samples are heated to 65°C for 5–10 min, and 0.1 volume of a 50% glycerol, 0.3% bromophenol blue, 0.1 M sodium phosphate buffer stock is added just before loading samples on the gel. Electrophoresis is at a constant voltage of 100 V with recirculating buffer. Gels can be stained with ethidium bromide or acridine orange.

cDNA Library Construction and Screening

cDNA libraries are prepared using 2.5 to 10 μg of poly(A)$^+$ RNA isolated from normal or regenerating goldfish retinas. The details of preparing cDNA libraries have already been described in this series (18) and are not repeated here. Briefly, RNA is reverse transcribed into cDNA with avian myeloblastosis virus (AMV) reverse transcriptase, linkered with *Eco*RI phosphorylated linkers, and then cloned into the *Eco*RI site of bacteriophage λ gt10. Libraries are plated and lifted onto nylon membranes as described (18).

To isolate clones encoding nAChRs from this library, we screened it with a mixed receptor probe. This probe consisted of nick-translated (19) cDNAs encoding the α subunits of the *Torpedo* electric organ nAChR, the rat muscle nAChR, and the rat neural nAChR α-4 subunit (20–22). We chose this diversity because it was not clear which nAChR genes would be expressed in the goldfish retina or how similar their nucleotide sequence would be to those clones already isolated. In retrospect it is clear that the neuronal nAChR genes are expressed in the retina and that these genes have a high nucleotide sequence similarity to their rat homologs (23–26). Therefore, it would have been sufficient to use a single rat neuronal nAChR subunit encoding cDNA to screen the goldfish library for retinal nAChR genes. Nick-translation reactions were also successfully performed using the BRL (Gaithersburg, MD) nick-translation system. Hybridization and washing conditions were as described in a previous volume in this series (18). Briefly, filters are hybridized in 0.75 M NaCl, 57 mM NaH$_2$PO$_4$, 5 mM EDTA (pH 7.4), 1% SDS, and 100 μg/ml denatured herring sperm DNA at 65°C and washed in 0.75 M NaCl, 75 mM sodium citrate (pH 7.0), 0.1% SDS at 65°C. These screening conditions are sufficient for isolating cDNAs encoding both the α and β subunits of neural nAChRs (23–26). Similar strategies have now been used to isolate glutamate receptor- and tubulin-encoding cDNAs from the goldfish library.

Sequence Analysis of cDNA Clones

To confirm that cDNAs hybridizing to the probe encode proteins with homology to the nAChR gene family, the DNA sequences must be determined. In addition, if one isolates a novel nAChR subunit cDNA, functional expression of the protein encoded by the cDNA is required to show that it represents a protein that participates in the formation of a ligand-gated ion channel. Since most of the members of the nAChR gene family have been isolated previously from rat and chick libraries (27), DNA sequencing is often sufficient for identifying a goldfish clone. We employed a unidirectional deletion strategy for sequencing cDNAs. Early work involved subcloning λ cDNA inserts into the single-stranded M13 phage (23–26). This phage is useful if single-stranded DNA is to serve as a template for DNA sequencing. More recently we have abandoned this phage and subcloned into the Bluescript phagemid vector (Stratagene, La Jolla, CA). The advantages of the latter vector include a color selection for recombinants, T3 and T7 promoters flanking the multiple cloning site for generating RNA probes, multiple unique restriction sites containing 5' and 3' overhangs for generating unidirectional deletions with exonuclease III (Exo II), and the ability to isolate single-stranded DNA using helper phage.

Phagemids containing cDNA inserts are grown in 500 ml of YT (15) supplemented with 40 μg/ml ampicillin (added after autoclaved medium cools). Phagemid DNA is isolated following standard protocols (15) and banded in CsCl (15). To sequence the cDNA insert, unidirectional deletions are created using exonuclease III. This strategy takes advantage of the fact the Exo III will digest DNA from a blunt or 5' overhang, while a 3' overhang protects the DNA from Exo III digestion. Therefore, one digests the recombinant phagemid DNA (10–25 μg) with two enzymes. The enzymes are chosen so that the one that generates a 5' overhang or a blunt end digests within the multiple cloning site of the phagemid closest to the insert cDNA, while the other enzyme digests between the latter enzyme site and the sequencing primer binding sequence. Neither enzyme should digest within the cDNA insert. After digestion the DNA is extracted with phenolchloroform and precipitated with sodium acetate (0.3 M) and 2 volumes of ethanol on ice for 10 min. The DNA is collected by centrifugation at 12,000 g for 10 min in a microfuge. The supernatant is removed by aspiration, the DNA pellet washed with 70% ethanol and centrifuged for 2 min, and the supernatant again removed by aspiration. The DNA pellet is dried under vacuum and resuspended in water at a concentration of 1 μg/μl. Ten microliters of this DNA is diluted into 98 μl of water and 12 μl of 10× Exo III buffer (0.66 M Tris-Cl, pH 8.0, 66 mM MgCl$_2$).

Twenty labeled tubes containing 7.5 μl of S1 reaction mixture (2.2 M

NaCl, 0.3 M potassium acetate, pH 4.5, 45% glycerol, 1.8 mM ZnSO$_4$, 60 units S1 nuclease) are placed on ice. The DNA solution is incubated at 37°C for 5 min and then a 2.5-μl aliquot is added to the first labeled tube. Six microliters of Exo III (~65 units/ml) is added to the DNA solution, which is vortexed and returned to the water bath. Aliquots (2.5 μl) are transferred to the S1 reaction buffer at 1- to 5-min intervals. We find under these conditions that Exo III digests DNA at an approximate rate of 150 nucleotides/ min. After taking all the time points the tubes are incubated at 30°C for 30 min. One microliter of S1 stop solution (0.3 M Tris base, 50 mM EDTA, pH 8.0) is added and the tubes heated to 70°C for 10 min. At this point 2-μl aliquots of the samples can be analyzed for size by electrophoresis in a 1% agarose gel. It is useful to also run linearized recombinant vector and parent vector lacking cDNA insert for comparison.

Appropriate time points (~200-bp intervals) are chosen for the fill-in reaction and ligation. One can either pool samples or keep time points separate. We prefer the latter since it is easier to identify deleted clones for sequencing. To each time point or 10μl of pooled sample, 1 μl of Klenow mixture (0.2 M MgCl$_2$, 10 mM Tris-HCl, pH 7.5, 0.3 U/μl of Klenow fragment of $E.$ $coli$ DNA polymerase I) is added, and the mixture is incubated for 5 min at 37°C. One microliter of 0.5 mM dNTP mix is added to each sample, and the mixtures are incubated for an additional 15 min at 37°C. Ligations are carried out by adding 40 μl of ligase mixture [50 mM Tris-HCl, 10 mM MgCl$_2$, 10 mM DTT, 50 μg/ml bovine serum albumin (BSA), 0.5 mM ATP, 7.5% polyethylene glycol (PEG) 8000, 0.05 units/μl of T4 DNA ligase] and incubating at room temperature for 2 hr. The ligation buffer lacking ligase, ATP, and PEG can be prepared as a 10× stock and stored at −20°C. Competent XL1-Blue (Stratagene) cells are transformed with 5 μl of this ligated material (see below).

Competent bacteria are prepared by diluting an overnight culture of XL1-Blue (Stratagene) bacteria 20- to 50-fold into YT broth and growing at 37°C with vigorous aeration until the absorbance at 660 nm is about 0.6. Cells are centrifuged at 5000 rpm at 4°C and resuspended in ice-cold 50 mM CaCl$_2$. Cells are incubated on ice for 20 min, pelleted as above, and resuspended in 0.1 volume of ice-cold 50 mM CaCl$_2$. These cells can be used immediately for transformation or stored for up to 1 week at 4°C. The cells are transformed by placing 0.2 ml of competent cells in a prechilled culture tube. Five to ten microliters of ligated DNA is added and incubated on ice for an additional 30 min. Cells are then heated at 42°C for 2 min and returned to room temperature; 1 ml of YT broth is added, and the cells are incubated at 37°C for 30 min with vigorous shaking. Cells are pelleted at 5000 rpm for 3 min, resuspended in 0.4 ml of YT broth, and plated on YT plates containing the appropriate antibiotic. Plates are placed at 37°C overnight, and the following morning colonies are picked from each plate and grown in 3 ml of medium with antibiotic. Minipreps of the phagemid DNA are performed (15), and the resulting DNA is cleaved with appropriate restriction enzymes to analyze the extent of the deletion on an agarose gel.

DNA sequencing is performed on either single- or double-stranded phagemid DNA. We prefer to sequence double-stranded DNA since isolation of this DNA is more reproducible than isolating the single-stranded phagemid. Miniprep DNA subjected to alkaline lysis and PEG precipitation is generally good for DNA sequencing reactions. There are a number of procedures recommended by various molecular biology manuals and companies for preparing DNA for sequencing reactions. If problems arise in sequencing DNA from the minipreps, we find that DNA purified over CsCl is also a good substrate for DNA sequencing reactions. Since DNA sequencing reactions are fairly standard and performed using manufacturer's directions and kits, they will not be repeated here. We have used Sequenase (United States Biochemical, Cleveland, OH) to sequence double-stranded DNA with good results. More recently we have used Taq polymerase and thermal cycling along with the Applied Biosystems (Foster City, CA) automated DNA sequenator to obtain DNA sequence data. Procedures are those recommended by the supplier.

RNase Protection Assays

RNase protection assays are used as a quantitative method for assaying specific mRNA species in the retina. In addition these assays are used to determine if heterogeneity exists in a specific mRNA species that may indicate alternative splicing or processing of the primary transcript. Radiolabeled antisense RNA probes for protection assays are generated by run-off transcription of linearized vectors (28) as described above. In this case, however, full-length probes are required, and so the alkaline hydrolysis step is omitted. We generally use restriction enzymes that keep our probes under 800 nucleotides in length. The reason for this is that as one makes longer probes more heterogeneity is observed in the transcription products. Restriction enzymes for linearizing the recombinant phagemid should generate a blunt or 5' overhang, since 3' overhangs have been reported to generate sense RNA. Probe sizes are confirmed on denaturing 8 M urea, 6% polyacrylamide gels.

Probe RNA ($\sim$100,000 cpm) is hybridized with cellular RNA (5–50 μg) in a 30-μl volume containing 50% formamide, 40 mM PIPES (pH 6.7), 0.74 M NaCl, and 1 mM EDTA (pH 8.0). Samples are first denatured for 5 min in a boiling water bath and then placed in a 55°C incubator. The sample tubes are placed on their sides to reduce evaporation. Samples are incubated for 3–20 hr (overnight is generally most convenient). Following hybridization, 170 μl of RNase digestion buffer (10 mM Tris-HCl, pH 7.2, 5 mM EDTA, 270 mM NaCl, 40 μg/ml RNase A, and 1000 units/ml RNase T1) is added, and samples are incubated for 1 hr at room temperature. The digestion is terminated by adding 150 μl of water, 25 μl of 10% SDS, and 25 μl of proteinase K stock (10 mg/ml) and incubating for an additional 30 min at room temperature. Samples are then extracted with an equal volume of phenol–chloroform–isoamyl alcohol (25:24:1, v/v) and the hybridized RNA precipitated by addition of 1 ml of ethanol. Samples are incubated on ice for 10 min and pelleted in a microfuge at 15,000 rpm for 10 min. The radioactive supernatant is carefully pipetted from the tube and disposed of in radioactive waste. The tube is briefly spun again and any remaining fluid removed. The precipitate is dried briefly under vacuum and resuspended in 3 μl of for-mamide containing just enough bromophenol blue to give color. The sample is vortexed thoroughly to resuspend the RNA and denatured for 5 min in a 90°C water bath just prior to loading on a 6% acrylamide, 8 M urea denaturing gel. The size of the protected fragments can be determined from an autoradio-graph of the dried gel using labeled DNA fragments run on the same gel as molecular weight markers.

Acknowledgments

We thank Dr. Roger Davis for valuable advice and for assistance in the preparation of Fig. 1. We also thank Dr. Davis and Frank Hoover for critically reading the manuscript. This work was supported by a grant from the Lucille P. Markey Charitable Trust to D.G.

References

1. D. G. Attardi and R. W. Sperry, *Exp. Neurol.* **7,** 46 (1963).
2. M. Murray, *Exp. Neurol.* **39,** 489 (1973).
3. H. R. Burrell, L. A. Dokas, and B. W. Agranoff, *J. Neurochem.* **31,** 289 (1978).
4. M. Murray and B. Grafstein, *Exp. Neurol.* **23,** 544 (1969).
5. A. M. Heacock and B. W. Agranoff, *Proc. Natl. Acad. Sci. U.S.A.* **73,** 828 (1976).
6. B. Grafstein, *in* "The Retina: A Model for Cell Biological Studies" (R. Adler and D. Farber, eds.), Part 2, p. 275. Academic Press, Orlando, Florida, 1986.
7. A. D. Springer and B. W. Agranoff, *Brain Res.* **128,** 405 (1977).
8. J. T. Schmidt and D. L. Edwards, *Brain Res.* **269,** 29 (1983).
9. R. Y. S. Lo and R. L. Levine, *J. Comp. Neurol.* **191,** 295 (1980).

10. A. D. Springer and S. M. Cohen, *Brain Res.* **225,** 23 (1981).

11. G. E. Landreth and B. W. Agarnoff, *Brain Res.* **118,** 299 (1976).

12. J. E. Johnson and J. E. Turner, *J. Neurosci. Res.* **8,** 315 (1982).

13. V. Hieber, B. W. Agranoff, and D. Goldman, *J. Neurochem.* **58,** 1009 (1992).

14. J. T. McCabe and D. W. Pfaff, *in* ''Methods in Neurosciences'' (P. M. Conn, ed.), Vol. 1, p. 98. Academic Press, San Diego, 1989.

15. T. Maniatis, E. G. Fritsch, and J. Sambrook, ''Molecular Cloning: A Laboratory Manual.'' Cold Spring Harbor Laboratory, Cold Spring Harbor, New York, 1982.

16. J. M. Chirgwin, A. E. Przybyla, R. J. MacDonald, and W. J. Rutter, *Biochemistry* **18,** 5294 (1979).

17. P. Chomczynski and N. Sacchi, *Anal. Biochem.* **162,** 156 (1987).

18. J. Boulter and P. D. Gardner, *in* ''Methods in Neurosciences'' (P. M. Conn, ed.), Vol. 1, p. 328. Academic Press, San Diego, 1989.

19. P. W. J. Rigby, M. Dieckmann, C. Rhodes, and P. Berg, *J. Mol. Biol.* **113,** 237 (1977).

20. M. Noda, H. Takahashi, T. Tanabe, M. Toyosato, Y. Furutani, T. Hirose, M. Asai, S. Inayama, T. Miyata, and S. Numa, *Nature (London)* **299,** 793 (1982).

21. J. Boulter, W. Luyten, K. Evans, P. Mason, M. Ballivet, D. Goldman, S. Stengelin, G. Martin, S. Heinemann, and J. Patrick, *J. Neurosci.* **5,** 2545 (1985).

22. D. Goldman, E. Deneris, W. Luyten, A. Kochhar, J. Patrick, and S. Heinemann, *Cell (Cambridge, Mass.)* **48,** 965 (1987).

23. K. Cauley, B. W. Agranoff, and D. Goldman, *J. Cell Biol.* **108,** 637 (1989).

24. K. Cauley, B. W. Agranoff, and D. Goldman, *J. Neurosci.* **10,** 670 (1990).

25. V. Hieber, J. Bouchey, B. W. Agranoff, and D. Goldman, *Nucleic Acids Res.* **7,** 5293 (1990).

26. V. Hieber, J. Bouchey, B. W. Agranoff, and D. Goldman, *Nucleic Acids Res.* **7,** 5307 (1990).

27. E. S. Deneris, J. Connoly, S. W. Rogers, and R. Duvoisin, *Trends Neurosci.* **12,** 34 (1991).

28. D. A. Melton, P. A. Krieg, M. R. Rebagliati, T. Maniatis, K. Zinn, and M. R. Green, *Nucleic Acids Res.* **12,** 7035 (1984).

[9] Receptor Protein–Tyrosine Kinases in the Nervous System, Cloning and Expression of *trk*B

David S. Middlemas

Introduction

As many receptors are members of large families related by amino acid sequence homology, several techniques for discovery of new members of these families based on sequence homologies have been developed. Subsequent expression allows for characterization of the receptors and identification of ligands for these orphan receptors (a receptor whose ligand is unknown). Cloning strategies based on homology have proved successful in identifying new members of the receptor protein-tyrosine kinase (PTK) family. Screening with degenerate oligonucleotides matching conserved regions in the kinase domains (1–4) and low-stringency screening of cDNA libraries with probes derived from subfamily members (5) have proved very successful. The use of degenerate oligonucleotides for direct screening of cDNA libraries laid the foundation for improved strategies that use the polymerase chain reaction (PCR) with primers comprised of degenerate oligonucleotides corresponding to conserved regions in the kinase domains (6–8). Alternatively, strategies that are based on enzymatic activity rather than sequence homology can be used to find family members that may not be closely related in amino acid sequence. For example, antibodies against phosphotyrosine have been used to successfully screen cDNA expression libraries to identify cDNA clones encoding active PTKs (9, 10).

As many receptors for growth factors are in the PTK family, it was postulated that there may be neuron-specific receptor PTKs. Earlier reports that expression of v-*src* in PC12 cells caused the cells to extent neurites (11) and evidence for phosphorylation of cellular proteins on tyrosine in PC12 cells treated with nerve growth factor (NGF) suggested there might be neuron-specific receptor PTKs (12). To this end, we screened a rate cerebellar library using degenerate oligonucleotides which correspond to the kinase domains of receptor PTKs. This approach had been used successfully to clone both protein-serine kinases (1) and protein-tyrosine kinases (2). Rat TrkB, a putative receptor protein-tyrosine kinase which is expressed predominantly in the nervous system, was identified using degenerate oligonucleotide probes

(3), whereas murine *trk*B was discovered using probes derived from the previously known human oncogene *trk* (13). In an intriguing development, *trk*B was found to express both a full-length receptor PTK and a truncated receptor containing both the extracellular domain and the transmembrane domain but lacking the kinase domain (3, 14). *trk*B is a member of the *trk* subfamily of PTKs, which so far includes *trk*(15), *trk*B, and *trk*C (16). The family is defined by strong sequence homology in the extracellular domains (35–50% amino acid sequence similarity). The finding that NGF is a ligand for Trk (17–20) prompted several laboratories to test whether members of the NGF/neurotrophin family are ligands for TrkB. The neurotrophins brain-derived neurotrophic factor (BDNF), neurotrophin-3 (NT-3) (21–24), and neurotrophin-5 (NT-5) (25) are ligands for TrkB.

Degenerate oligonucleotides used as primers for the PCR offer an enhanced strategy for using sequence homology to clone new members of extended gene families (6–8). Advantages of a PCR strategy over the use of radiolabeled degenerate oligonucleotides to directly screen a cDNA library include the ability to use more degenerate primers and the lower background of identification of nonkinase cDNAs. However, one may still wish to screen cloned PCR products using assays based on a degenerate olgonucleotide in the amplified region. Therefore, the protocols described by Hanks and others (1–4) should still prove useful.

In this chapter, methods for using PCR for rapid cloning of cDNAs into both bacterial and mammalian expression vectors are presented. The bacterially expressed proteins are suitable for raising specific antisera, whereas expression in mammalian cells allows expression of functional receptors. Having identified an orphan receptor, it is often important to express this receptor in an active form for further characterization. In the case of receptor PTKs, it is possible to screen directly for ligands with an autophosphorylation assay using antireceptor antibodies and cell lines expressing the orphan receptor. Following immunoprecipitation of the receptor, autophosphorylation can be monitored most readily by antiphosphotyrosine Western blotting (26).

Methods

A strategy for the introduction of new restriction sites into a *trk*B cDNA for cloning into a bacterial expression vector was used. This strategy requires the design of oligonucleotide primers for PCR which allow introduction of restriction sites at the ends of the PCR product which can correspond to any region of the template cDNA clone (Fig. 1).

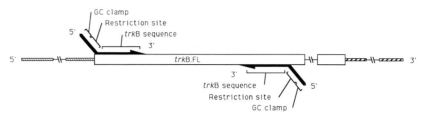

FIG. 1   Schematic diagram indicating the design of PCR primers.

Primer Design

At the 5′ end of the oligonucleotide, a GC clamp (4 base pairs of G or C in random order) is used. On the 3′ side of this GC clamp in the primer is placed a restriction enzyme site followed by a region which corresponds to the cDNA being cloned. The GC clamp facilitates recognition of the cleavage site by the restriction enzymes. If synthesis of the PCR products is incomplete (resulting in ragged ends), the restriction sites may still be intact because they are separated by 4 base pairs from the ends of the PCR product. Alternatively, the GC clamp may prevent breathing at the ends of the PCR product which could result in poor recognition by restriction enzymes. Regardless of the mechanism, the clamp allows efficient recognition of cleavage sites by restriction enzymes. The reverse primer is designed in the same fashion, but the use of a different restriction enzyme site is desirable to allow unidirectional ligation into an expression vector. The predicted melting temperature (T_m) of each of the primers (counting only the region which corresponds to the cDNA to be cloned) for the cDNA should roughly match. For primers of about 35 bases or less, the simple summation of 4°C for each GC pair and 2°C for each AT pair suffices (27). The temperature for the PCR reaction annealing should be 4°C below the predicted T_m of the primers. Primer with T_m values between 48 and 60°C work well. The design of the primer sequence should take into account any convenient matching sequences between potential restriction enzyme sites and the cDNA clone. For example, a primer at the 3′ untranslated region of a cDNA does not have to be precisely located. Therefore, a shorter primer can be designed if a region with a suitable restriction site or part of a restriction site is included in the primer. Yield and purity of oligonucleotide synthesis decrease with increasing length of the primer being synthesized; in addition, less expense is incurred by minimizing the length of the oligonucleotides synthesized.

The advantage of using the PCR to introduce precisely located restriction sites is illustrated by the cloning of *trk*B into a bacterial expression vector (Fig. 2). Pines and Hunter (28) modified a pGEM vector by introduction of a ribosomal binding site (RBS) to the 5′ side of an *Nco*I site and found enhanced expression of human cyclin A in bacterial with this modified vector. The *Nco*I site contains an ATG sequence which allows the placement of a start codon at a precise distance from the ribosomal binding site. The PCR was used to introduce an *Nco*I site into the 5′ end of a *trk*B cDNA.

Primers were designed for expression of the extracellular domain of a TrkB as a bacterial nonfusion protein. The predicted N-terminal sequence of the TrkB receptor is CPMSCK after cleavage of the signal peptide. The primer used for expression would result in the sequence MACPMSCK. The extra alanine residue on the amino terminal was required in order to introduce the *Nco*I restriction site. As the N-terminal cysteine may be critical for the structure of the N-terminal end of TrkB, we opted for adding an extra residue as opposed to replacing the cysteine with alanine. A 27-mer,

5′ GCG-CCC-ATG-GCC-TGC-CCC-ATG-TCC-TGC 3′

was designed to encode the amino terminus of the bacterial protein. The C-terminal primer,

5′ GGC-CAA-GCT-TAC-TCC-CGA-TTG-GTT-TGG-TC 3′

corresponds to the sequence of the extracellular region next to the transmembrane domain in addition to containing a *Hin*dIII restriction site sequence. (The *Nco*I and *Hin*dIII restriction sites are underlined.) As many transmembrane proteins are glycosylated, one should keep in mind that bacterial expression of a protein does not result in glycosylation. In addition, the disulfide bonds that form in a protein expressed intracellularly in bacteria and purified may be different than the natural disulfide bonds. As a result, the protein may not be folded properly. Nevertheless, this strategy for raising antisera did prove successful for the TrkB receptor.

Conditions for PCR

The conditions for PCR are basically those developed by Cetus (Emeryville, CA) using Taq polymerase. Add to a 0.5-ml microcentrifuge tube, 16 μl dNTPs stock (1.25 mM each of dATP, dTTP, dGTP, and dCTP), 10 μl of 10× Taq buffer [500 mM KCl, 100 mM Tris-HCl, pH 8.3, 15 mM MgCl$_2$, and 0.1% (w/v) gelatin], 5 μl of both of the primer solutions (10 μM each in water), 0.5 μl Taq polymerase [5 units (V) = μl, Perkin-Elmer/Cetus, Nor-

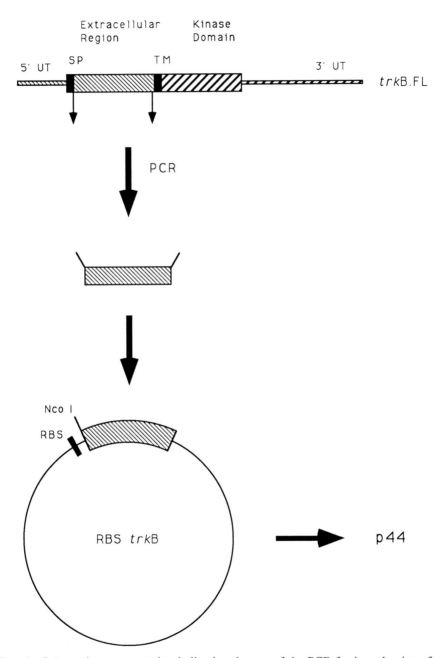

F_{IG}. 2   Schematic representation indicating the use of the PCR for introduction of
specific restriction sites into a cDNA for cloning into a bacterial expression vector.

walk, CT] 5μl cDNA solution containing 0.5 μg *trk*B.FL cDNA, and 58.5 μl water, which makes a total reaction volume of 100 μl. The final concentrations of dNTPs are 200 μ*M* each, and the final concentrations of primers are 0.5 μ*M* each. The solution is then overlayed gently with 50–75 μl mineral oil (Sigma, St. Louis, MO). The PCR conditions for each cycle are 1.5 min at 92°C (denaturation), 1.5 min at 53°C (hybridization), and 1.0 min at 72°C (extension). A total of 35 to 45 cycles are used, followed by storage at 6°C.

Polymerase chain reactions using 35 cycles and room temperature storage following the reaction are usually satisfactory. If one encounters problems such as nonspecific PCR products, options include raising the annealing temperature. If a PCR product is not observed, lowering the annealing temperature of the reaction should be tried first.

Cloning into Expression Vector

The 1.2-kb PCR product is resolved by agarose gel electrophoresis (0.8% agarose) in TAE buffer (40 m*M* Tris–acetate, 1 m*M* EDTA) containing ethidium bromide (0.5 μg/ml) for visualization of the DNA (27). The eluted PCR product is purified using NA45 paper according to the manufacturer's instructions (Schleicher and Schuell, Keene, NH). The PCR product is then digested with appropriate restriction enzymes, in this case *Nco*I and *Hin*dIII, for 2 hr using OnePhorAll buffer (Pharmacia, Piscataway, NJ) followed by chloroform–phenol extraction and ethanol precipitation (27). The PCR product is ligated using T4 ligase into the RBS pGEM vector previously digested with *Nco*I and *Hin*dIII and purified (27). The product of the ligation reaction can then be used to transform the DH5α strain of *Escherichia coli,* which are subsequently grown under ampicillin selection on plates. Several colonies should be picked and grown under ampicillin selection, and then plasmid DNA should be prepared for restriction enzyme digestion analysis. In this case, a plasmid (pRBS *trk*B) containing the *trk*B insert was then used for transformation of BL 21(DE3) cells carrying the pLys S plasmid (29) for expression of the *trk*B extracellular domain. The transformed BL 21 cells are grown under chloramphenicol (25 μg/ml) and ampicillin (50 μg/ml) selection. Multiple colonies are picked for further analysis.

Expression of T7 RNA polymerase is inducible by isopropyl-β-D-thiogalactopyranoside (IPTG) in the BL 21 (DE3) strain. As the T7 RNA polymerase is inducible by IPTG, expression of *trk*B, which is under the control of a T7 RNA polymerase promoter, is induced by IPTG. In addition, rifampicin specifically blocks host (*E. coli*) RNA polymerase, allowing exclusive expression of genes under the control of T7 RNA polymerase (30). Since *E. coli* RNA has a short half-life, rifampicin blocks essentially all host protein synthesis to allow expression only of proteins under the control of a T7 RNA polymerase promoter.

pLys S is a plasmid carrying a T7 lysozyme gene and also confers chloramphenicol resistance. As T7 lysozyme inhibits basal T7 RNA polymerase, it increases the tolerance of BL 21(DE3) cells to potentially toxic proteins under the control of the T7 RNA promoter, which is discussed in detail by Studier *et al.* (29). In addition, the expression of lysozyme aids in the lysis of cells during purification of proteins. Mild treatments that disrupt the inner membrane can result in lysis of the cells containing T7 lysozyme. This eliminates the need to add exogenous lysozyme to breakdown *E. coli* cell walls during lysis of cells for the subsequent purification of proteins.

Expression of p44^{trkB}

Test inductions are performed to confirm protein expression in BL 21 cells. In this case, the cells transformed with pRBS *trk*B should express p44^{trkB}. Twenty microliters of an overnight culture of pRBS *trk*B-transfected BL 21 cells is seeded into 3 ml Luria broth (LB) (27) with ampicillin and chloramphenicol and grown to a cell density giving an optical density (OD) of 0.5 absorbance units (AU) at 37°C (requires about 3 hr). Four 400-μl aliquots are taken, and two are treated with IPTG (400 μM). IPTG induces the expression of the exogenous protein. If p44^{trkB} is produced in the presence of IPTG and rifampicin, it confirms that is the result of transformation of the cells with the RBS *trk*B plasmid. These cultures are grown an additional 3 hr at 37°C, and then two of the cultures, one induced with IPTG and one uninduced, are treated with rifampicin (200 μg/ml) for 15 min. The four cultures are then pelleted by centrifugation in a microcentrifuge. At this point 10 μl of a solution of Tran^{35}Slabel (ICN, Cosa Mesa, CA) at a concentration of about 1.0 μCi/μl in LB is added to the pellets, vortexed, and incubated for 30 min at room temperature. Then 25 μl of sodium dodecyl sulfate–polyacrylamide gel electrophoresis (SDS–PAGE) sample buffer is added. After heating at 100°C for 5 min, the samples are submitted to SDS–PAGE (15% acrylamide gel) and subjected to autoradiography.

The bacteria expresses a protein which migrates with an M_r of about 44,000 on SDS–PAGE (Fig. 3). p44^{trkB} is made in the presence of rifampicin in the IPTG-induced cells, which indicates it is due to transcription of the pRBS *trk*B plasmid by T7 polymerase. The proteins with M_r values of about 29,000 and 27,000 that are synthesized in the presence of rifampicin in the IPTG-induced cells are β-lactamase precursor and β-lactamase (30). Since the *amp*^{R} gene is downstream of the T7 promoter, the β-lactamase coding sequence is encoded 3' to *trk*B on a polycistronic RNA and is therefore translated.

To determine whether the protein is soluble or contained in inclusion bodies, inclusion bodies are isolated. Twenty microliters of an overnight culture of pRBS *trk*B-transfected BL 21 cells is seeded into 3 ml LB with ampicillin and chloramphenicol and grown to an OD of 0.5 AU at 37°C. Two 400-μl aliquots are taken, induced with 400 μM IPTG, and grown for an additional 3 hr at 37°C. The cells are pelleted by centrifugation in a microcentrifuge. One sample is lysed in 40 μl sample buffer for SDS–PAGE. The second pellet is subjected to a freeze–thaw cycle in 120 μl lysis buffer (50 mM Tris-HCl, 100 mM NaCl, and 1 mM EDTA buffered to pH 8.0); the pellets incubated 15 min at room temperature, cooled to 4°C, and sonicated twice for 15 sec. After centrifugation in a microcentrifuge, the pellet is washed with 120 μl Triton wash buffer (50 mM Tris-HCl, 100 mM NaCl, 1 mM EDTA, 0.5% Triton X-100, and 10 mM EDTA buffered to pH 8.0) and centrifuged again. The pellet is then taken up in 40 μl sample buffer for SDS–PAGE. After electrophoresis on a 15% gel, the gel is stained with Coomassie Brilliant Blue R250. If both lanes contain equal amounts of expressed protein, in this case p44^{trkB}, the protein is in the insoluble fraction (inclusion bodies).

Preparation of Antigen

An overnight culture of pRBS *trk*B-transfected BL 21 cells in 1 ml LB is seeded into 400 ml LB containing chloramphenicol and ampicillin. The culture should be grown to 0.5 OD, which required 3–4 hr at 37°C. At this point, the culture is induced with 400 μM IPTG and grown for an additional 3 hr at 37°C. The cells are pelleted by centrifugation at 9000 g. The cells are then subjected to one freeze–thaw cycle in 120 ml lysis buffer (50 mM Tris-HCl, 100 mM NaCl, and 1 mM EDTA buffered to pH 8.0); the cells are incubated 15 min at room temperature, cooled to 4°C, and sonicated twice for 15 sec. After centrifugation at 14,000 g, the pellet is washed with 120 ml Triton wash buffer (50 mM Tris-HCl, 100 mM NaCl, 1 mM EDTA, 0.5% Triton X-100, and 10 mM EDTA buffered to pH 8.0) and centrifuged at 14,000 g. The pellet is then taken up in sample buffer for SDS–PAGE gel electrophoresis. The protein can be quantitated by serial dilution of the protein followed by analysis on SDS–PAGE gel. The gel is stained with

FIG. 3   Test inductions of BL 21(DE3) strain containing pLys S transformed with pRBS trkB. From the left: lane 1, uninduced control; lane 2, uninduced control treated with rifampicin; lane 3, IPTG induced; lane 4, IPTG induced and treated with rifampicin.

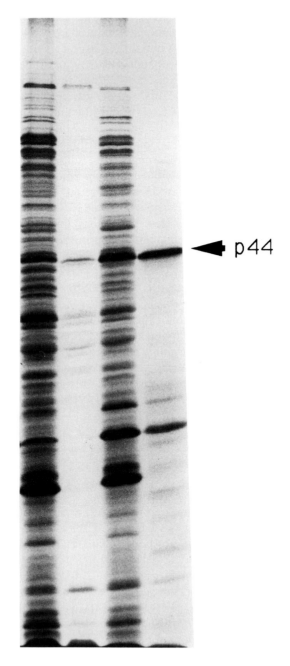

Coomassie Brilliant Blue R250. Lanes containing a serial dilution of antigen protein are contrasted to lanes containing serially diluted protein standards (Bio-Rad, Richmond, CA) in order to quantitate the concentration of protein.

Solubilized pellet containing 200 μg of p44^{trkB} is loaded on a 15% polyacrylamide gel that is 2 mm thick and 15 cm wide. After SDS–PAGE, the proteins are visualized as opaque bands by treatment with 250 mM KCl at 4°C. The p44^{trkB} band is excised by cutting the gel with a razor blade. The gel slices are placed in a 1.5-ml microcentrifuge tube which has holes in the bottom. The holes are made by piercing the tube with an 18-gauge hypodermic needle. The tube loaded with gel slices and which has holes in the bottom is placed on top of a second microcentrifuge tube. A gel slurry is collected in the bottom tube by centrifugation at full speed for 15–30 sec in a microcentrifuge. The slurry is combined with an equal volume of either Freund's complete adjuvant (first injection) or Freund's partial adjuvant (subsequent boosts) and injected into 4-pound female New Zealand White rabbits subcutaneously at four injection sites. The boosts are done at 1-month intervals. Procedures for bleeding rabbits and preparing serum are described in detail by Harlow and Lane (31).

Generation of Mammalian Lines Expressing TrkB Receptor

To generate stable expression of *trk*B in variety of mammalian cell lines, we opted for a retroviral mediated gene transfer system (Fig. 4). Advantages may include the ability to infect difficult to transfect cell lines such as neuroblastomas. As Northern blot analysis indicated that Rat-2 cells do not express *trk*B, this cell line was chose for the initial characterization and testing of anti-TrkB antisera. The pLNL-SLX CMV vector, which contains cloning sites on the 3' side of the human cytomegalovirus promoter and a *neo*^{R} selectable marker (32), was used. This vector is derived from LNL-XHC (33). The entire *trk*B coding region including a predicted signal peptide was cloned into pLNL-SLX CMV using the approach discussed above. The PCR reaction was carried out using a 29-mer with a predicted T_m of 60°C which contains a *Bam*HI site (5' CCG-GCT-CGA-GCT-AGC-CTA-GGA-TGT-CCA-GG 3') and a 28-mer with a predicted T_m of 60°C which contains an *Xho*I site (5' GCG-CGG-ATC-CAT-GGC-GCG-GCT-CTG-GGG-C 3'). (The restriction sites are underlined.) PCR conditions used were identical with those described above except that the annealing temperature was 58°C and only 35 cycles were used. The 2.4-kb PCR product was purified by agarose gel electrophoresis, digested with *Bam*HI and *Xho*I restriction enzymes, and ligated into the pLNL-SLX CMV vector, which had also been digested with *Bam*HI and *Xho*I. The pSLX *trk*B construct was transfected into the DH5 strain of *E. coli* and purified for transfection using CsCl gradients (27).

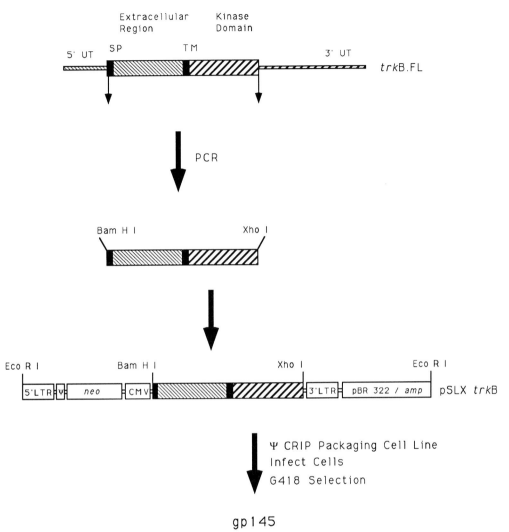

FIG. 4   Schematic illustration depicting the strategy for expression of TrkB receptor in Rat-2 fibroblasts using a retroviral gene transfer system.

Establishment of Cell Lines with pSLX trkB

The ΨCRIP amphotrophic packaging cell line is used to package the pSLX *trk*B construct into viral particles (34). Alternatively, the ecotrophic packaging ΨCRE cell line can be used (34). The protocol described by Sharfmann *et* *al.* with slight modifications is used for production of viruses and subsequent infection of recipient cell lines (32). ΨCRIP cells (1×10^6) are plated in 10% calf serum–Dulbecco's modified Eagle's medium (10% CS/DMEM) in a 10-cm dish. After 12–18 hr at 37°C, 20 μg DNA is transfected using a standard calcium phosphate precipitation method (27). Briefly, 50 μl of 2.5 *M* CaCl$_2$ is mixed with 20 μg plasmid in 200 μl water, supplemented with 200 μl of 2× HBS buffer adjusted to pH 7.12 (280 m*M* NaCl, 50 m*M* HEPES, and 1.5 m*M* Na$_2$HPO$_4$), and incubated a room temperature for 20 min. The suspension is then added to the ΨCRIP cells and incubated 16 hr at 37°C. The cells are washed with phosphate-buffered saline (PBS) twice and fed with 10 ml of 10% CS/DMEM. An alternative method for transfection using calcium phosphate precipitation of DNA has also been used with success (35).

Twenty-four hours later, the ΨCRIP cells are washed twice with VE (versenate buffer) and fed with 10% CS/DMEM. The recipient Rat-2 fibroblasts are plated at 1×10^6 per 10-cm plate at this time. Twenty-four hours later, the medium from the ΨCRIP cells which contains recombinant retrivirus is filtered and added to the Rat-2 cells. Polybrene is added to a concentration of 8 μg/ml, which improves the infection efficiency. Twenty-four hours later, the cells are split at several dilutions into 10% CS/DMEM containing G418 (400 μg/ml). The G418 selection takes 10–14 days. Medium should be changed every 3 days. 10–20 colonies are picked and assayed for expression using Western blotting analysis. In this case, Western blotting with the anti-TrkB antiserum should detect TrkB receptor. Northern analysis to confirm expression of mRNA should also be done. A Rat-2 LacZ pool of cells was also established as a control by infection of Rat-2 cells with recombinant virus produced by a ΨCRIP line that packages a pLNL-SLX CMB β-galactosidase construct (32).

It is often desirable to establish a retroviral packaging cell line for producing recombinant virus. The procedure described in detail by Scharfmann *et al.* (32) involves transfection of ΨCRE cells, harvesting virus, and infecting ΨCRIP cells. The infected cells are selected in the presence of G418 (400 μg/ml). Single colonies are isolated and expanded, then assayed for viral titer on NIH 3T3 or Rat-2 cells. The medium from these lines can be filtered and frozen at −80°C for infections at later times. In a variation on this theme, colonies of ΨCRIP cells transfected with pSLX *trk*B above were isolated by selection with G418 and assayed for virus on Rat-2 cells. A transfected line, PST 10 (packaging pSLX *trk*B), was isolated which has a titer of slightly greater than 1×10^3 neomycin-resistant colonies/ml. If infected ΨCRIP packaging cell lines are isolated, titers of greater than 5×10^4 can be obtained.

Western Blotting Analysis

Confluent cells in 10-cm culture dishes are lysed with 400 μl of SDS-PAGE gel sample buffer and boiled, and then 40 μl is loaded per lane for SDS-PAGE. Prestained Rainbow markers (Amersham, Arlington Heights, IL) are used for molecular weight standards to eliminate an unnecessary staining step. After electrophoresis, the proteins are transferred to Immobilon-P (Millipore, Bedford, MA), using a semidry rapid blotter (Acrylictech, San Diego, CA) using the protocol of Kyhse-Anderson (36). Western blotting is performed with the method of Glenney with slight modifications (37). The membrane is blocked by shaking at room temperature for 30 min in MTA buffer [5% nonfat dried milk, 0.2% Tween 20, 3 mM NaN$_3$ in PBS (27)]. The membrane is then incubated with antibodies for 12–16 hr at room temperature at various dilutions of antibody in order to determine titers. Shorter incubation periods (2–12 hr) are possible with many antibodies. Ideally, one needs to do a time course for this step to optimize the time required to complete the Western analysis. The membrane is then washed with shaking at room temperature 6 times for 5 min with wash buffer (150 mM NaCl, 10 mM sodiuim phosphate, 0.2% Tween 20 adjusted to pH 7.5). The membranes are then incubated with ^{125}I-labeled protein A (Amersham, Arlington Heights, IL) at a concentration of 10 μCi/20 ml in MTA for 1 hr with shaking at room temperature. The membrane is washed 6 times for 5 min as before in wash buffer. After brief drying, the membrane is covered with Saran Wrap and autoradiographed.

Both of the antisera raised against the extracellular domain of the TrkB receptor, anti-TrkB 5049 and anti-TrkB 5050, detect a 145-kDa protein in the RST (rat-2 and pSLX *trk*B) cells (Fig. 5). The highest expression of gp145^{trkB} was observed in the lines RST 1 in the first series of infections and RST 15 in the second series of infections. The titer for Western blotting with anti-TrkB 5049 is about 1:4000, and the titer of anti-TrkB 5050 is slightly less than 1:4000. Both antisera work for immunoprecipitation, although anti-TrkB 5050 works slight better for immunoprecipitation than anti-TrkB 5049.

Autophosphorylation Assay

Confluent plates of RST 1 cells are treated with either BDNF of NT-3 in 10% CS/DMEM for 5 min at 37°C. The plates are then washed twice with 4°C TD (Tris-buffered saline) and lysed in 1 ml RIPA (31) buffer at 4°C. TrkB recep-tor is incubated with 5 μl anti-TrkB 5050 for 2 hr followed by incubation with 20 μl protein A coupled to Sepharose beads (Repligen, Cambridge, MA) with rotation at 1 hr at 4°C. The beads are then pelleted

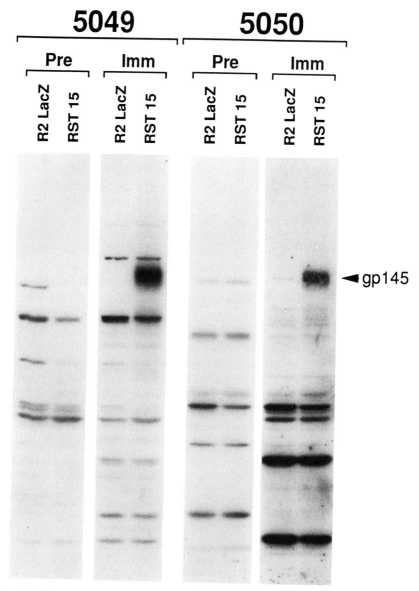

FIG. 5 Western blotting of proteins isolated from RST 15 cells with anti-TrkB antibodies (1:400. (Left) Western blotting with 5049 anti-TrkB, both preimmune (Pre) and immune (Imm). (Right) Western blotting with 5050 anti-TrkB. Each serum was assayed against protein from Rat-2 LacZ (control) and RST 15 (gp145^{trkB}) cells.

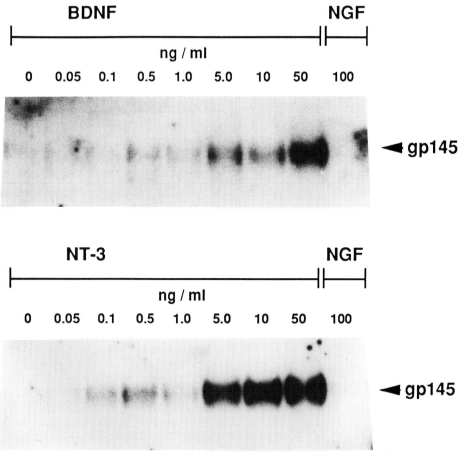

FIG. 6   Antiphosphotyrosine Western blot of immunoprecipitated TrkB receptor from RST 1 cells treated with BDNF, NT-3, and NGF for 5 min at 37°C at various concentrations.

by centrifugation for 30 sec in a microcentrifuge. The beads are washed with 1 ml RIPA buffer at 4°C briefly and centrifuged. This washing procedure is repeated 3 more times. The pellet is then taken up in 20 μl of 2× sample solvent for SDS-PAGE. After resolution on a 7.5% polyacrylamide gel, the proteins are transferred as described above. Antiphosphotyrosine Western blotting assays are performed using the detailed protocols developed by Kamps (26).

Both BDNF and NT-3, but not NGF, stimulate phosphorylation of gp145[trkB] on tyrosine in both the RST 1 and RST 15 cell lines, presumably due to autophosphorylation (Fig. 6). Although NT-3 works at slightly lower concentrations than BDNF in this experiment, it should be kept in mind that this is a kinetic experiment with data taken at one time point only (5 min). This experiment does not address the affinities of BDNF and NT-3 for the TrkB receptor directly. Indeed, there is evidence that NT-3 stimulates TrkB autophosphorylation more rapidly than BDNF (21).

The expression of receptor PTKs coupled with antiphosphotyrosine Western blotting can allow for rapid screening of potential ligands, either purified or in conditioned medium, for orphan receptor PTKs. In addition, antiphosphotyrosine Western blotting can also be used for the identification of substrates for receptor PTKs. TrkB antisera may prove invaluable for immunohistochemical localization of TrkB protein expression in the nervous system.

Acknowledgments

I thank Drs. Arnon Rosenthal and Karoly Nikolics at Genentech for providing purified recombinant BDNF and NT-3. This work was carried out in Dr. Tony Hunter's laboratory, whom I thank for suggestions and advice, in addition to support. Drs. Bart Sefton and Mark Kamps provide antiphosphotyrosine antibodies.

References

1. S. K. Hanks, *Proc. Natl. Acad. Sci. U.S.A.* **84,** 388 (1987).
2. R. A. Lindberg and T. Hunter, *Mol. Cell. Biol.* **10,** 6316 (1990).
3. D. S. Middlemas, R. A. Lindberg, and T. Hunter, *Mol. Cell. Biol.* **11,** 143 (1991).
4. S. K. Hanks and R. A. Lindberg, *in* "Methods in Enzymology" (T. Hunter and B. M. Sefton, eds.), Vol. 200, p. 525. Academic Press, San Diego, 1991.
5. M. H. Kraus and S. A. Aaronson, *in* "Methods in Enzymology" (T. Hunter and B. M. Sefton, eds.), Vol. 200, p. 546. Academic Press, San Diego, 1991.
6. A. F. Wilks, R. R. Kurban, C. M. Hovens, and S. J. Ralph, *Gene,* **85,** 67 (1989).
7. C. Lai and G. Lemke, *Neuron* **6,** 691 (1991).
8. A. F. Wilks, *in* "Methods of Enzymology" (T. Hunter and B. M. Sefton, eds.), Vol. 200, p. 533. Academic Press, San Diego, 1991.
9. R. A. Lindberg, D. P. Thompson, and T. Hunter, *Oncogene* **3,** 629 (1988).
10. R. A. Lindberg and E. B. Pasquale, *in* "Methods of Enzymology" (T. Hunter and B. M. Sefton, eds.), Vol. 200, p. 557. Academic Press, San Diego, 1991.
11. S. Alema, P. Caselbore, E. Agostoni, and F. Tato, *Nature (London)* **316,** 557 (1985).
12. P. A. Maher, *Proc. Natl. Acad. Sci. U.S.A.* **85,** 6788 (1988).
13. R. Klein, L. F. Parada, F. Coulier, and M. Barbacid, *EMBO J.* **8,** 3701 (1989).
14. R. Klein, D. Conway, L. F. Parada, and M. Barbacid, *Cell (Cambridge, Mass.)* **61,** 647 (1990).

15. D. Martin-Zanca, R. Oskam, G. Mitra, T. Copeland, and M. Barbacid, *Mol. Cell. Biol.* **9,** 24 (1989).
16. F. Lamballe, R. Klein, and M. Barbacid, *Cell (Cambridge, Mass.)* **66,** 967 (1991).
17. D. Kaplan, D. Martin-Zanca, and L. F. Parada, *Nature (London)* **350,** 158 (1991).
18. R. Klein, S. Jing, V. Nanduri, E. O'Rourke, and M. Barbacid, *Cell (Cambridge, Mass.)* **65,** 189 (1991).
19. B. L. Hempstead, D. Martin-Zanca, D. R. Kaplan, L. F. Parada, and M. V. Chao, *Nature (London)* **350,** 678 (1991).
20. D. L. Kaplan, B. L. Hempstead, D. Martin-Zanca, M. V. Chao, and L. F. Parada, *Science* **252,** 554 (1991).
21. D. Soppet, E. Escandon, J. Maragos, D. S. Middlemas, S. W. Reid, J. Blair, L. E. Burton, B. R. Stanton, D. R. Kaplan, T. Hunter, K. Nikolics, and L. F. Parada, *Cell (Cambridge, Mass.)* **65,** 895 (1991).
22. S. P. Squinto, T. N. Stitt, T. H. Aldrich, S. David, S. M. Bianco, P. Masiakowski, M. E. Furth, D. M. Valenzuela, P. S. DiStefano, and G. D. Yancopoulos, *Cell (Cambridge, Mass.)* **65,** 885 (1991).
23. R. Klein, V. Nanduri, S. Jing, F. Lamballe, P. Tapley, S. Bryant, C. Cordon-Cardo, K. R. Jones, L. F. Reichardt, and M. Barbacid, *Cell (Cambridge, Mass.)* **66,** 395 (1991).
24. D. J. Glass, S. H. Nye, P. Hantzopoulos, M. J. Macchi, S. P. Squinto, M. Goldfarb, and G. D. Yancopoulos, *Cell (Cambridge, Mass.)* **66,** 405 1991).
25. L. R. Berkemeier, J. W. Winslow, D. R. Kaplan, K. Nikolics, D. V. Goeddel, and A. Rosenthal, *Neuron* **7,** 857 (1991).
26. M. P. Kamps, *in* "Methods in Enzymology" (T. Hunter and B. M. Sefton, eds.), Vol. 201, p. 101. Academic Press, San Diego, 1991.
27. J. Sambrook, E. F. Fritsch, and T. Maniatis, "Molecular Cloning: A Laboratory Manual." Cold Spring Harbor Laboratory Press, Cold Spring Harbor, New York, 1989.
28. J. Pines and T. Hunter, *Cell (Cambridge, Mass.)* **58,** 833 (1989).
29. F. W. Studier, A. H. Rosenberg, J. J. Dunn, and J. W. Dubendorff, *in* "Methods in Enzymology" (D. V. Goeddel, ed.), Vol. 185, p. 60. Academic Press, San Diego, 1990.
30. S. Tabor and C. C. Richardson, *Proc. Natl. Acad. Sci. U.S.A.* **82,** 1074 (1985).
31. E. Harlow and D. Lane, "Antibodies: A Laboratory Manual." Cold Spring Harbor Laboratory, Cold Spring Harbor, New York, 1988.
32. R. Scharfmann, J. H. Axelrod, and I. M. Verma, *Proc. Natl. Acad. Sci. U.S.A.* **88,** 4626 (1991).
33. M. A. Bender, T. D. Palmer, T. E. Gelinas, and A. D. Miller, *J. Virol.* **61,** 1639 (1987).
34. O. Danos and R. C. Mulligan, *Proc. Natl. Acad. Sci. U.S.A.* **85,** 6460 (1988).
35. C. Chen and H. Okayama, *Mol. Cell. Biol.* **7,** 2745 (1987).
36. J. Kyhse-Anderson, *J. Biochem. Biophys. Methods* **10,** 203 (1984).
37. J. Glenney, *Anal. Biochem.* **156,** 315 (1986).

[10] Transfection with Nerve Growth Factor Receptor Complementary DNA

Hiroshi Matsushima and Emil Bogenmann

Introduction

Nerve growth factor (NGF) is an essential neurotrophic factor responsible for survival and differentiation of sympathetic and peripheral sensory neurons (1). Recently, cholinergic neurons of the basal forebrain and cerebellar neurons of mammals were shown to be NGF target cells. Expression of NGF and NGF binding sites have been found in glia cells and several nonneural tissues including kidney, testis, muscle, and lymphoid cells (2, 3). Although NGF was thought to be a factor required for neuronal cell survival and differentiation, it was recently demonstrated that NGF acts as mitogen in NIH 3T3 fibroblasts transfected with the high-affinity NGF receptor gene, and the NGF signal cascade may also play an important role in the regulation of the inflammatory processes (4, 5). These findings suggest that NGF has various biological functions depending on the cell type.

The various functions of NGF in a variety of different cell types are mediated by binding to membrane-associated cell surface receptors (NGFR). A low-affinity receptor component (gp75^{NGFR}) (6) and a high-affinity receptor (gp140$^{proto\text{-}trk}$) (7) have been identified, and their genes have been cloned from various species (8, 9). The function(s) of each of the receptor molecule has not completely been worked out (10). Expression of the cDNA encoding gp140$^{proto\text{-}trk}$ in NIH 3T3 cells results in the assembly of low- and high-affinity binding sites, and NGF treatment activates immediate early genes (i.e., c-*fos*) in the absence of gp75^{NGFR} expression (4, 7). Long-term NGF treatment generates a transformed phenotype, suggesting that the gp140$^{proto\text{-}trk}$ activation induces a mitogenic stimulus. On the other hand, expression of gp75^{NGFR} cDNA in a mutant PC 12 cell line also generated low- and high-affinity binding sites; NGF treatment was followed by c-*fos* gene activation, but no neurite outgrowth could be observed (11). The same authors also suggest that a functional NGF signal cascade requires the concomitant expression of gp75^{NGFR} and gp140$^{proto\text{-}trk}$ genes (12).

We, therefore, investigated the role of gp75^{NGFR} and gp140$^{proto\text{-}trk}$ in cell growth and differentiation when expressed in neuronal cell lines. We expressed the corresponding human cDNAs in cell lines which lack a functional NGF cascade, and special attention was given to the role the extracellular matrix (ECM) may play in the NGF-regulated pathway of cell differentiation.

Methods in Neurosciences, Volume 12

Transfection of gp75^{NGFR} cDNA into a Human Neuroblastoma Cell Line

Human Neuroblastoma (NB), a neural crest-derived tumor, is the most common solid neoplasm in childhood. The tumors show various degrees of cellular differentiation, and spontaneous regression may occur (13). The *MYCN* oncogene has been implicated in the progression of the tumor (14). Most NB cell lines isolated from patients with advanced disease do not express a functional NGF/NGFR cascade. We have isolated an NB cell line (HTLA230) from a patient with stage IV disease (15). The cell line shows high N-*myc* gene amplification, expresses neuron-specific genes, such as the NF68 and SCG-10 gene, and stains positive with the A2B5 and HNK-1 antibodies. The cell line, however, does not express either the gp75^{NGFR} or the gp140$^{proto-trk}$; thus, no functional NGF/NGFR cascade is present.

We transfected the human gp75^{NGFR} cDNA expression vector (pMVE-1, kindly provided by Dr. Chao) (6) into HTLA230 cells by either electroporation or calcium phosphate precipitation. Neomycin (G418)-resistant clones were selected on R$_{22}$C1F matrix, an *in vitro* synthesized ECM (16). Positive transfectants were selected by an immunological rosette assay followed by five consecutive rounds of fluorescence-activated cell sorting (FACS) using the mouse anti-human gp75^{NGFR} antibody (ME-20.4) (17). With this combination of procedures we were able to select clones which showed more than 90% of the cells positive for the surface-bound gp75^{NGFR} molecules compared to background staining of HTLA230 cells. Northern blot analysis of one clone (98-3) demonstrated a message for the gp75^{NGFR} which was slightly larger than that seen in human melanoma (A875) cells. Immunoprecipitation of [^{35}S]methionine-labeled cells using the ME-20.4 antibody revealed a major band of 70–75 kDa, which corresponds in size to the glycosylated form of gp75^{NGFR} present in A875 cells (18).

Semiquantitative determination of the number of gp75^{NGFR} molecules present in 98-3 cells by immunoprecipitation using serial dilutions of A875 membrane preparations demonstrated that approximately 3×10^4 receptors/cell were expressed at the cell surface. Similar numbers were obtained by Scatchard plot analysis using ^{125}I-labeled NGF. Low-affinity binding sites (K_d 10^{-9} M, 3×10^4 receptors/cell) and high-affinity binding sites (K_d 10^{-11} M, $1–2 \times 10^3$ receptors/cell) were determined in 98-3 cells, but no specific binding was seen with HTLA230 parental cells. These numbers, however, need to be interpreted with caution, since FACS analysis of antibody-stained cells showed a broad fluorescence histogram and, therefore, ^{125}I-labeled NGF binding just represents an average number per cell. Also, we have not yet been able to determine whether high-affinity receptors are expressed on all cells or only in a subset of the clonally derived transfected cell line. We concluded, however, that transfection of the gp75[NGFR] cDNA into NB cells can generate low- and high-affinity receptors (19).

Next we tested the function of the transfected receptor gene and we the onset of early response genes (c-*fos* gene) as an indicator. Treatment of 98-3 cells with NGF (0.5 μg/ml) showed activation of the c-*fos* protooncogene within 40 min, not seen with parental cells. The level of induction was superinduced in the presence of the protein synthesis inhibitor anisomycin. Serum-starved 98-3 cells in which 80% of the cells were arrested in G_0/G_1 phase of the cell cycle also showed a higher level of c-*fos* induction than nonsynchronized cells, suggesting a cell cycle-dependent expression of a functional NGF/NGFR cascade (19).

Neuronal differentiation is characterized by outgrowth of neurites and expression of neuron-specific marker genes. When we treated 98-3 cells for 8–10 days with NGF using cells grown either on plastic or on an ECM, these experiments demonstrated unequivocally that transfectants were capable of morphologically differentiating by extending a network of long neurites, but only when grown on the ECM (Fig. 1). Furthermore, differentiated cells lacked detectable *MYCN* protein as determined by immunofluorescence and did not incorporate [³H]thymidine, suggesting that NGF-responsive NB cells became postmitotic (20). Therefore, NGF-induced neuronal differentiation requires extrinsic factors present within the ECM as well as a functional NGF–receptor complex.

Our observations that 98-3 cells morphologically differentiated only when grown on an ECM suggested to us that specific components of the matrix are required. We tested the $R_{22}C1F$ matrix, which is composed of 90% collagen (types I and III), 10% glycoproteins, and small amounts of glycosaminoglycans, for the ability to promote cell differentiation in 98-3 NB cells induced by NGF. Trypsin digestion of the ECM removes glycoproteins and core proteins of proteoglycan and guanidinium chloride extracts matrix proteoglycans (21). Parental and transfected 98-3 cells were grown on complete or trypsinized matrix in the continuous presence of NGF, and the number of morphologically differentiated cells was determined 8 days later. Biochemical alterations of the ECM did not affect cloning efficiency nor cell growth of either cell line. Parental HTLA230 cells demonstrated the same undifferentiated morphology under all culture conditions. However, the study showed that trypsin-extracted matrix lacks the components to promote NB cell differentiation, whereas combined trypsin and guanidinium treatment, on the other hand, restores the ability of the ECM to stimulate NGF- as well as retinoic acid (RA)-induced morphological differentiation (22). Similar observations were obtained with other cell types, such as skeletal muscle cells induced to differentiate *in vitro* (21). Further biochemical analysis of the components necessary for differentiation demonstrated that the ratio of

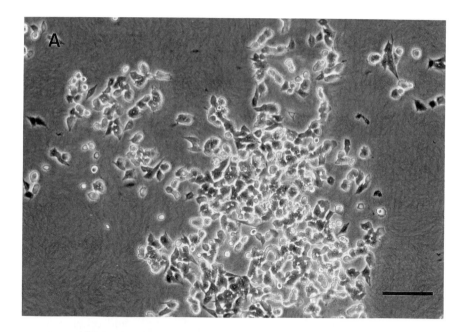

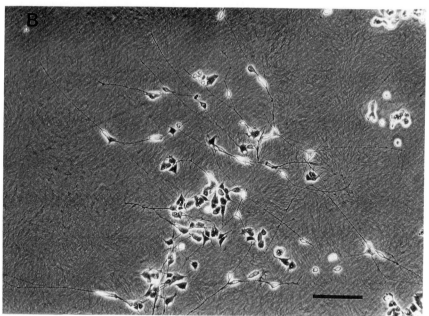

FIG. 1 Phase-contrast micrograph of 98-3 cells grown for 8 days on ECM in the absence (A) or presence (B) of NGF (0.5 μg/ml). Note the undifferentiated morphology of the cells in the absence of NGF, whereas in the presence of NGF many cells with long bifurcated dendrites are seen, suggesting differentiation of 98-3 cells. Bar, 50 μm.

hyaluronic acid to chondroitin sulfate present in the ECM is crucial for neuronal differentiation.

These experimental observations, however, are intriguing since gp75^{NGFR} is thought to be the low-affinity receptor for NGF and gp140$^{proto\text{-}trk}$ could not be detected by Northern analysis in NGF-treated 98-3 cells. One, therefore, has to speculate that expression of the cDNA in cells with a neuronal background can reconstitute a functional NGF–receptor complex by an unknown mechanism. Further biochemical studies are needed to investigate the role of gp75^{NGFR} in NB cells.

Transfection of gp140$^{proto\text{-}trk}$ into Neuronal Cell Lines

Previous reports demonstrated that the *trk* protooncogene (23), originally identified in a human colon carcinoma cell line, binds NGF with high affinity while other members of this gene family bind different neurotrophic factors (24, 25). Expression of gp140$^{proto\text{-}trk}$ in NIH 3T3 cells in the presence of NGF was shown to stimulate cell growth, and other reports suggest that coexpression of gp75^{NGFR} and gp140$^{proto\text{-}trk}$ is required for a functional NGF cascade. We investigated similar questions and expressed the gp140$^{proto\text{-}trk}$ cDNA (kindly provided by M. Barbacid) in various cell lines of neuroectodermal origin which lack a functional NGF cascade to test its role in neuronal cell growth and differentiation *in vitro*.

A rat central nervous system-derived cell line (B104) (26) which expresses low levels of gp75^{NGFR} was initially used, and gp140$^{proto\text{-}trk}$ stable transfectants (B104-*trk*) were established. Immediate early genes, such as the c-*fos* gene, were activated on NGF treatment, and surface expressions of the gp140$^{proto\text{-}trk}$ receptor was demonstrated by FACS analysis using an anti-gp140$^{proto\text{-}trk}$ antibody directed against the extracellular domain of the receptor (kindly provided by Dr. Eager). A growth analysis of B104-*trk* cells in the presence or absence of NGF (200 ng/ml) was performed with cells seeded at low cell density (Fig. 2). B104-*trk* cells grown in the presence of NGF showed growth stimulation compared to control cells; Clonal analysis of B104-*trk* cells in presence or absence of NGF was performed at low (1%) serum concentration (Fig. 3), and the results confirmed the earlier findings. The number of clones, their morphology which was dominated by round cells, and the size of the clones in the presence of NGF were significantly different when compared to parental B104 or nontreated B104-*trk* cells.

The data obtained with B104-*trk* cells grown at low cell density are in agreement with other reported observations. However, Northern blot analysis of long-term NGF-treated B104-*trk* cells also demonstrated the expression of marker genes of neuronal differentiation, such as transin, previously dem-

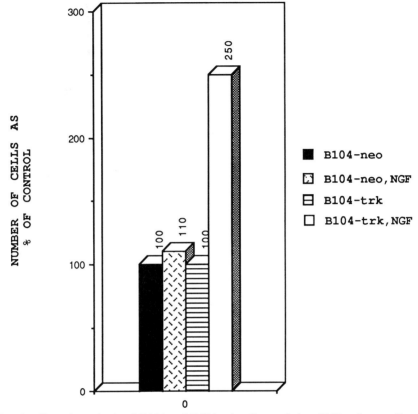

FIG. 2 Growth analysis of B104 and B104-*trk* cells seeded at 2000 cells per 60-mm dish and grown in the absence or presence of NGF (0.1 μg/ml) for 6 days. Cell counts are expressed as a percentage of untreated control cells. Note the 2.5-fold increase in cell number with B104-*trk* in the presence of NGF.

onstrated in mature neurons (27). It is worth noting that B104-*trk* cells morphologically differentiate after 4–6 days treatment with NGF, and cells with long processes, not present in untreated B104-*trk* cells, were seen. These morphological alterations are in agreement with the change in gene expression observed by Northern analysis.

Growth analysis, however, demonstrated that B104-*trk* cells in the presence of NGF proliferate, despite the expression of neuronal differentiation markers. Therefore, reconstitution of the NGF–receptor complex in B104 cells can activate the pathway of differentiation but it is not coupled with growth control. Alternatively, these cells, which are tumorigenic variants of

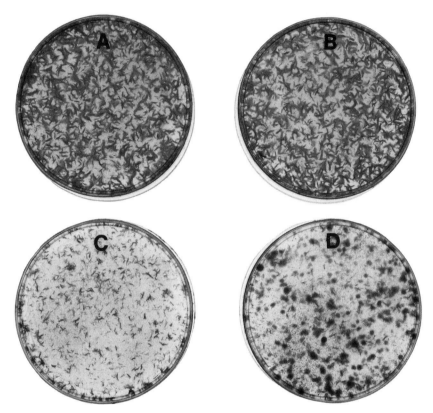

FIG. 3   Clonal analysis of B104-*trk* cells was performed with cells grown in low (1%) fetal bovine serum (FBS) (C, D) in the absence (C) or presence (D) of NGF (0.1 µg/ml) and cells grown in 10% FBS (A, B) without (A) or with (B) NGF. Note that when cells are grown under high serum conditions, there is no obvious growth effect visible after 8–10 days in culture. However, cells grown at clonal density and with added NGF showed a substantial growth advantage over untreated B104-*trk* cells.

normal embryonic brain cells, may represent an aberrant phenotype owing to transformation events.

Similar experiments, as mentioned above, were also performed with a human melanoma cell line (A875) known to express high levels of gp75^{NGFR} but to lack gp140$^{proto-trk}$ gene expression. Stable transfectants showed a typical c-*fos* response on NGF treatment, and, more importantly, long-term NGF treatment significantly reduced cell growth compared to untreated control cells (Fig. 4). The cells did not change their morphology in the presence of NGF when grown on either plastic or an ECM. It remains to be determined whether treated A875-*trk* cells express differentiation-specific marker genes, but so far no pigmented melanoma cells could be found.

With these experimental data, we demonstrated that expression of gp140$^{proto\text{-}trk}$ in neuronal cell lines expressing the endogenous gp75^{NGFR} gene reconstitutes a functional NGF signal cascade. The role of gp75^{NGFR} receptor molecule in the NGF cascade remains to be determined.

To investigate the question whether gp75^{NGFR} is necessary for neuronal differentiation, we used HTLA230 cells which lack expression of gp75^{NGFR} and gp140$^{proto\text{-}trk}$ genes. Stable transfectants (HTLA230-trk) with gp140$^{proto\text{-}trk}$ cDNA were established using electroporation followed by selection with G418 (1 mg/ml). Immunofluorescence analysis showed surface expression of the transfected receptor gene, and NGF treatment was followed by transient activation of the c-fos gene, suggesting a functional response in the absence of gp75^{NGFR} similar to that seen in NIH 3T3 cells. Long-term treatment with

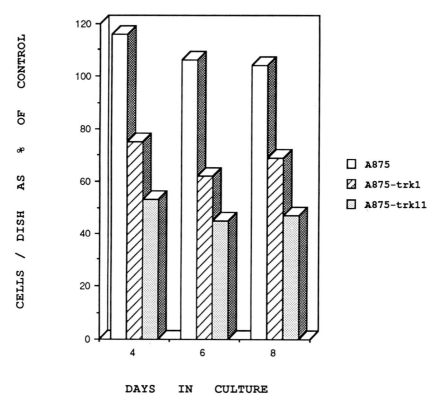

FIG. 4 Cells of a human melanoma (A875) cell line were transfected with the gp140$^{proto\text{-}trk}$ cDNA and stable transfectants grown in the absence or presence of NGF (0.1 μg/ml). Note the growth reduction in A875-trk clones 1 and 11 in the presence of NGF compared to NGF-treated A875 parental cells. Cell counts are expressed as a percentage of untreated control cells.

NGF showed dramatic growth reduction over a 10-day culture period (Fig. 5). NGF-treated HTLA230-*trk* cells also showed profound morphological changes within 2 days, and a network of dendrites indicated that these malignant cells underwent differentiation (Fig. 6). Furthermore, Northern blot analysis of long-term NGF-treated cells clearly showed activation of differentiation markers (i.e., GAP-43) (28), and down-regulation of the *MYCN* gene could be demonstrated (Fig. 7). These observations prompted us to speculate that gp140$^{proto-trk}$ expression in NB cells is sufficient to activate cell differentiation, which results in morphologically and molecularly differentiated as well as growth-arrested cells. We furthermore speculate that reconstitution of the high-affinity NGF receptor in NB cells has "tumor

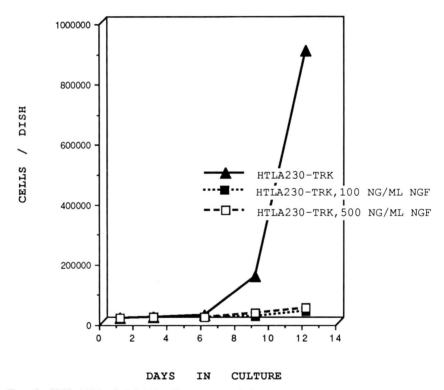

DAYS IN CULTURE

FIG. 5   HTLA230-*trk* (18-10) cells were seeded onto R$_{22}$C1F matrix at 10,000 cells per 60-mm dish and either left untreated or treated with 0.1 or 0.5 μg/ml of NGF. Note the nearly complete growth arrest of HTLA230-*trk* cells in the presence of NGF compared to untreated cultures.

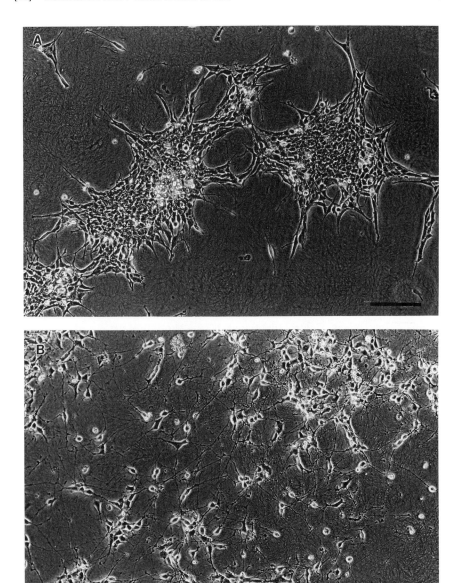

FIG. 6   HTLA230-*trk* (18-10) cells were grown for 8 days on R$_{22}$C1F matrix in the absence (A) or presence (B) of NGF (0.1 μg/ml). Note the undifferentiated cell morphology in (A) compared to the extended outgrowth of neurites seen with cells in (B). Bar, 50 μm.

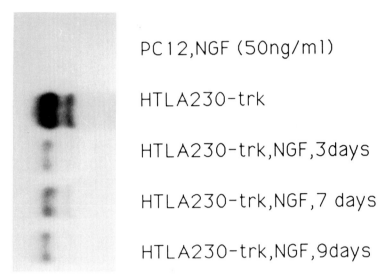

PC12,NGF (50ng/ml)

HTLA230-trk

HTLA230-trk,NGF,3days

HTLA230-trk,NGF,7 days

HTLA230-trk,NGF,9days

Fig. 7   HTLA230-*trk* (18-10) cells were grown on $R_{22}C1F$ matrix for 2 days followed by treatment without or with NGF (0.1 μg/ml), and RNA was isolated at indicated time points. Control (untreated) HTLA230-*trk* RNA was isolated at day 6. Note the high levels of *MYCN* expression in HTLA230-*trk* cells and the rapid down-regulation of it in treated cells.

suppressor'' activity, and current experiments explore this aspect using immunosuppressed animals inoculated with the transfected HTLA230-*trk* cells.

Out data, however, cannot yet explain the previously made observations which demonstrated that gp75[NGFR] expression in HTLA230 cells or mutant PC 12 cells can generate a functional response in the absence of gp140[proto-*trk*] expression. One may have to speculate that additional ''neuronal'' components are necessary to generate high-affinity binding with gp75[NGFR] or that other members of the *trk* may also be involved, although no such messages could be found by Northern blot analysis in HTLA230 cells. Alternatively, the gp75[NGFR] receptor, which has been shown to have substantial homology with other membrane-bound molecules, such as the tumor necrosis factor receptors (TNFR I and II) (29) as well as the CD30 molecule, the gene for which was recently isolated from a patient with Hodgkin's lymphoma (30), may have a yet unknown function. Since TNF is involved in programmed cell death (apoptosis), one could speculate that a similar function is exerted by gp75[NGFR] and that expression of the gene in tumor cells such as PC 12 and HTLA230 cells may aberrantly activate differentiation rather than apoptosis.

Summary

Our experimental observations provided the following conclusions.

1. Transfection of the gp75^{NGFR} cDNA into HTLA230 cells which lack gp75^{NGFR} and gp140$^{proto-trk}$ gene expression results in the functional assembly of a NGF–receptor complex. Activation with NGF leads to differentiation but only when cells are grown in appropriate culture conditions including an extracellular matrix.

2. Functional expression of gp140$^{proto-trk}$ in neuronal cells expressing the endogenous gp75^{NGFR} demonstrate that differentiation is activated on treatment of cells with NGF. Cell growth may or may not be affected by the presence of high-affinity receptor molecules, depending on the nature of the cell lines used.

3. Expression of gp140$^{proto-trk}$ in a human neuroblastoma cell line which lack both gp75^{NGFR} and gp140$^{proto-trk}$ expression generates a functional response which leads to differentiation and growth arrest, therefore suggesting that gp140$^{proto-trk}$ expression is sufficient for a functional NGF response.

Acknowledgments

We would like to thank Dr. M. V. Chao, Department of Cell Biology and Anatomy, Cornell University, New York for the pMVE-1 construct; Drs. M. Barbacid and V. B. Eager, Department of Molecular Biology, Bristol-Myers Squibb Pharmaceutical Research Institute, Princeton, NJ for the gp140$^{proto-trk}$ expression vector and the anti-gp140$^{proto-trk}$ antibody. We would like to thank D. Smith for typing the manuscript. This work was supported by public funds (NIH/NS 25795-04).

References

1. R. Levi-Montalcini, *Science* **237,** 1154 (1987).
2. H. Sariola, M. Saarma, K. Saimio, U. Arumäe, J. Palgi, A. Vaahtokari, I. Thesleff, and A. Karavanoy, *Science* **254,** 571 (1991).
3. M. V. Chao, *in* "Handbook of Experimental Pharmacology" (M. B. Sporn and B. Roberts, eds.), Vol. 95, p. 135. Springer-Verlag, Heidelberg, 1990.
4. C. Cordon-Cardo, P. Tapley, S. Jing, V. Nanduri, E. O'Rourke, F. Lamballe, K. Kovary, R. Klein, K. R. Jones, L. F. Reichardt, and M. Barbacid, *Cell (Cambridge, Mass.)* **66,** 173 (1991).
5. H. Matsuda, M. D. Coughlin, J. Bienenstock, and J. A. Denburg, *Proc. Natl. Acad. Sci. U.S.A.* **85,** 6508 (1988).

6. D. Johnson, A. Lanahan, C. R. Buck, A. Sehgal, C. Morgan, E. Mercer, M. Bothwell, and M. Chao, *Cell (Cambridge, Mass.)* **47**, 545 (1986).

7. R. Klein, S. Jing, V. Nanduri, E. O'Rourke, and M. Barbacid, *Cell (Cambridge, Mass.)* **65**, 189 (1991).

8. T. H. Large, G. Weskamp, J. C. Helder, M. J. Radeke, T. P. Misko, E. M. Shooter, and L. F. Reichardt, *Neuron* **2**, 1123 (1989).

9. M. Barbacid, F. Lamballe, D. Pulido, and R. Klein, *Biochim. Biophys. Acta* **1072**, 115 (1991).

10. M. Bothwell, *Cell (Cambridge, Mass.)* **65**, 915 (1991).

11. B. L. Hempstead, L. S. Schleifer, and M. V. Chao, *Science,* **243**, 373 (1989).

12. B. L. Hempstead, D. Martin-Zanca, D. R. Kaplan, L. F. Parada, and M. V. Chao, *Science* **350**, 678 (1991).

13. R. C. Seeger, S. E. Siegel, and N. Sidell, *Ann. Intern. Med.* **97**, 873 (1982).

14. R. C. Seeger, G. M. Brodeur, H. Sather, A. Dalton, S. E. Siegel, K. Y. Wong, and D. Hammond, *N. Engl. J. Med.* **315**, IIII (1985).

15. H. Matsushima and E. Bogenmann, *Int. J. Cancer* **51**, 250 (1992).

16. P. A. Jones, T. Scott-Burden, and W. Gevers, *Proc. Natl. Acad. Sci. U.S.A.* **76**, 353 (1979).

17. A. H. Ross, P. Grob, M. Bothwell, D. E. Elder, C. S. Ernst, N. Marano, B. F. D. Christ, C. C. Slemp, M. Herly, B. Atkinson, and H. Koprowski, *Proc. Natl. Acad. Sci. U.S.A.* **81**, 6681 (1984).

18. P. M. Grob, A. H. Ross, H. Koprowski, and M. Bothwell, *J. Biol. Chem.* **260**, 8044 (1985).

19. H. Matsushima and E. Bogenmann, *Mol. Cell. Biol.* **10**, 5015 (1990).

20. H. Matsushima and E. Bogenmann, *Prog. Clin. Biol. Res.* **366**, 227 (1991).

21. T. Scott-Burden, E. Bogenmann, and P. A. Jones, *Exp. Cell Res.* **156**, 527 (1986).

22. H. Matsushima and E. Bogenmann, *Int. J. Cancer* **51**, 727 (1992).

23. D. Martin-Zanca, R. Oskam, G. Mitra, T. Copeland, and M. Barbacid, *Mol. Cell. Biol.* **9**, 24 (1989).

24. R. Klein, V. Nanduri, S. Jing, F. Lamballe, P. Tapley, S. Bryant, C. Cordon-Cardo, K. R. Jones, L. F. Reichardt, and M. Barbacid, *Cell (Cambridge, Mass.)* **66**, 395 (1991).

25. F. Lamballe, R. Klein, and M. Barbacid, *Cell (Cambridge, Mass.)* **66**, 967 (1991).

26. D. Schubert, S. Heinemann, W. Carlisle, H. Tarikas, B. Kimes, J. Patrick, J. H. Steinbach, W. Culp, and B. L. Brandt, *Nature (London)* **249**, 224 (1974).

27. C. M. Machida, K. D. Rodland, L. Matrisian, B. E. Magun, and G. Ciment, *Neuron* **2**, 1587 (1989).

28. L. R. Karns, N. Shi-Chung, J. A. Freeman, and M. C. Fishaman, *Science* **236**, 597 (1987).

29. S. Mallett and A. N. Barclay, *Immunol. Today* **12**, 220 (1991).

30. H. Durkop, U. Latza, M. Hummel, F. Eitelbach, B. Seed, and H. Stein, *Cell (Cambridge, Mass.)* **68**, 421 (1992).

[11] Strategy to Analyze Thymic Transcription of the α3 Nicotinic Neuronal Acetylcholine Receptor Subunit

Mirta Mihovilovic, Christine Hulette, and John R. Gilbert

Introduction

The analysis of normal and variant transcripts is of importance in characterizing the expression of genes and gaining insight into the characteristics of translational isoforms. This chapter discusses an approach which we have utilized to study the transcription of the α3 neuronal nicotinic acetylcholine receptor (nAChR) subunit in human thymus. The emphasis is on the techniques employed. The chapter is divided into two sections: isolation of thymic cDNA clones and their characterization through sequencing and polymerase chain reaction (PCR) analysis (Section I), and utilization of ribonuclease protection assays (RPAs) for the study of normal and variant forms of thymic transcripts (Section II).

Myasthenia gravis (MG) is an autoimmune disorder in which the neuromuscular acetylcholine receptor is a well-characterized target of the disease (1–4). Thymic abnormalities are found in 80 to 90% of the patients (5, 6). Thymectomy ameliorates the symptoms of the disease (7, 8), indicating an important role for the thymus in the triggering and/or maintenance of this condition. Since 1977 (9, 10), several laboratories have reported the expression in thymus of nAChR(s) or nAChR-like antigens (11–13).

The putative expression of a thymic cholinergic antigen, when taken in the context of the histological (14–17) and lymphocytic abnormalities (14, 15) seen in MG thymi, provides support for the hypothesis that a thymic cholinergic antigen may play a role in the process of triggering and maintaining the autoimmune attack in MG. Thus, it is important to determine whether metabolic deregulation of a cholinergic antigen at transcriptional and/or translational levels may play a role in the process of the intrathymic antigen presentation in MG.

Our attempts to characterize cholinergic thymic antigens led us to identify thymic transcripts that encode for the α3 subunit of a neuronal nAChR (18), indicating that a neuronal nAChR is expressed in thymus. This receptor could represent one of the thymic cholinergic antigens previously reported (13). The function of such thymic neuronal nAChR is, however, unknown.

It is possible that it may participate in the transduction or modulation of signals received through autonomic innervation. The thymus receives both sympathetic and parasympathetic innervation (19–22). This innervation is crucial for normal thymus development (23, 24) and influences thymic function (25). The identification of a number of neuroactive substances in thymus including adrenaline, vasoactive intestinal peptide, Met-enkephalin, and acetylcholine (19–22) indicates that a tight neuroregulatory control exists in thymus. In this context, the microenvironment of the thymus, playing a crucial role in T cell maturation (26), can be viewed as regulated by a network of neuroactive sustances and their corresponding receptors. Thus, metabolic deregulation of a thymic neuronal nAChR may have important implications for thymic function including the process of intrathymic T cell maturation. The methodology presented in this chapter is the result of our investigation of transcription in normal and MG thymi.

Our strategy utilizes characterized thymic cDNA clones to investigate the expression of their corresponding transcripts. The process of cDNA clone characterization is expedited by PCR analysis of the clones. The primers that are employed are derived from a thymic cDNA clone known to carry the sequence corresponding to the expected normal transcript. Deletion and insertion cDNA variants are easily recognized through PCR analysis. Normal and variant clones are then used as a source of probes for direct transcriptional analysis in the tissue of choice. The analysis, in this case, is performed through ribonuclease protection assays.

Isolation of cDNA Clones and Their Characterization through Sequencing and Polymerase Chain Reaction Analysis

Isolation of Thymic cDNA Clones: Preliminary Strategy

General Considerations

The isolation of cDNA clones is frequently the first step in the process that leads to the identification of transcripts that are expressed in a given tissue. The successful isolation of specific cDNA clones is dependent on the production of a high-quality cDNA library containing the desired cDNA clones and the availability of a suitable probe for detection. It is advisable to initiate screening with a well-characterized probe of known origin if one is available. A probe can be selected either for its specificity toward a specific transcript or for its potential to hybridize to a family of transcripts. With the advent of the PCR it has become possible to isolate isoforms of clones, members of gene families, or completely unknown genes which share sequence motifs through the preparation of unique or degenerate primers from published sequences (27). In general, the use of PCR-generated probes and of available homologous probes is complementary.

Before committing time to the production and screening of a cDNA library it should be determined whether the selected probe detects transcripts in the tissue of choice. Northern blot analysis is adequate for the detection of transcripts representing as little as 0.001% of the poly(A)$^+$ RNA population. The use of cRNA probes increases the sensitivity further, allowing Northern blot analysis to approach the sensitivity of S1 nuclease and ribonuclease protection methods (28). The use of Northern blot analysis offers several advantages. First, it provides information on the tissue distribution, size, number, and abundance of the transcripts which are recognized by the probe. Second, it helps to optimize experimental conditions for library screening. In addition, the blots themselves can be saved and reused.

In the experimental approach outlined below, preliminary Northern blot analyses using a mouse neuromuscular α-1 subunit probe indicated that transcripts which hybridize to the nAChR probe were present in human thymus (Fig. 1). To increase the chance of detecting human transcripts utilizing a homologous mouse probe (see below), blotted human RNA transcripts were targeted with antisense cRNA probes derived from the corresponding cDNA clones. Antisense and sense probes were produced through run-off RNA synthesis employing linearized plasmids that contain bacteriophage RNA promoter sequences flanking the insert site (29). A series of vectors that allow the production of sense and antisense cRNA are commercially available. The procedures described here have been optimized employing the Bluescript SK+ vector (Stratagene, La Jolla, CA), but they can be used with any vector that carries bacteriophage T7, T3, or SP6 RNA promoters or with PCR products containing promoter sequences that have been incorporated during the amplification step (30).

To detect transcripts and/or search for cDNA clones that code for poorly characterized peptides, either degenerate oligonucleotides (derived from partial peptide sequences) or homologous cDNA clones can be used as probes. Knowing that nAChRs are oligomeric proteins composed of two or more homologous subunits (31), we searched for a cDNA-derived probe that could hybridize to different members of the nAChR family. A mouse neuromuscular α-1 cDNA clone (generous gift of Drs. J. Boulter, S. Heinemann from The Salk Institute, La Jolla, CA, and J. Patrick from Baylor College, Houston, TX), which has a high degree of coding region sequence conservation among different mammalian species (32) and which shares highly conserved sequence domains with other members of the nAChR family (31–33), was chosen to generate a probe for preliminary Northern blot analyses (Fig. 1) and for subsequent library screening (18).

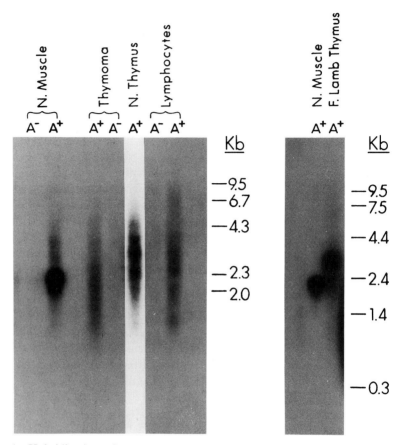

FIG. 1 Hybridization of a mouse probe that codes for evolutionarily conserved domains of the $\alpha 1$ subunit of the neuromuscular nAChR to transcripts of muscular and thymic origin. A HincII/PstI fragment of the carboxy terminus of the clone pMAR $\alpha 15$ (32) that codes for the $\alpha 1$ subunit of the neuromuscular nAChR was subcloned into the Bluescript SK+ vector (Stratagene). An antisense cRNA was produced by RNA run-off synthesis and used to probe RNA species that had been electrophoresed and blotted onto GeneScreen Plus (NEN Boston, MA) nylon membranes. Blots were washed at 55°C employing 2× SSC, 0.1% SDS. Human muscle RNA: 2 μg/lane; all other lanes: 10 μg/lane. A^+ and A^-, respectively, stand for poly(A)$^+$ and poly(A)$^-$ RNA preparations; N, normal; F, fetal. The photographic composite shows that the mouse probe hybridizes to both muscle and thymic RNA species that can be differentiated by their molecular weights. Both human and fetal lamb RNA species are recognized by the mouse probe. Normal human thymus appears to express higher amount of RNA species that hybridize with the $\alpha 1$ probe than MG thymoma. The normal thymus lane has been photographed with higher aperture to visualize details of the Northern blot.

An $\alpha 1$ cDNA subclone corresponding to a *Hinc*II/*Pst*I fragment of the carboxy terminus of the mouse $\alpha 1$ clone pMAR $\alpha 15$ (32) was generated through subcloning into the Bluescript SK+ vector. The subcloned cDNA encodes for highly conserved sequences of the four transmembrane domains of the mouse $\alpha 1$ subunit extending downstream from position 15 of the first putative transmembrane domain. Probes generated from this subclone are expected to cross-hybridize under low-stringency conditions with different members of the nAChR family.

The process of constructing and screening cDNA libraries is not discussed in this chapter. Many detailed protocols have been described, including an integrative strategy on the topic (34).

Practical Considerations on Tissue Handling

Success in detecting RNA transcripts expressed in a given tissue depends on the quality of the RNA preparations that are utilized. For human tissue it is crucial that measures are taken to avoid ribonuclease degradation of tissue transcripts. Thus, as soon as the tissue is available it is immediately frozen in liquid nitrogen and stored at $-80°C$. No changes in RNA yields have been observed on storage for up to 6 years. In addition, frozen tissues are easily powdered under liquid nitrogen employing a stainless steel tissue mallet, which facilites the process of homogenization in buffers containing guanidine salts.

Practical Considerations on Handling and Purification of RNA

All glassware employed is baked. Materials that cannot be baked are treated with 1–2 M NaOH solutions and rinsed with sterilized water followed by rinsing with RNase-free water [i.e., diethyl pyrocarbonate (DEPC)-treated water or irrigation water, Abbott Laboratories, Chicago, IL]. All solutions are made with RNase-free water and further treated for at least 1 hr with 2% DEPC, followed by autoclaving which inactivates the remaining DEPC.

For total RNA preparations we routinely employ an initial homogenization in buffer containing 6 M guanidine isothiocyanate (35). Depending on the nature of the tissue, however, other chaotropic agents and conditions can be considered. We have found that for human brain tissue guanidine hydrochloride-based extractions of RNA (36) give excellent results. In addition, many commercial poly(A)$^+$ RNA purification kits yield poly(A)$^+$ preparations of good quality for immediate use, but not for storage.

Homogenization of tissues in guanidine isothiocyanate-containing buffer is routinely performed at a tissue to buffer ratio of 1 to 20 (w/v). For metabolically active tissues this ratio is increased up to 1 to 40. For fat-rich tissues the detergent concentration in the homogenization buffer is empirically increased. We have found it helpful to prepare a homogenization buffer containing 20% N-lauryl sarcosine which we add in 0.5-ml aliquots to 25 ml of the fat-rich homogenization mixture during the process of homogenization. The addition of detergent-rich solution continues until the fat droplets have been emulsified.

For poly(A)$^+$ RNA isolation we utilize affinity chromatography over oligo(dT)-cellulose (37). Electrophoretic analysis is performed in formaldehyde-containing agarose gels (38).

Practical Overview on the Production of cRNA Probes

cRNA probes are produced through RNA run-off synthesis employing as templates linearized cDNA clones that contain a bacteriophage RNA promoter sequence flanking the cloned cDNA sequence. Linear DNA templates for cRNA run-off synthesis are usually generated by employing a unique restriction site located at one end of the cloned insert, although is also possible to restrict within the cDNA itself. In either case, the linearized clone should be flanked by a bacteriophage RNA promoter sequence that will permit the synthesis of antisense cRNA. The linearized cDNA clone should contain either a 5' end extension or a blunt end to avoid ectopic DNA priming mediated by a 3' end extension. Ectopic priming during cRNA synthesis results in the synthesis of stretches of complementary cRNA. The latter interacts with its complementary cRNA, not only decreasing the effective concentration of the probe but producing misleading artifacts. If a 3' end extension cannot be avoided, the linerized plasmid should be blunt ended before it is used as template for cRNA synthesis.

Procedures for Production of cRNA Probes

Linearization of Plasmids

Digest 10 μg of the cDNA clone with the restriction enzyme of choice. At the end of the digestion check the linearization of the plasmid electrophoretically (38) utilizing 2% of the sample. Once it is determined that linearization has occurred, purify the restricted DNA by extraction with equal volumes of phenol–chloroform and chloroform. After chloroform extraction, precipitate the DNA by the addition of 1/10 volume of 3 M sodium acetate and 2.5 volumes of ethanol. Incubate the sample for 20 min at $-80°C$ and pellet the DNA by centrifugation at 10,000 g for 30 min at 4°C.

Blunt Ending of 3' End Extended cDNAs

The exonuclease activity of T4 DNA polymerase is exploited to produce a blunt-ended cDNA, while its polymerase activity, in the presence of deoxynucleotides (dNTPs), ensures that there is no further cDNA hydrolysis of the linearized cDNA.

    10 × T4 DNA polymerase buffer
      0.330 M Tris–acetate, pH 8.0
      0.660 M Potassium acetate
      0.100 M Magnesium acetate
      5 mM Dithiothreitol (DTT)
    1 × Tris/EDTA (TE) buffer
      10 mM Tris-Cl, pH 7.5
      2 mM EDTA

Resuspend the linearized cDNA clone in 50 μl of water. Add 10 μl of 10× T4 DNA polymerase buffer, 4 μl of a solution that contains each of the deoxynucleotides triphosphates (dATP, dCTP, dGTP, and dTTP) at a concentration of 2.5 mM, and water to give a final volume of 100 μl. Start the reaction by adding 1.25 units of enzyme/μg of cDNA. Incubate for 5 min at 37°C, then inactivate the enzyme by treatment at 65°C for 10 min. Purify the blunt-ended DNA by extracting with equal volumes of phenol–chloroform and chloroform. Recover the DNA through ethanol precipitation in the presence of 0.300 M sodium acetate utilizing 2.5 volumes of ethanol (see above).

Production of Radiolabeled cRNA Probes

The general radiolabeling procedure generates about 200 ng of *in vitro* synthesized cRNAs. It works equally well with T3, T7, and SP6 RNA polymerases. Radiolabeling is achieved by utilizing [α-P^{32}]UTP.

    10× RNA polymerase buffer
      0.400 M Tris-Cl, pH 8.0
      80 mM MgCl$_2$
      20 mM Spermidine
      0.500 M NaCl

To 1 μg of linearized cDNA resuspended in 0.2× TE add 2.5 μl of 10× RNA polymerase buffer, 1 μl each of 10 mM ATP, CTP, and GTP, 1 μl of 0.5 mM UTP, 1 μl of 0.750 M DTT, 5 μl of 400–800 Ci/mmol, 10 mCi/ml [α-^{32}P]UTP, 25 units of human placental RNase inhibitor (Amersham, Arlington Heights, IL), and RNase-free water to yield a final volume of 25 μl. Start the reaction by adding 15 units of the RNA polymerase (Stratagene, La Jolla, CA) and incubate for 30 min at 37°C.* At the end of the incubation, hydrolyze

---

* cRNA probes with higher specific activities are produced by decreasing the amount of "cold" UTP and proportionally increasing the amount of radiolabeled UTP in the reaction.

the DNA template by incubation of the mixture for 15 min at 37°C in the presence of 20 units of RNase-free DNase (Stratagene, La Jolla, CA), adding the DNase directly to the reaction mixture.

Determine the extent of ^{32}P incorporated into the cRNA through trichloroacetic acid (TCA) precipitation as follows: Transfer a 1-μl aliquot of the reaction mixture into 100 μl of a salmon sperm DNA solution containing 0.500 mg/ml of sonicated DNA in 50 mM EDTA, pH 8.0, and add 5 ml of ice-cold 10% TCA. Incubate for 10 min on ice. Filter through GF/C glass fiber filters (Whatman, Maidstone, England). Wash the filters three times with 5 ml of 5% TCA, and three times with 5 ml of ethanol. Calculate the yield of cRNA synthesis from the amount of [α-^{32}P]UTP incorporated into the TCA precipitated material, taking into account the specific activity of the radiolabeled UTP present in the original reaction mixture and assuming that 1 mol of UTP is incorporated for every 4 mol of synthesized cRNA.

To purify the cRNA probes nonincorporated nucleotides are removed through filtration employing cutoff filtration units. Currently a number of units are commercially available. We prefer a unit that will fit in a small bench-top centrifuge. Ultrafree-MC units, as provided by the manufacturer (Millipore, Bedford, MA), have given excellent results. Wash the mixtures containing the cRNA probes 4 times with 0.4 ml of TE, monitoring the filtration of the free radiolabeled UTP in the eluate after each wash.

For cRNA probes to be used in Northern blot analysis no further purification is necessary. (See section on RPAs for electrophoretic purification of cRNA probes). Check the quality of the cRNA probe electrophoretically employing formaldehyde denaturing gels (38). After electrophoresis treat the gel for 30 min in 7% TCA, press the TCA-treated gel between 10 layers of filter paper for 45 min, and autoradiograph the TCA-treated flattened gel (XAR5 EK film, Kodak, Rochester, NY).

Procedure for Northern Blot Hybridization Employing cRNA Probes

Northern blot analysis used in conjunction with cRNA probes can be a highly sensitive procedure. We have found that the use of poly(A)$^+$ RNA, nylon membranes, UV cross-linking of the RNA species to the nylon support, and washes of the blots at temperatures equal or higher to that used during hybridization optimizes the sensitivity of the analysis. Since cRNA–RNA hybrids are extremely stable, conditions of very high stringency can, if necessary, be used to eliminate residual nonspecifically bound probe. High-stringency washes will, however, accelerate the degradation and release of RNA from the blots, reducing the number of times a given blot can be reused.

> Prehybridization medium
>> A 50% formamide-containing medium is recommended. Once made, it can be aliquoted and stored at 4°C.

For 500 ml of medium mix:

250 ml of deionized formamide

10 ml of 50× Denhardt's solution (see below)

100 ml of 50% dextran sulfate (Pharmacia, Piscataway, NJ)

50 ml of 10% sodium dodecyl sulfate (SDS)

25 ml of 1 M sodium phosphate buffer, pH 7.0

29 g of NaCl.

After a homogeneous solution is obtained, add 50 mg of denatured salmon sperm DNA (Sigma, St. Louis, MO). Dilute to 500 ml with water.

50× Denhardt's solution

1% Ficoll type 400 (Pharmacia)

1% polyvinylpyrrolidone (Sigma)

1% bovine serum albumin, fraction V (Sigma)

Stocks of salmon sperm DNA

Solutions (5.0–10.0 mg/ml) are prepared by suspension of the DNA in TE, followed by sonication of the suspension with 5 min × 20 kHz bursts delivered at a power output of 40 W (Sonicator Model 185E, Heat Systems Inc., Plainview, NY). Prior to mixing into the prehybridization medium, the DNA is further denatured by boiling for 7 min.

Prehybridization and Hybridization Conditions

Both prehybridization and hybridization are performed at 55°C. Prehybridization is carried out for at least 2 hr, followed by hybridization for 16 to 20 hr employing approximately 20 to 40 × 10⁶ dpm of cRNA probe (25 to 50 ng/ 10 ml of hybridization solution).

cRNA probes are treated with yeast tRNA (GIBCO/BRL, Gaithersburg, MD) prior to hybridization as follows. Mix the cRNA probe with an aqueous solution of yeast tRNA so the final tRNA concentration in the hybridization medium will be 0.100 mg/ml (usually 1.0 mg of tRNA per blot), boil the mixture for 7 min, and slowly cool it to 55°C. Add 0.5 ml of prewarmed prehybridization solution and transfer this mixture to the hybridization container.

Washing Conditions

The blots are washed at 55°C. To avoid nonspecific binding of the probe, care should be taken that the temperature does not drop below 55°C, particularly in the first wash.

10× NaCl/sodium citrate buffer (SSC)

0.150 M NaCl

15 mM sodium citrate

Start with a quick 5 min wash at low stringency (2× SSC, 0.1% SDS). Perform successive washes for 20 min, increasing the stringency after two washes. The following empirical set of stringencies provide a good baseline for determining the best wash conditions for a given probe: low stringency, 2× SSC, 0.1% SDS; medium stringency, 1× SSC, 0.1% SDS; high stringency, 0.1× SSC, 0.1% SDS. Terminate the washing when radiation measured with a hand-held Geiger counter falls below 0.4 mrads/hr. Expose the blot to a sensitive photographic film (XAR5, EK, Kodak).

Remarks

Preliminary Northern blot analyses dictated the tissue of choice for library production and the type of probe best suited for library screening (Fig. 1). For our studies, a normal thymic library was produced; it was cloned in the EcoRl site of the λ ZAPII vector (Stratagene) and consisted of 3×10^6 clones with an average insert size of 1.0 kb. This library was screened with the *Hinc*II/*Pst*I cDNA radiolabeled mouse α1 probe (see above). Screening of 250,000 clones resulted in the isolation of 14 cDNA clones.

Sequencing and PCR Analysis of Isolated cDNA Clones

Practical Considerations for Sequencing of cDNA Clones

The identification of homologous cross-reactive clones is ultimately verified through sequencing. We sequence our isolated clones directly from double-stranded DNA phagemid preparations. This avoids the need of subcloning and generates, in the process, a set of primers spaced through the complete cDNA sequence under analysis. In addition, the sequencing of both strands of the double-stranded template allows for simultaneous sequence verification.

The Bluescript phagemids are recovered through *in vivo* excision from the λ ZAPII cDNA clones (39). Sequencing is performed employing alkali-denatured miniplasmid preparations of the excised phagemid as templates, the Bluescript T3 and T7 promoter sequences as priming sites, and a Sequenase-based (USB, Cleveland, OH) standard Sanger protocol (40). Numerous variations for denaturation of DNA templates are equally effective, among them boiling of the template followed by quick −70°C freezing (41).*

The initial 3′ and 5′ end sequences were compared against the GenBank

---

* Commercially available PCR-based sequencing kits that allow direct DNA sequencing from λ-derived bacteriophages, cosmids, as well as genomic DNA have become available. These kits are ideal for obtaining preliminary sequencing data. Definitive sequence information is, however, best obtained by employing M13 or plasmid-cloned DNA.

database (42). Three cDNA clones which encoded for the α-3 neuronal nAChR subunit were chosen for further sequencing, with attention centering on clone 72NTBSZ which had the longest 5' end extension. This clone, however, lacked the complete sequence encoding the putative signal peptide as well as 8 amino acids of the mature subunit. To select a clone that carried these 5' end sequences, a PCR study utilizing the α1-reactive clones (see below) was carried out employing an 18-mer α3 primer complementary to positions 72 through 90 of the 72NTBSZ clone and a primer corresponding to either the T3 or T7 promoter plasmid flanking sequence (18). The cDNA clones which produced the largest PCR-amplified products were selected for 5' end sequencing. Clone 61NT-2-18 was shown to carry the complete coding sequence for the human α3 subunit.

Strategy for PCR Analysis of α1 Homologous cDNA Clones

Sequence analysis of numerous cDNA clones is time consuming. To expedite the process of cDNA characterization, each of the 14 isolated α1-reactive

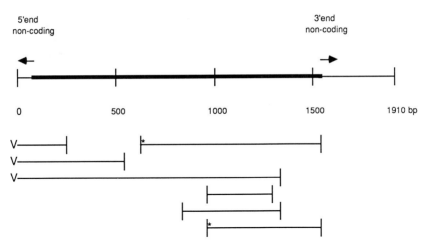

FIG. 2   Schematic representation of primer pairs utilized in the PCR analysis of α1 nAChR subunit-reactive cDNA clones. Primers developed during the sequencing of cDNA clones encoding the normal α3 subunit of a neuronal nAChR were utilized for PCR analysis of 14 α1-reactive thymic clones. The segmented bar represents the α3 clone 61NT-2-18, which carries the complete coding sequence of the α3 nAChR subunit. The PCR-amplified normal products obtained with different primer pairs are represeneted by bars. Of 14 clones investigated, 13 coded for the α3 nAChR subunit; of these, 2 clones produced larger PCR-amplified products than expected. An asterisk at the right of a bar highlights the PCR-amplified products that were larger for the 2 variant clones. V: a vector primer flanking the cDNA sequence was utilized in conjunction with an α3 nAChR subunit primer in the PCR analysis.

clones was subjected to a PCR analysis. Primer pairs developed for sequencing purposes were employed to prime each isolated cDNA clone. Primer combinations that were employed are shown in Fig. 2. Using this strategy 1 of 14 clones was ruled out as representing an $\alpha 3$ cDNA, and 2 variant $\alpha 3$ cDNAs clones were detected (Fig. 3).

I)

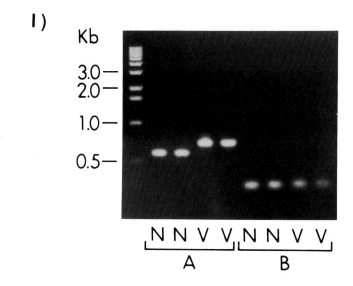

II)

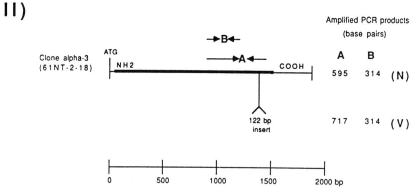

FIG. 3   PCR analysis of thymic clones, showing data for two PCR reactions conducted for four $\alpha 3$ cDNA clones (I). The primer pairs used (A and B) are also schematically represented (II). Two variant (V) clones produced a PCR-amplified product about 130 bp longer than its normal (N) counterparts when primers that extend between exon 5 and exon 6 of the $\alpha 3$ nAChR subunit were used.

Procedures for PCR Analysis

We have employed a general baseline PCR protocol successfully for the amplification of cDNAs and genomic DNA derived sequences. The main feature is the original enrichment of PCR products under high-stringency conditions, followed by subsequent amplification under lower stringency conditions. The result is a protocol that is highly specific and that provides high yields of DNA sequences of interest.

> 10×PCR buffer
>> 0.100 M Tris-Cl, pH 8.4
>> 15 mM $MgCl_2$
>> 0.500 M KCl
>> 0.10% gelatin

PCR Mixture

For each set of primers prepare one reaction mixture combining the following reagents in the order given. Defining n as the number of reactions to be run, mix $n + 1$ amounts of the following reagents:

> 2.5 μl of 10× PCR buffer
> 2 μl of a deoxynucleotide mixture 2.5 mM in dATP, dCTP, dGTP, and dTTP
> 5 to 25 pmol of each primer (usually 50 ng of each 18-mer primer)
> 1 unit of Taq polymerase
> Water to provide for a final volume of 25 μl

PCR Reaction

Aliquot the PCR mixture into tubes, and add either 1 to 5 ng of DNA vector 125 to 250 ng of human genomic DNA per PCR tube. Overlay with 20 μl of mineral oil and start the PCR cycling. If the volume of TE buffer exceeds 5 μl and poor yields of the PCR products are obtained, compensate for the EDTA concentration present in the TE buffer by adding $MgCl_2$.

PCR Cycling

Denature at 94°C for 7 min. Proceed with 15 cycles as follows: anneal at 55°C for 3 min; extend at 72°C considering that 1000 bp is incorporated per minute of incubation, but setting the extension time at 1 min minimum; denature at 94°C for 3 min. Follow with 15 additional cycles as indicated: anneal at 37°C for 3 min; extend as above; denature at 94°C for 3 min. Analyze the PCR products electrophoretically, using 1 to 2% agarose gels (38).

Occassionally we have found that primers with melting temperatures of 45°C or lower have to be annealed at temperatures no higher than 50°C. Annealing at 55°C to 60°C is sometimes necessary to eliminate secondary PCR products.

Remarks

PCR analysis of cDNA clones facilitates the process of cDNA characterization (Fig. 3) and dictates the approach to follow for direct transcriptional analysis (see below). The variant α3 cDNA clones were further characterized through sequencing and were shown to carry 122 extra nucleotides. Sequence comparison of the variant human clones with the highly homologous α3 genomic sequences of chicken and rat (43) indicated that the 122-bp fragment was spliced between exon 5 and 6 of the α3 nAChR subunit sequence.

Utilization of Ribonuclease Protection Assays for the Study of Normal and Variant Forms of Thymic Transcripts

General Considerations

Two basic methods are used for the investigation of transcript expression: amplification by PCR of reverse-transcribed RNA (44) and protection of single-stranded cDNA (45, 46) or cRNA (29) species from nuclease attack by their corresponding transcripts. In most instances the two methods complement each other, but questions of sensitivity, prior knowledge of the system under study, and focus of the particular study should be considered before choosing one over the other.

Transcriptional analysis of known point mutations and transcripts generated through alternative splicing could be carried out equally well with either method. Of the two methods, amplification by PCR of reverse-transcribed RNA is more sensitive but also more prone to technical artifacts such as primer–dimer formation and mispriming arising through PCR amplification (47).

Methods based on the protection of single-stranded cDNA from S1 nuclease attack (45, 46) or of cRNA probes from RNase attack (29) have been widely used and because radiolabeled cRNA probes are easily synthesized from vectors carrying bacteriophage promoter sequences that flank cloned cDNA sequences (see Section I). Ribonuclease protection assays (RPAs) appear to produce fewer technical artifacts (48). They have gradually, over the past several years, begun to supercede S1 nuclease protection assays as the method of choice in many applications involving mapping and quantitation of transcripts.

Production of normal and variant α3 cRNA probes was straightforward, and RPAs were chosen to study the expression of the α3 thymic transcripts. A *Sac*I/*Msc*I fragment that extends from exon 5 to exon 6 of the α3 (normal and variant) cDNAs were subcloned into a *Sac*I/*Eco*RV-restricted Bluescript SK+ vector. These subclones provided the templates for the production of normal and variant cRNAs. The T3 and T7 promoter sequences of the vector were utilized to direct the synthesis of sense and antisense cRNA probes, respectively.

Procedures for Ribonuclease Protection Assays

Purification of cRNA Probes

cRNA probes are produced as described in Section I,1. To avoid background signals in the RPA, however, it is advisable to purify them electrophoretically. A purification protocol for typical cRNA probes with lengths of 100 to 700 bp is given below.

> RPA loading solution
> > 80% deionized formamide
> > 2 mM EDTA, pH 8.0
> > 0.1% bromophenol blue
> > Store in aliquots at −20°C.
> 10× TBE buffer
> > 0.900 M Tris base
> > 0.900 M boric acid
> > 20 mM EDTA
> Probe elution solution
> > 0.500 M ammonium acetate
> > 2 mM EDTA
> > 0.1% SDS

To 25 μl of the nonpurified, DNase-treated, radiolabeled cRNA probe add 100 μl of loading solution. Warm for 10 min at 70°C and subject to electrophoresis in 5% acrylamide (acrylamide to bisacrylamide ratio 30.0:0.8), 1 × TBE, 7.5 M urea, 0.75 mm thick gels at 300 V for approximately 1 hr.

For calibration of the gel we run a commercially available RNA ladder (GIBCO/BRL, Gaithersburg, MD). The excised ladder lanes are silver stained (silver stain kit, Bio-Rad, Richmond, CA), dried, and corrected for gel swelling by proportional size reduction applied to the dried gel. This procedure has the advantage that it utilizes well-characterized markers without further treatment. Alternatively, 5′ end radiolabeled RNA markers produced employing polynucleotide kinase (49) or radiolabeled cRNA markers produced employing well-characterized cDNA clones available to the laboratory can be used.

After the electrophoresis is complete, lift one glass plate of the electrophoretic assembly and cover the gel held in the second glass plate with Saran Wrap. Position a photographic film (XAR5 EK, Kodak) so that the edges of the glass plate match the edges of the film. Expose for 1 min. Slide the exposed film underneath the glass plate holding the gel. Flush the edges of the glass plate and the autoradiograph. Excise the gel band with a razor blade following the contour of the autoradiograph and transfer it to a sterile tube. Cover the excised gel band with 350 μl of probe elution solution. Incubate at 37°C for 4 to 16 hr and recover the purified cRNA probe in the elution solution.

Hybridization of cRNA Probes to RNA Transcripts

For most applications, the optimal hybridization conditions are the most stringent. This is particularly true when dealing with a family of homologous transcripts and/or with repetitive transcribed sequences. Typically a 70% formamide-based hybridization solution and temperatures of 42°C or higher are recommended. The method we present has worked equally well with the normal cRNA probe at temperatures ranging from 42 to 70°C and formamide concentrations at 70%.

In our study the α3 variant clones contain a 122-bp Alu element. It was found that higher stringencies were necessary for the detection of the variant transcript forms (55°C or higher) (Fig. 4). Lower stringencies (45°C) resulted almost exclusively in the limited protection of the 122-bp Alu sequence present in the variant cRNA probe (not shown). At lower stringencies, highly homologous Alu sequences present in many partially spliced thymic transcripts appear to compete with the α3 variant thymic transcripts for binding to the cRNA probe.

> Stock hybridization solution
> 80% deionized formamide
> 40 mM PIPES-Cl, pH 6.7
> 0.400 M NaCl
> 1 mM EDTA.
> Stored under nitrogen at −20°C

For each RNA preparation to be tested prepare two tubes. One tube of the pair will be treated with ribonuclease; the other will receive an identical solution without the nuclease. In addition, prepare two blank tubes containing yeast tRNA. As a negative control run the corresponding sense cRNA probe.

Mix, in a volume of 4.5 μl, either total or poly(A)$^+$ RNA with the antisense cRNA probe, utilizing about a 4 to 10 times molar excess of the probe. (We typically mix 30 pg of a 300-bp probe with a specific activity of 1000 dpm/pg and 2 to 5 μg of poly(A)$^+$ RNA.) Add 30 μl of stock hybridization solution. Treat for 10 min at 90°C and hybridize at the chosen temperature (42–70°C) from 18 to 22 hr. If the mixture of RNA and cRNA probe exceeds the recommended volume of 4.5 μl, precipitate the nucleic acids with 2.5 volumes of ethanol in the presence of 0.300 M sodium acetate, dry the pellet and resuspend in 4.5 μl of IXTE.

Ribonuclease Treatment of cRNA–RNA Hybrids

The procedure described has been used successfully with RNase A, which cleaves after pyrimidines, RNase T1, which cleaves after guanine nucleotides, and with combinations of both.

Ribonuclease treatment buffer
10 mM Tris-Cl, pH 7.5
1 mM EDTA
0.300 M NaCl
Stock RNase solutions
Aqueous solutions of RNase A (Sigma) at 200 Kunitz units/ml and RNase T1 (Sigma) at 500 units/ml are boiled for 10 min and stored at 4°C for use within 1 week. For long-term storage, RNase solutions are kept at −20°C.

After hybridization is completed, equilibrate the tubes at room temperature, quickly spin the contents, and add to each duplicate tube either 350 μl ribonuclease treatment buffer or a stock RNase solution that has been diluted 1/500 in this buffer. Incubate for 30 min at 37°C.

It may be necessary to determine empirically the best conditions for RNase treatment in terms of the concentration of the RNase (RNase stock solutions diluted 1/50 to 1/5000), the length of the incubation (15 to 60 min), and the temperature (4 to 37°C) of the treatment. For A–U-rich RNA sequences, it may be necessary to use only RNase T1 in the assay. Owing to transient strand separation of A–U-rich RNA–RNA hybrid sequences, RNase A may degrade a perfectly matched A–U-rich region.

Preparation of Samples for Electrophoretic Analysis

Proteinase K/tRNA/SDS solution
Prepare fresh by mixing:
1 volume of a 10.0 mg/ml aqueous solution of proteinase K (GIBCO/BRL) (stock solution kept at −20°C)
0.4 volume of a 5.0 mg/ml aqueous solution of yeast tRNA (GIBCO/BRL) (stock solution kept at −20°C)
1 volume of 10.0% SDS

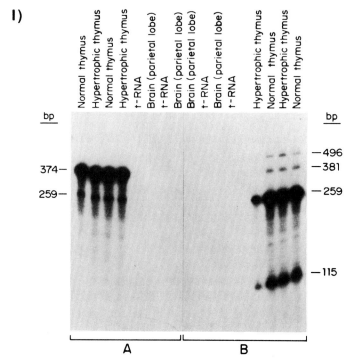

I)

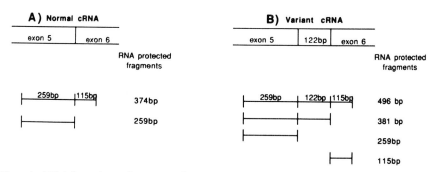

II)   CHARACTERISTICS OF THE cRNA PROBES EMPLOYED AND OF THE cRNA FRAGMENTS PROTECTED IN THE ASSAY

FIG. 4   RPA based on the protection of normal and variant α3 cRNA probes by thymic transcripts. *SacI/MscI* fragments of the normal and the 122 bp variant clone carrying 259 bp of exon 5 and 115 bp of exon 6 (II) were subcloned into the *SacI/EcoRV*-restricted Bluescript SK+ vector (Stratagene). Sense and antisense radiolabeled cRNAs were obtained by T3 and T7 promoter-directed run-off cRNA synthesis. Protection by RNA of the radiolabeled probes was monitored electrophoretically. Data for a RPA employing antisense α3 probes which were hybridized for 20 hr at

At the end of the RNase treatment, eliminate the RNases and at the same time provide carrier tRNA by adding 24 μl of the proteinase K/tRNA/SDS solution. Incubate for 30 min at 37°C. Add 400 μl of phenol–chloroform–isoamyl alcohol (5:4.8:0.2, v/v), mix, and then spin at 10,000 g for 5 min at room temperature. Transfer 375 μl of the aqueous phase to a fresh tube. Precipitate the RNA in solution by adding 2.5 volumes of ethanol, incubating at −70°C for 25 min. Recover the precipitate by centrifuging at 10,000 g for 30 min. Eliminate the ethanol by vacumm-drying the pellets for 5 min.

Resuspend the pellets in 10 μl of RPA loading solution. Warm for 10 min at 70°C and analyze electrophoretically employing 5% acrylamide 1 × TBE/ 7. 5 M urea gels as indicated above for cRNA probe purification. Run the gels at 300 V for 60 to 90 min. At the end of the run separate the plates of the electrophoretic assembly, leaving the gel supported by one glass plate. Lift the gel from the glass plate employing a sheet of filter paper (GB002, Schleicher and Schuell, Keene, NH), wrap the gel and its paper support in Saran Wrap, and expose it to a high-sensitivity photographic film (XAR5 EK, Kodak) for at least 12 hr.

Final Remarks

Figure 4 shows data depicting the cRNA species that are protected from RNase A digestion by thymic transcripts isolated from both normal and MG hypertrophic thymuses. The level of protection of the cRNA probes by normal and MG-derived thymic transcripts are equivalent, indicating that transcription of the normal and variant forms of the α3 nAChR subunit in the MG hypertrophic thymus does not differ from that of its normal counterpart. RPAs do not, however, show the topological differences in the distribution of α3 transcripts that may exist in different thymic tissues. *In situ* hybridization analyses are under way and should provide this information.

60°C is presented in (I). The eight lanes at left represent results obtained with the normal cRNA probe, and the lanes at right represent results obtained with the variant cRNA probe. Lane 5 from the right is blank. Except for lane 4 from the right, which represents data employing 2 μg of poly(A)[+] RNA, all others represent protection by 8 μg of poly(A)[+] RNA. The sequence characteristics of the cRNA probes and of the RNA-protected species electrophoretically resolved are schematized in (II). The results show that the normal transcript (RPA 374-bp fragment) is the dominant species in all thymuses tested (data for thymomas are not shown). The variant transcript appears to be a minor species (RPA 496 bp). Three partially spliced forms are also detected (RPA 259 bp, 381 bp, and 115 bp), representing the exon 5 sequence, the exon 5 sequence spliced to the 122-bp sequence, and the exon 6 sequence.

In contrast to hypertrophic thymus, we have found that thymomas show decreased levels of $\alpha 3$ transcript expression (50, 51), a finding that was suggested by our preliminary Northern analysis of "cholinergic" thymic transcription (Fig. 1).

At present, the role that nAChR(s) play in thymus and the number of different nAChR or nAChR-related proteins expressed in this organ are unknown. We are seeking to characterize these "cholinergic" proteins. Our approach relies on the screening of thymic cDNA libraries employing available homologous probes, the characterization of the isolated clones through sequencing, and the production of cDNA-derived probes with which to investigate thymic cholinergic transcripts and their corresponding translation products. The latter are studied through the production of antibodies to recombinant peptides expressed by the characterized cDNA clones.

In this chapter, we have presented a strategy for the characterization of isolated cDNA clones and the analysis of transcriptional variants. This strategy uses sequencing primers for the PCR analysis of isolated cDNA clones, permitting rapid detection of insertion or deletion cDNA variants. Selected variant clones, along with their normal counterparts, are then employed as sources of probes for transcriptional analysis.

Acknowledgments

This work was supported by PO1-NS-26630 (M.A.P.V.), a Program Project Grant from the National Institutes of Health, a Clinical Research Grant (M.M.) from the Muscular Dystrophy Association (MDA), and MO1-RR-30, National Center for Research Resources, General Clinical Research Centers Program, from the National Institutes of Health.

References

1. J. M. Lindstrom, M. E. Seybold, V. A. Lennon, S. Whittingham, and D. D. Duane, *Neurology* **26**, 1054 (1976).
2. D. B. Drachman, S. de Silva, D. Ramsay, and A. Pestronk, *Ann. N.Y. Acad. Sci.* **505**, 90 (1987).
3. A. Vincent, P. J. Whiting, M. Schluep, F. Heldenreich, B. Lang, A. Roberts, N. Willcox, and J. Newsom-Davis, *Ann. N.Y. Acad. Sci.* **505**, 106 (1987).
4. F. M. Howard, Jr., V. A. Lennon, J. Finley, J. Matsumoto, and L. R. Elveback, *Ann. N.Y. Acad. Sci.* **505**, 526 (1987).
5. B. Castleman and E. H. Norris, *Medicine* **20**, 27 (1949).
6. G. D. Levine and J. Rosai, *Hum. Pathol.* **9**, 495 (1978).

7. A. J. Oosterhuis, *Ann. N.Y. Acad. Sci.* **377,** 678 (1981).

8. C. W. Olanow, A. S. Wechsler, and A. D. Roses, *Ann. Surg.* **196,** 113 (1982).

9. I. Kao and D. B. Drachman, *Science* **196,** 74 (1977).

10. W. K. Engel, J. L. Trotter, D. E. McFarlin, and C. L. McIntosh, *Lancet 1,* 1310 (1977).

11. H. Wekerle, R. Hohfeld, U.-P. Ketelsen, J. R. Kalden, and I. Kalies, *Ann. N.Y. Acad. Sci.* **277,** 455 (1981).

12. M. Schluep, N. Willcox, A. Vincent, G. H. Dhoot, and J. Newsom-Davis, *Ann. Neurol.* **22,** 212 (1987).

13. T. Kirchner, S. Tzartos, F. Hoppe, B. Schalke, H. Wekerle, and H. K. Muller-Hermelink, *Am. J. Pathol.* **130,** 268 (1988).

14. M. Bofill, G. Janossy, N. Willcox, M. Chilosi, L. K. Trejdosiewics, and J. Newsom-Davis, *Am. J. Pathol.* **119,** 462 (1985).

15. M. Chilosi, A. Iannucci, L. Fiore-Donati, G. Tridente, M. Pampanim, G. Pizzolo, M. Ritter, M. Bofill, and G. Janossy, *J. Neuroimmunol.* **11,** 191 (1986).

16. W. Savino, H. Emonard, and J.-A. Grimaud, *J. Neuroimmunol.* **13,** 223 (1988).

17. M. J. Kornstein, J. J. Brooks, A. O. Anderson, A. I. Levinson, R. P. Lisak, and B. Zweiman, *Am. J. Pathol.* **117,** 184 (1984).

18. M. Mihovilovic and A. D. Roses, *Exp. Neurol.* **111,** 175 (1991).

19. J. M. Williams and D. L. Felten, *Anat. Rev.* **199,** 531 (1981).

20. D. L. Felten, S. Y. Felten, S. L. Carlson, J. A. Olschowka, and S. Livnat, *J. Immunol.* **135,** 755s (1985).

21. K. Bulloch, *in* "The Neuro–Immune–Endocrine Connection" (C. S. Cotman, R. E. Brinton, A. Galaburda, B. McEwen, and D. M. Schneider, eds.), p. 34. Raven, New York, 1987.

22. F. Magni, F. Bruschi, and K. Marianne, *Brain Res.* **424,** 379 (1987).

23. K. Bulloch, *in* "Neural Modulation of Immunity" (R. Guillemin, M. Cohen, and T. Melnechuk, eds.), p. 110. Raven, New York, 1985.

24. K. Bulloch and W. Pomerantz, *J. Comp. Neurol.* **228,** 57 (1984).

25. T. L. Rozzman and W. H. Brooks, *J. Neuroimmunol.* **10,** 59 (1985).

26. L. A. Jones, L. T. Chin, and A. M. Kruisbeek, *Thymus* **16,** 195 (1990).

27. M. J. McPherson, K. M. Jones, and S. J. Gurr, *in* "PCR: A Practical Approach" (M. J. McPherson, P. Quiake, and G. R. Taylor, eds.), p. 171. IRL Press, Oxford Univ. Press, Oxford, New York, and Tokyo, 1991.

28. R. Krumlauf, *in* "Methods in Molecular Biology" (E. J. Murray, ed.), Vol. 7, p. 307. Humana Press, Clifton, New Jersey, 1991.

29. D. A. Melton, P. A. Krieg, M. R. Rabagliati, T. Maniatis, K. Zinn, and M. P. Green, *Nucleic Acids Res.* **12,** 7035 (1984).

30. F. Samson, J. E. Lee, W.-Y. Hung, T. G. Potter, M. Herbstreith, A. D. Roses, and J. Gilbert, *J. Neurosci. Res.* **27,** 441 (1990).

31. E. S. Deneris, J. Boulter, J. Connolly, E. Wada, K. Wada, D. Goldman, L. W. Swanson, J. Patrick, and S. Heinemann, *Clin. Chem.* **35,** 731 (1989).

32. J. Boulter, W. Luyten, K. Evans, P. Mason, M. Ballivet, D. Goldman, S. Stengelin, G. Martin, S. Heinemann, and J. Patrick, *J. Neurosci.* **5,** 2545 (1985).

33. J. Boulter, K. Evans, D. Goldman, G. Martin, D. Treco, S. Heinemann, and J. Patrick, *Nature (London)* **319,** 368 (1986).

34. J. Boulter and P. D. Gardner, *in* "Methods in Neurosciences" (P. M. Conn, ed.), Vol. 1, p. 328. Academic Press, San Diego, 1989.

35. J. M. A. Chirgwin, A. E. Przybyla, R. J. MacDonald, and W. J. Rutter, *Biochemistry* **18,** 5294 (1979).

36. R. Ilaria, D. Wines, S. Pardue, S. Jamison, S. R. Ojeda, J. Snider, and M. R. Morrison, *J. Neurosci. Methods* **15,** 165 (1985).

37. H. Aviv and P. Leder, *Proc. Natl. Acad. Sci. U.S.A.* **69,** 1408 (1972).

38. J. Sambrook, E. F. Fritsch, and T. Maniatas, *in* "Molecular Cloning: A Laboratory Manual," pp. 6.9, 7.43. Cold Spring Harbor Laboratory, Cold Spring Harbor, New York, 1989.

39. J. M. Short, J. Fernandez, J. A. Sorge, and W. Huse, *Nucleic Acids Res.* **16,** 7583 (1988).

40. F. Sanger, S. Nicklen, and A. R. Coulson, *Proc. Natl. Acad. Sci. U.S.A.* **74,** 5463 (1977).

41. J.-L. Casanova, C. Pannetier, C. Jaulin, and P. Kourilsky, *Nucleic Acids Res.* **18,** 4028 (1990).

42. J. Devereux, P. Haeberli, and O. Smithies, *Nucleic Acids Res.* **12,** 387 (1984).

43. P. Nef, C. Oneyser, C. Alliod, S. Couturier, and M. Ballivet, *EMBO J.* **7,** 595 (1988).

44. E. S. Kawasaki and A. M. Wang, *in* "PCR Technology: Principles and Applications for DNA Amplification" (H. A. Erlich, ed.), p. 89. Stockton, New York, London, Tokyo, Melbourne, and Hong Kong, 1989.

45. A. J. Berk and P. A. Sharp, *Cell (Cambridge, Mass.)* **12,** 721 (1977).

46. T. J. Ley, N. P. Anagnou, G. Pepe, and A. W. Nienhuis, *Proc. Natl. Acad. Sci. U.S.A.* **79,** 4775 (1982).

47. R. K. Sakai, *in* "PCR Technology: Principles and Applications for DNA Amplification" (H. A. Erlich, ed.), p. 7. Stockton, New York, London, Tokyo, Melbourne, and Hong Kong, 1989.

48. J. Sambrook, E. F. Fritsch, and T. Maniatis, *in* "Molecular Cloning: A Laboratory Manual," p. 7.71. Cold Spring Harbor Laboratory, Cold Spring Harbor, New York, 1989.

49. C. C. Richardson, *in* "The Enzymes" (P. D. Boyer, ed.), 3rd Ed., Vol. 14, p. 299. Academic Press, New York, 1981.

50. M. Mihovilovic, C. Hulette, J. Mittelstaedt, and A. D. Roses, *Abstracts Soc. for Neurosci.* **17,** 1507 (1992).

51. M. Mihovilovic, C. Hulette, J. Mittelstaedt, C. Austin, and A. D. Roses, *Ann N.Y. Acad. Sci* (1993).

[12] Absolute Quantitation of γ-Aminobutyric Acid A Receptor Subunit Messenger RNAs by Competitive Polymerase Chain Reaction

Dennis R. Grayson, Patrizia Bovolin, and Maria-Rita Santi

Introduction

In recent years, it has become increasingly clear that many neurotransmitter receptors include families of protein subunits which are assembled in hetero-oligomeric structures, the combinations of which participate in the pharmacological and functional specificity of the receptorial response. This receptor structural diversity is well represented in the superfamily of ligand-gated ion channels (1), whose members contain four membrane-spanning domains, as well as numerous G-protein-coupled metabotropic receptors characterized by the presence of seven membrane-spanning domains (2). The ligand-gated ionotropic receptors include those for the amino acid neurotransmitter receptors glutamate, glycine, and γ-aminobutyric acid (GABA), each of which is encoded by a complex and tightly regulated family of homologous genes. For example, molecular cloning studies have shown that at least 15 different subunit mRNAs ($\alpha 1$, $\alpha 2$, $\alpha 3$, $\alpha 4$, $\alpha 5$, $\alpha 6$, $\beta 1$, $\beta 2$, $\beta 3$, $\gamma 1$, $\gamma 2S$, $\gamma 2L$, $\gamma 3$, δ, and ρ) of the $GABA_A$ receptor exist in the rodent central nervous system (CNS), with individual subunits showing varying degrees of homology to each other and to other members of the ligand-gated ion channel superfamily (3–5). Transient transfection studies have shown that particular combinations of receptor subunits when coexpressed display distinct pharmacological properties (4, 6, 7). Moreover, each receptor subunit mRNA is differentially expressed in distinct and overlapping sets of neurons and glia throughout the adult CNS and during development.

There are multiple problems associated with the application of conventional molecular biological approaches to the study of homologous mRNA families. Using as our example the $GABA_A$ receptor, the high degree of sequence homology existing within members of the α, β, and γ subunit families complicates the interpretation of hybridization data obtained using full-length cDNA probes, as they cross-hybridize with mRNAs encoding other receptor subunits from the same family (8–12). Moreover, the quantitation of mRNA levels by Northern analysis will be problematic, particularly if the receptor subunit mRNAs are present in low copy numbers in the CNS structure under investigation. Because many of the $GABA_A$ receptor subunit mRNAs are present in low abundance in selected brain regions, we sought to develop an application of the polymerase chain reaction (PCR) that is sensitive and would allow us to measure and compare each of the receptor subunit mRNAs expressed in discrete central nervous system structures of interest.

Relative Quantitation

To determine the relative abundances of the various mRNAs encoding subunits of the $GABA_A$ receptor, we prepared a series of amplification primers which were specific to each of the receptor subunit cDNAs (13, 14). The primers were designed to amplify nonhomologous portions of the cDNAs extending from within the C-terminal intracellular loop (see Refs. 3 and 4 for recent reviews) to the 5'-most portion of the 3' untranslated regions. In these regions, individual sequences differ substantially so that subunit-specific mRNA amplification is achieved.

Total cellular RNA is isolated from the CNS structure of interest, and cDNA is generated by reverse transcription. We have performed numerous reverse transcription reactions in the presence of trace amounts of [^{32}P]dCTP with differing amounts of RNA from various brain regions and have found that the extent of ^{32}P incorporation is reproducible from experiment to experiment. Moreover, the reaction kinetics are pseudo-first-order with respect to RNA concentration (P. Bovolin, unpublished data, 1990). The PCR is performed with increasing amounts of starting template in the presence of [^{32}P]dCTP for different numbers of amplification cycles. Amplification products are separated from primers by electrophoresis, and the incorporated radioactivity is measured after cutting the ethidium bromide-stained bands from the gel.

The correlation between the amount of amplified product and the amount of reverse-transcribed template depends critically on the abundance of the targeted template in the RNA sample. The analysis is most sensitive to differences in RNA levels when the incorporated radioactivity values are within the exponential range. Further increasing either the number of cycles or the amount of starting template does not allow for the detection of differences in the amount of amplified product because of the so-called plateau effect (15). For this reason, it is necessary to perform similar titration experiments for each set of specific primers when analyzing RNA from different brain regions.

A limitation to the above approach is that the information obtained is relative and does not allow for a direct comparison of mRNA levels corresponding to different receptor subunits. This occurs because the amplification rates of different target templates using different primer pairs will also vary. For this reason and because of a limited amount of space, we have chosen to present procedures we are currently using to determine absolute amounts of receptor subunit mRNAs using competitive PCR analysis (16–21) of reverse-transcribed RNA. The reader is referred to some recently published protocols for obtaining relative GABA$_A$ receptor subunit mRNA levels (13, 22) and mRNA levels encoding other proteins (23).

Competitive PCR with Internal Standards

Overview

To quantitate the different GABA$_A$ receptor subunit mRNA levels, a series of internal standards were synthesized that use the same primer pairs which are used to amplify the mRNA-derived template. Each standard template was cloned into a dual bacteriophage Sp6/T7 RNA polymerase promoter-containing plasmid (pGEM 1) so that cRNA synthesized *in vitro* can be added to the unknown sample prior to the reverse transcriptase reaction. In this way, each subsequent manipulation (reverse transcription, PCR amplification, chloroform extraction, etc.) affects both templates equally and is therefore internally controlled. The internal standard template differs from the target mRNA template minimally, for example, 3 of 304 bp for the $\alpha 1$ subunit and 2 of 340 bp for the $\alpha 5$ subunit (see Fig. 2 for these two examples of site-directed mutations). The internal standards contain a restriction enzyme cleavage site (*Bgl*II) midway through the target sequence which allows one to differentiate between the amplification products arising from the mRNA and those arising from the added cRNA in a postamplification step. Because a single primer pair (per subunit cDNA) is used to amplify both templates (i.e., mRNA and cRNA) and because the two templates are identical in size and virtually identical in sequence, they are amplified at the same rate. This approach has been used successfully to quantitate absolute amounts of mRNAs encoding 13 different subunits of the GABA$_A$ receptor in primary cultures of cerebellar granule neurons and cerebellar glia prepared from neonatal rat pups (21).

Figure 1 shows a schematic overview of competitive PCR analysis, each step of which is discussed in some detail in the sections that follow. cRNA is synthesized *in vitro* from the linearized internal standard, and known amounts are added to constant amounts of total RNA isolated from the sample of interest. Both templates are reverse transcribed and coamplified with the subunit-specific primer pair in the presence of trace amounts of an

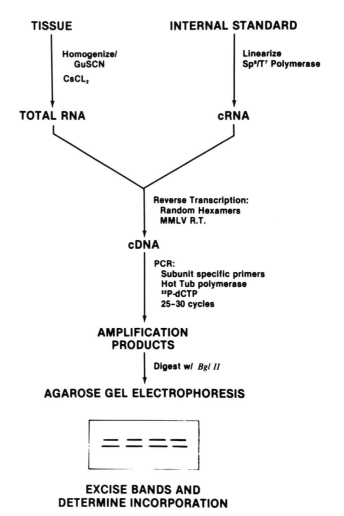

FIG. 1 Schematic overview of the steps involved in using internal standards for competitive PCR analysis of RNA.

isotopically labeled dNTP. Following PCR, each amplification reaction is chloroform extracted, and aliquots are digested with *Bgl*II overnight. The digestion products are separated by agarose gel electrophoresis, and the amount of each product is determined by measuring the extent of isotope incorporation into the respective amplification products. Because the *Bgl*II restriction site has been directed to the middle of the amplified sequence, only two bands need to be excised from the gel and counted. The data are then plotted as the counts incorporated into the standard amplification product divided by the counts incorporated into the corresponding subunit mRNA amplification product as a function of the amount of internal standard cRNA added initially to the sample (see, e.g., Fig. 4). The point of equivalence represents the point at which the ratio of counts is unity and corresponds to the concentration of standard template which is present in the unknown sample.

Basic Methodology

Oligonucleotides

We routinely synthesize amplification primers on an Applied Biosystems (Foster City, CA) Model 381A DNA synthesizer using phosphoramidite chemistry leaving the terminal dimethoxytrityl group intact. The oligonucleotides are removed from their columns using concentrated NH_4OH (30% NH_3 content, Aldrich Chemical Co., Milwaukee, WI), and the various protecting groups are eliminated by incubation at room temperature for 24 hr. All primers are purified by reversed-phase chromatography using oligonucleotide purification cartridges following the recommendations of the manufacturer (Applied Biosystems). Although we found that this extra purification step greatly reduced our overall yield of material, the improved specificity and yield of PCR product were well worth it. Amplification primers that were either repeatedly lyophilized or butanol extracted (24) and lyophilized gave rise to significant amounts of amplification artifacts that were not present when the cartridge-purified primers were used.

The concentration of the purified oligonucleotide is calculated from the absorbance of an aliquot at 260 nm. We calculate an extinction coefficient for each primer as follows: $\varepsilon_{260} = \Sigma$ [(#A's × 15.4) + (#C's × 7.3) + (#G's × 11.7) + (#T's × 8.8)] ml/μmol (25). PCR primer stocks are prepared by diluting the concentrated oligonucleotide to 20 μM with nuclease-free water (Promega, Madison, WI) using positive displacement pipettes (Microman, Gilson, Middletown, WI).

The $GABA_A$ receptor–primer pairs were designed to allow amplification of 300 to 340 bp and to include cDNA sequences with the lowest degree of homology between the various receptor subunits. Moreover, each primer pair contained a comparable G/C content (50%) in order to minimize variability in the hybridization efficiency at the annealing temperature (21). This last point is perhaps more a matter of convenience as it allows for a single annealing temperature for each template to be analyzed. For cloning purposes, restriction sites are built onto the ends of the amplification primers with an additional

α_1 GABA$_A$ RECEPTOR SUBUNIT

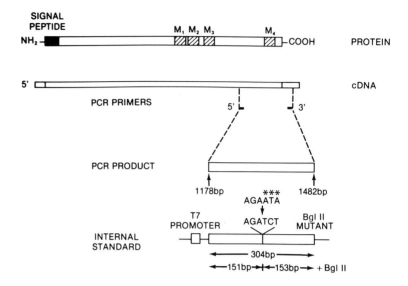

α_5 GABA$_A$ RECEPTOR SUBUNIT

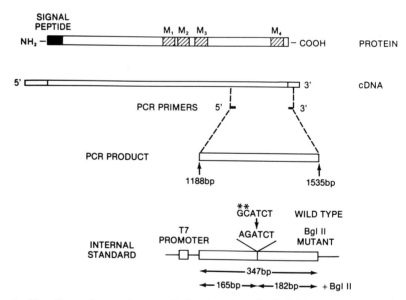

FIG. 2 Site-directed mutations made in the $\alpha 1$ and $\alpha 5$ GABA$_A$ receptor subunit cDNA templates to introduce a *Bgl*II restriction site in the center of the sequence. [Reprinted with permission from P. Bovolin, M. R. Santi, M. Memo, E. Costa, and D. R. Grayson, *J. Neurochem.* **59,** 62 (1992). (Raven Press).]

3 bp to 5' to the restriction site (i.e., 5'-XXXGAATTC-24bp-3', where X is A, C, G, or T). As we have now developed a large number of internal standards, we routinely add an *Eco*RI site to the 5' amplimer and a *Hin*dIII site to the 3' amplimer so that each primer is cloned in the same orientation. Occasionally, an internal *Eco* or *Hin*d site exists within the amplification target sequence, in which case we substitute with another site (e.g., *Bam*HI) so as to maintain the cloning orientation.

Synthesis of Internal Standards

Internal standard templates are generated by site-directed mutagenesis using PCR overlap extension (26). Although each standard is designed so as to introduce a *Bgl*II restriction site midway between the amplification primers, we have also successfully used the same strategy to create 30 to 40 bp deletion templates as well. Single-stranded internal primers are designed and synthesized so that the restriction site would be introduced with only a minimal number of base substitutions. In each case this involves changing only from 2 to 3 of the 6-bp recognition sequence. For example, approximately midway between the GABA$_A$ receptor α1 amplification primers was the sequence AGAATA (see Fig. 2), which required only a 3-bp change to AGATCT. The internal mutating primers are designed such that there is a 24-bp overlap of the primary PCR products (i.e., the products of the two initial amplifications). The percent (G + C) content of this overlap is designed such that the annealing temperature is the same as that of the outside primers. (We have designed our primers such that each is 24 bp in length with an approximate G + C content of 50%. Using these primers, we have found that an annealing temperature of 60°C results in fairly specific amplification with the absence of artifactual DNA bands).

The internal standards can be synthesized either from a linearized cDNA template or from reverse-transcribed cDNA if the clone of interest is not available. Different concentrations of heat-denatured linear starting cDNA template (from 1 to 100 ng) are incubated with 1 μM of each primer pair (internal + outside), and PCR is performed with Hot Tub DNA polymerase (Amersham, Arlington Heights, IL) buffered as indicated below in the competitive PCR procedure. The primary PCR is performed using 35 amplification cycles, each cycle consisting of 94°C/45 sec, 60°C/1 min, and 72°C/1 min (with a final 72°C extension of 15 min) using a DNA Thermal Cycler (Perkin-Elmer/Cetus, Norwalk, CT). Material from the primary PCR is phenol–chloroform extracted (1:1 v/v) and purified by chromatography on PCR purification columns (Qiagen, Chatsworth, CA) as described by the manufacturer. The yield of product following purification is determined either by visual inspection of the ethidium bromide-stained bands following gel electrophoresis or by using the fluorescent dye Hoescht 33258 (27) and a DNA Mini-Flurometer (Hoefer, San Francisco, CA).

For the secondary PCR, increasing and equivalent amounts of the two primary amplification products are pooled, and the PCR is then performed using only the two outside primers. The product from the second PCR is purified on PCR purification columns (as before) and aliquots are then digested with *Bgl*II to ensure the presence of the restriction site.

Following digestion of the secondary PCR product with *Eco*RI and *Hind*III, the material is extracted from low melting point agarose (Bethesda Research Laboratories, Gaithersburg, MD) and cloned into the corresponding sites of pGEM 1 (Promega-Biotech), or any convenient RNA polymerase promoter-containing plasmid used in the laboratory, using standard cloning methodology (25). Minipreps of the transformed ligation mixes can be screened by digesting with *Eco*RI and *Hind*III to verify the size of any recombinants and in parallel with *Eco*RI and *Bgl*II to verify the presence of the targeted restriction site. We have sequenced each of our internal standard templates using a Genesis 2000 automated sequencer (Du Pont–NEN, Boston, Ma) to subsequently verify the location of the mutations.

RNA Isolation

The usual precautions for handling RNA [sterile disposable glass, diethyl pyrocarbonate (DEPC)-treated solutions, gloved hands, etc; see Ref. 25, Section 7.3] are taken to prevent RNase degradation. Dissected brain tissue is frozen immediately after removal on dry ice. The frozen tissue is homogenized in 5 M guanidium isothiocyanate, 100 mM Tris-HCl, pH 7.4, and 1 mM EDTA, and total RNA is isolated following $CsCl_2$ ultracentrifugation as previously described (28). The RNA pellet is resuspended in 500 μl nuclease-free water, phenol–chloroform (1:1, v/v/) extracted, and ethanol (2.5 volumes) precipitated from 0.3 M sodium acetate, pH 5.2. The yield of total RNA is determined by measuring the optical density (OD) of an aliquot of the precipitated stock at 260/280 nm.

The OD measurement is critical for the comparison of mRNA levels from, for example, different preparations or different brain regions. Preliminary experiments have shown a strong correlation between the amount of RNA (isolated as described above) as determined by OD readings and the yield of radioactivity incorporated into cDNA during RNA reverse transcription (25). Total RNA prepared using other published protocols (29), as well as various commercially available RNA isolation kits (RNAzol, Tel-Test, Inc.; Fast Track, Invitrogen, San Diego, CA; QuickPrep, Pharmacia LKB, Piscataway, NJ), will also work well with the competitive PCR technique but may suffer by the degree to which the material is contaminated by either DNA or protein. In our hands, RNA prepared by ultracentrifugation, as described above, always resulted in RNA free of any genomic DNA contamination. It is always advisable to compare RNA recovery (yield) by additional means such as measuring the extent of incorporation following reverse transcription or measuring the levels of an additional RNA which should remain constant in each sample (we are currently measuring the amount of 18 S rRNA in our samples; both β-actin and cyclophilin are commonly in use for this purpose, but these also change developmentally and so can sometimes be misleading). Once the concentration of the newly isolated RNA has been determined, it is convenient to partition the sample and freeze multiple aliquots (e.g., 25–30 μg each) for future use.

cRNA Synthesis

cRNA is prepared from the linearized standard templates such that the corresponding sense strand is produced. The following discussion is applicable to any RNA polymerase promoter-containing vector system but will require modification depending on the particular choice. The internal standards are linearized with *Sph*I (BRL) which cuts 432 bp downstream of the *Hin*dIII 3' cloning site (except for the β3 subunit standard which contains an internal *Sph*I site). Any 3' overhangs left by the linearizing restriction enzyme are routinely "filled in" so as to minimize opposite strand synthesis with 5 units/μg template of the Klenow fragment of DNA polymerase I (Promega).

cRNA is synthesized starting from 2 to 5 μg of the linearized template and T7 RNA polymerase as described in a technical manual provided by the manufacturer (Promega). The reaction contains 2 to 5 μg template, 40 mM Tris-HCl, pH 7.5, 6 mM MgCl$_2$, 2 mM spermidine, 10 mM NaCl, 10 mM dithiothreitol (DTT), 100 units of RNasin (Promega), 0.5 mM each rNTP, and from 30 to 40 units of T7 RNA polymerase in a 100-μl reaction volume. All reagents are added to the template at room temperature, and the reaction is initiated by the addition of enzyme. The transcription reaction is incubated at 37°C for 120 min, at which time from 2 to 5 units of RQ1 RNase-free DNase (Promega) is added to digest the template DNA. This reaction is kept at 37°C from 30 to 60 min to ensure complete digestion of residual DNA template. Although the amount of plasmid DNA present in the dilute amounts of RNA used in our experiments is likely to be very low, any DNA contamination present will coamplify with the added cRNA during the PCR (see below, Potential Sources of Error).

Following DNase digestion, the reaction is extracted one time with TE-saturated phenol–chloroform and precipitated from 3.75 M ammonium acetate by adding 2 volumes of ice-cold ethanol. The precipitation is carried out at $-70°C$ for a minimum of 30 min or overnight at $-20°C$. The RNA is then pelleted in a refrigerated ($4°C$) microcentrifuge for 15 min at 12,000 g and resuspended in 100 μl of 1 M ammonium acetate is repeated as before, and following centrifugation the pellet is washed with 500 μl of ice-cold 70% ethanol, dried under reduced pressure, and resuspended in 20 μl of nuclease-free water. Approximately 25% of this material is used to determine the absorbance at 260/280 nm. Typically, we obtain from 25 to 35 μg of cRNA using the above conditions as determined by this measurement.

Synthesis of cDNA

As indicated in Fig. 1, all manipulations for the test RNA sample and the internal standard cRNA are subsequently carried out in the same tube. Because the absolute abundance of the mRNA of interest is not known in the test sample, the quantitation is carried out in two steps. The first involves a rather broad titration of the unknown sample using concentrations of cRNA that span a wide concentration range (e.g., from 1 pg to 1 ng), while the second series spans a more restricted concentration range such that the standard cRNA is both above and below the amount present in the unknown sample. These two ranges roughly correspond to those shown in Fig. 4, where the more restricted range is shown in the insets. The more points that are used in the second titration experiment, the more accurate the regression line will be for determining the equivalence point.

Various amounts of cRNA prepared from the internal standard template are added to a constant amount of total RNA isolated from the source of interest (usually 0.5 μg or less depending on the abundance of the transcript being measured). The RNA–cRNA mixtures are reverse transcribed with cloned Moloney murine leukemia virus (MMLV) reverse transcriptase using random hexamer primers. Two hundred units of MMLV reverse transcriptase (BRL) is added to the RNA in 50 mM Tris-HCl, pH 8.3, 74 mM KCl, 3 mM MgCl$_2$, 1 mM random hexamers [pd(N)$_6$, Pharmacia] in a volume of 20 μl. The reaction mixture is incubated at room temperature for 10 min, then at $37°C$ for 1 hr, heat-denatured at $95°C$ for 5 min, and ethanol-precipitated from 0.3 M sodium acetate by adding 2 volumes of ice-cold ethanol.

Competitive PCR Amplification

Aliquots containing from 25 to 50% of the reverse-transcribed material are amplified with Hot Tub Polymerase (Amersham) in a Thermal Cycler (Perkin-

Elmer/Cetus). We have found, in side by side tests, that use of Hot Tub Polymerase (Amersham) reproducibly results in higher yields (by as much as 2- to 4-fold) and is more specific (i.e., yielding a single band after agarose gel electrophoresis rather than multiple bands) when compared to a comparable number of units of AmpliTaq DNA polymerase (Perkin-Elmer/Cetus). Since the amplifications are internally controlled, the yield of product is not so important. However, because the amplification products are separated in a postamplification step, the specificity of the process is critical to the interpretation of the data.

Prior to amplification, the cDNA is heated at 95°C for 5 min and quick chilled on ice to denature the starting template. The reaction mixture contains cDNA, 1 μM of subunit-specific primer pairs, 200 μM dNTPs, 1.5 mM MgCl$_2$, 50 mM Tris-HCl, pH 9.0, 20 mM ammonium sulfate, and 2.5 units of Hot Tub Polymerase in a 100-μl volume. Trace amounts of [^{32}P]dCTP (1–2 μCi/ sample) are added to the reaction for subsequent quantitation. The mixture is overlaid with 50 μl of mineral oil and amplified between 25 and 30 cycles. Although the particular cycling parameters used will depend on the personal bias of the investigator, we have found the following to work well: a denaturation step (94°C, 45 sec), an annealing step (60°C, 1 min), and an elongation step (72°C, 1 min) with a final 15-min (72°C) elongation. Following the final amplification cycle, samples are cooled to 4°C until use. Mineral oil is removed from the samples by extracting with 100 μl of chloroform and a 2-min microcentrifugation. Amplification products arising from both the mRNA and cRNA templates are analyzed as described below.

Postamplification Separation of Products

As indicated above, we have chosen to construct our internal standards such that postamplification digestion with a restriction enzyme allows us to differentiate the templates of origin. Although we have found that this works well, there are several other possibilities the reader may wish to consider. The technique of overlap extension PCR can also be used to generate deletions, as well as insertions, into various internal standard templates (26). The advantage of using a template that differs in size is that it eliminates the need for a postamplification digestion step. Using overlap extension PCR, we have generated templates with internal deletions of approximately 10% of the target sequence (30–35 bp). However, we found that in order to separate the amplification products, it is necessary to resolve the products on rigid agarose gels (2.0%) and to electrophorese for longer periods of time (2 hr or more). Because the separation of products is somewhat more convenient with the restriction site-containing standards, this system was adopted for the mRNAs we wanted to measure. Alternatively, large deletions (30–50%

of the target sequence) in the internal standard template would probably work as well as using a restriction site to differentiate the amplification products.

It has also been shown that an internal standard prepared by linking 12 different amplification primers in tandem is coamplified at the same rate as the mRNA templates being measured (16). The advantage of this approach is that a single internal standard is used to quantitate multiple mRNAs. However, it is important to verify that the internal standard is amplified at the same rate as each template being measured, and the construction of the template can be a technical limitation. More recently, competitor DNA fragments that differ in size have been generated by amplifying DNA from an evolutionarily distant species under low-stringency annealing conditions (30). Competitor templates generated in this way would also use the homologous primer pairs and could therefore serve as appropriate PCR internal standards.

We have previously illustrated the feasibility of coamplifying distinguishable templates using a single primer pair and determining their origin in a postamplification step (20). Following PCR and chloroform extraction, 51-μl aliquots of each PCR are digested with 30 units of BglII (overnight) in 50 mM Tris-HCl, pH 8.0, 10 mM MgCl$_2$, and 100 mM NaCl in a 60-μl volume. The digestion is allowed to go overnight (minimum digestion time for >95% digestion is 4 hr) at 37°C. The following day, 6.6 μl of a 10× loading dye [where 1× is 0.025% (w/v) bromophenol blue, 0.025% (w/v) xylene cyanol, 5% (v/v) glycerol] is added, and 20-μl aliquots are analyzed on 1.8% (w/v) agarose gels in 0.5× TBE mM Tris–borate, 1 mM EDTA).

The following illustrates the use of competitive PCR analysis to determine the amounts of two GABA$_A$ receptor subunit mRNAs in RNA isolated from adult rat cortex. Known concentrations (see legend to Fig. 3) of cRNA derived from each internal standard were added to constant amounts of total cortical RNA. The mixtures were reverse transcribed in parallel and PCR amplified in the presence of [^{32}P]dCTP (as described above). Following amplification, aliquots were digested (as above) with BglII and analyzed by agarose gel electrophoresis in triplicate. Figure 3 is a photograph of the ethidium bromide-stained gel showing the amplification products arising from the mRNA template (α1, 304 bp; α5, 340 bp) and the corresponding digestion products arising from the respective cRNA (α1, 151 + 153 bp; α5, 165 + 175 bp).

To quantitate the amount of product corresponding to the amplified mRNA and cRNA, the ethidium bromide-stained bands are removed and the extent of incorporation determined by counting. A corresponding gel slice is excised from an adjacent lane with a PCR control sample (no added template) in order to determine that background radiation. Results from the above com-

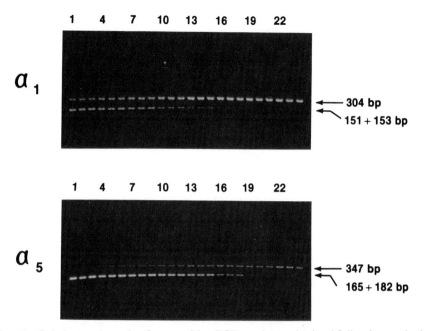

FIG. 3   Gel electrophoresis of competitive PCR products obtained following analysis of the $\alpha 1$ and $\alpha 5$ GABA$_A$ receptor subunit mRNA levels. Increasing amounts of cRNA were added to 1 μg of total RNA isolated from adult rat cortex. The mixtures were reverse transcribed and amplified in the presence of trace amounts of [^{32}P]dCTP as described in the text. Aliquots were digested with *Bgl*II in triplicate and analyzed by gel electrophoresis. The amounts of cRNA added to 1 μg total RNA were 1 pg (lanes 1–3), 10 pg (lanes 4–6), 50 pg (lanes 7–9), 100 pg (lanes 10–12), 200 pg (lanes 13–15), 400 pg (lanes 16–18), 600 pg (lanes 19–21), and 1 ng (lanes 22–24).

petitive PCR analyses are shown in Fig. 4, where the data are presented as the counts incorporated into the amplified cRNA standard divided by the counts incorporated into the corresponding subunit mRNA amplification product versus the known amount of internal standard cRNA added to the test sample. The insets Fig. 4 show expansion of the scale near the point of equivalence (i.e., the point at which the ratio of counts is unity). A comparison of the two plots presented in Fig. 4 demonstrates the different levels of the corresponding receptor subunit mRNA present in RNA isolated from adult rat cortex. Specifically, 1 μg of total cortical RNA contains 260 pg (2.5 fmol) specific to the $\alpha 1$ receptor subunit mRNA and 22 pg (0.19 fmol) of the $\alpha 5$ mRNA. Using this approach we have been able to quantitate receptor subunit mRNA levels which differ by as much as 200-fold in the same RNA sample (20, 21).

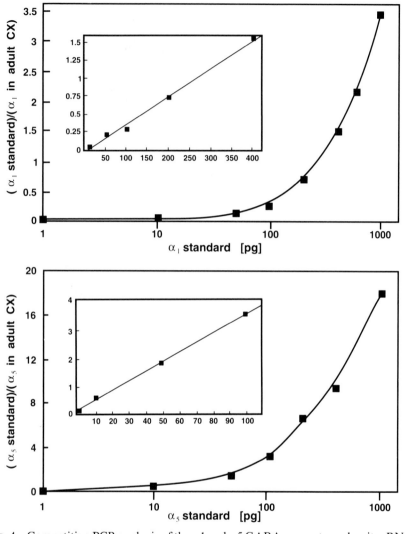

FIG. 4   Competitive PCR analysis of the $\alpha1$ and $\alpha5$ GABA$_A$ receptor subunit mRNAs in RNA from adult rat cortex. The amplification products shown in Fig. 3 were excised from the gel, and the amount of radioactivity incorporated into each band was determined by Cerenkov counting. As discussed in the text, the amount of each subunit mRNA in the adult cortical sample is determined from the point of equivalence of each standard curve.

Potential Sources of Error

The most common and potentially most difficult to correct source of error involves reagent contamination (i.e., primer, dNTP, internal standard). This can generally be avoided by rigorous use of either positive displacement pipettes (Gilson Microman series, Rainin, Woburn, MA) or aerosol filter pipette tips (e.g., Aerosol Resistant Tips, Continental Laboratory Products, San Diego, CA) when preparing reagent stocks. Reagent contamination can be particularly costly when it involves either enzyme stocks or primer preparations. It is also worthwhile to prepare multiple aliquots of stock solutions for multiple users (other procedures for minimizing PCR-product carryover can be found in Ref. 31). Finally, internal standard cRNA dilutions should be prepared, aliquoted, and frozen so as to minimize the possibility of introducing RNase into the stock standard template. To check for contamination we routinely include a control which lacks added template and another which lacks added primers.

Another potential source of error arises if the isolated RNA contains residual genomic DNA, the effect being an overestimate of the amount of mRNA template present. In general, this will only be a problem when the amount of specific mRNA present in the sample is very low (<1 to 5 copies per cell) or if the corresponding gene is present at multiple copies per genome. We have estimated the amount of DNA contamination of RNA isolated by the guanidium isothiocyanate–CsCl$_2$ procedure (28) to be approximately 1% or 10 ng (based on the amount of fluorescent Hoescht 33258 binding to 1 μg RNA). If the genomic structure is known, this error can be avoided completely by designing the amplification primers such that they span an intron.

The presence of residual internal standard-containing plasmid which may still be present in the cRNA samples used for competitive PCR represents another potential artifact that would lead to an underestimation of the concentration of template in the unknown sample. This would occur if the RQ1 digestion of the cRNA produced in the *in vitro* transcription reaction were incomplete. The RQ1 DNase digestion time we have suggested is longer than that recommended by the manufacturer for this reason. If this type of artifact is suspected, one can redigest an aliquot of the cRNA with the DNase (as before) and then perform side by side PCR comparisons to determine if the second digestion had an effect. In general, this source of error may only be a significant problem when attempting to measure very rare mRNAs.

Conclusions

We have described a competitive PCR assay that we are using to quantitate the mRNA levels of each of the GABA$_A$ receptor subunits that have thus far been cloned. The approach uses an internal standard template to generate cRNA, known amounts of which are added to constant amounts of total RNA. The mixtures are reverse transcribed, and subunit-specific primer pairs are used to coamplify cDNA encoding the subunit of interest and cDNA derived from the internal standard template. The amplification products arising from each template are then separated in a postamplification step. As each step of the procedure (from the reverse transcription through the PCR amplification) is carried out in the same tube, the reactions are internally controlled. Data obtained from this approach are absolute and so can be used to compare the abundances of any mRNA of interest. We are currently using this approach to quantitate various subunits of the glutamate receptor (GluR1–R6; Ref. 32), the metabotropic glutamate receptor (33), and the more recently cloned N-methyl-D-aspartate (NMDA)-selective glutamate receptor (34). This will allow for the first time a direct comparison of the absolute amounts of mRNAs encoding families of inhibitory and excitatory neurotransmitter receptor subunits.

Acknowledgments

The authors would like to thank Jon Baldvins and Jim Kirby for excellent technical assistance. We thank Drs. Michel Khrestchatisky and Alan Tobin (Dept. of Biology, University of California, Los Angeles) for the rat $\alpha1$, $\alpha2$, $\alpha5$, and $\beta1$ GABA$_A$ receptor subunit cDNAs which were used in the construction of the corresponding internal standards. The $\alpha5$ GABA$_A$ (35) receptor subunit mRNA is identical in sequence to the $\alpha4$ cDNA previously published from another laboratory (36). This work was supported by a grant from the Fidia Research Foundation, Washington, D.C. A complete listing of the amplification primer sequences and the primers used in constructing the internal standards for each GABA$_A$ receptor subunit mRNA that we are measuring ($\alpha1$, $\alpha2$, $\alpha3$, $\alpha4$, $\alpha5$, $\alpha6$, $\beta1$, $\beta2$, $\beta3$, $\gamma1$, $\gamma2S$, $\gamma2L$, $\gamma3$, δ, and ρ) is available on request.

References

1. H. Betz, *Neuron,* **5,** 383 (1990).
2. P. R. Schofield, B. D. Shivers, and P. H. Seeburg, *Trends Neurosci.* **13,** 8 (1990).
3. R. W. Olsen and A. J. Tobin, *FASEB J.* **4,** 1469 (1990).
4. P. H. Seeburg, W. Wisden, T. A. Verdoorn, D. B. Pritchett, P. Werner, A. Herb, H. Lüddens, R. Sprengel, and B. Sakmann, *Cold Spring Harbor Symp. Quant. Biol.* **55,** 29 (1990).
5. H. Lüddens and W. Wisden, *Trends Pharmacol. Sci.* **12,** 49 (1991).
6. G. Puia, S. Vicini, P. H. Seeburg, and E. Costa, *Mol. Pharmacol.* **39,** 691 (1991).

7. T. A. Verdoorn, A. Draguhn, S. Ymer, P. H. Seeburg, and B. Sakmann, *Neuron* **4,** 919 (1990).

8. K. M. Garrett, N. Saito, R. S. Duman, M. S. Abel, R. A. Ashton, S. Fujimori, B. Beer, J. F. Tallman, M. P. Vitek, and A. J. Blume, *Mol. Pharmacol.* **37,** 652 (1990).

9. P. Montpied, B. M. Martin, S. L. Cottingham, B. K. Stubblefield, E. I. Ginns, and S. M. Paul, *J. Neurochem.* **51,** 1651 (1988).

10. P. Montpied, E. I. Ginns, B. M. Martin, D. Stetler, A.-M. O'Carrol, S. J. Lolait, L. C. Mahan, and S. M. Paul, *FEBS Lett.* **258,** 94 (1989).

11. P. Montpied, A. L. Morrow, J. W. Karanian, E. I. Ginns, B. M. Martin, and S. M. Paul, *Mol. Pharmacol.* **39,** 157 (1991).

12. P. Montpied, E. I. Ginns, B. M. Martin, D. Roca, D. H. Farb, and S. M. Paul, *J. Biol. Chem.* **266,** 6011 (1991).

13. M. Memo, P. Bovolin, E. Costa, and D. R. Grayson, *Mol. Pharmacol.* **39,** 599 (1991).

14. M. Memo, P. Bovolin, E. Costa, and D. R. Grayson, *in* "Neurotransmitter Regulation of Gene Expression" (E. Costa, ed.), Fidia Research Foundation Symposium Series 8. Thieme Medical Publ., New York, 1991.

15. M. A. Innis and D. H. Gelfand, *in* "PCR Protocols" (M. A. Innis, D. H. Gelfand, J. J. Sninsky, and T. J. White, eds.), p. 3. Academic Press, San Diego, 1990.

16. A. M. Wang, M. V. Doyle, and D. F. Mark, *Proc. Natl. Acad. Sci. U.S.A.* **86,** 9717 (1989).

17. A. M. Wang and D. F. Mark, *in* "PCR Protocols" (M. A. Innis, D. H. Gelfand, J. J. Sninsky, and T. J. White, eds.), p. 70. Academic Press, San Diego, 1990.

18. G. Gilliland, S. Perrin, K. Blanchard, and H. F. Bunn, *Proc. Natl. Acad. Sci. U.S.A.* **87,** 2725 (1990).

19. G. Gilliland, S. Perrin, and H. F. Bunn, *in* "PCR Protocols" (M. A. Innis, D. H. Gelfand, J. J. Sninsky, and T. J. White, eds.), p. 60. Academic Press, San Diego, 1990.

20. P. Bovolin, M.-R. Santi, M. Memo, E. Costa, and D. R. Grayson, *J. Neurochem.* **59,** 62 (1992).

21. P. Bovolin, M.-R. Santi, G. Puia, E. Costa, and D. R. Grayson, *Proc. Natl. Acad. Sci. U.S.A.* **89,** 9344 (1992).

22. K. J. Buck, R. A. Harris, and J. M. Sikela, *BioTechniques* **11,** 636 (1991).

23. J. Golay, F. Passerini, and M. Introna, *PCR, Methods Appl.* **1,** 144 (1991).

24. M. Sawadago and M. W. Van Dyke, *Nucleic Acids Res.* **19,** 674 (1991).

25. J. Sambrook, E. F. Fritsch, and T. Maniatis, "Molecular Cloning: A Laboratory Manual." Cold Spring Harbor Laboratory, Cold Spring Harbor, New York, 1989.

26. S. N. Ho, H. D. Hunt, R. H. Horton, J. K. Pullen, and L. R. Pease, *Gene* **77,** 51 (1989).

27. Y. Kubota, *Bull. Chem. Soc. Jpn.* **63,** 758 (1990).

28. J. M. Chirgwin, A. E. Przybyla, R. J. MacDonald, and W. J. Rutter, *Biochemistry* **18,** 5294 (1979).

29. P. Chomczynski and N. Sacchi, *Anal. Biochem.* **162,** 156 (1987).

30. K. Überla, C. Platzer, T. Diamanstein, and T. Blankenstein, *PCR, Methods Appl.* **1,** 136 (1991).

31. S. Kwok, *in* "PCR Protocols" (M. A. Innis, D. H. Gelfand, J. J. Sninsky, and T. J. White, eds.), p. 142. Academic Press, San Diego, 1990.
32. R. Dingeldine, *Trends Pharmacol. Sci.* **12,** 360 (1991).
33. M. Masu, Y. Tanabe, K. Tsuchida, R. Shigemoto, and S. Nakanishi, *Nature (London)* **349,** 760 (1991).
34. K. Moriyoshi, M. Masu, T. Ishii, R. Shigemoto, N. Mizuno, and S. Nakanishi, *Nature (London)* **354,** 31 (1991).
35. M. Khrestchatisky, A. J. MacLennan, M. Chiang, W. Xu, M. B. Jackson, N. Brecha, and A. J. Tobin, *Neuron* **3,** 745 (1989).
36. D. B. Pritchett and P. H. Seeburg, *J. Neurochem.* **54,** 1802 (1991).

Section II

Receptor Subclasses

[13] Angiotensin II Receptor Subtypes in the Central Nervous System

MaryAnne E. Hunt, and James K. Wamsley

Introduction

The identification of the enzyme renin (originally believed to be a pressor agent) just prior to the turn of the twentieth century, and the eventual elucidation of its role as the catalyst for the conversion of angiotensinogen to angiotensin, supplied the impetus for the intensive study of the neuropeptidergic system. Continued research produced conclusive evidence of a homeostatic role for renin–angiotensin in normal and hypertensive states. Isolation of the decapeptide angiotensin I (AI) and the octapeptide angiotensin II (AII) ultimately led to the development of pharmacological agents which, because of their ability to inhibit formation of angiotensin or antagonize activity at AII receptor sites, proved to be highly effective antihypertensives. Saralasin ([Sar[1], Val[5], Ala[8]] angiotensin), introduced in 1971, is probably the best known of these peptidergic compounds (see Refs. 1 and 2). Unfortunately, the peptidergic antihypertensives and the angiotensin-converting enzyme (ACE) inhibitors were plagued with unpleasant side effects and had a brief duration of action. These drawbacks motivated researchers to develop effective nonpeptidergic AII antagonists. Discovery of such compounds eluded scientists until 1982 when Furukawa *et al.* (3, 4) patented S83078 and S8308, imidazole analogs with AII antagonist properties. A variety of related compounds followed, each slightly more specific for the AII receptor than its predecessor. The search culminated (at least so far) in the development of highly selective AII antagonists (5a,b,c, 6a,b) such as DUP753 (Losartan) (2-*n*-butyl-4-chloro-5-hydroxymethyl-1-[(2[1]-(1*H*-tetrazol-5-yl)-biphenyl-4-yl) methyl]imida-zole, potassium salt) and PD123177 {1-(4-amino-3-methylphenyl)-methyl-5-diphenylacetyl-4,5,6,7-tetrahydro-1*H*-imidazo[4,5-*c*]pyridine-6-carboxylic acid. Losartan possessed the added bonus of being orally active and showed none of the partial agonist activity commonly associated with therapeutic agents such as Saralasin (5a, 6a).

As early as 1976, evidence of a separate central nervous system (CNS) renin–angiotensin system (RAS) emerged, leading to a preponderance of confirmational data in the literature (7–11). Autoradiographic localization of ^{125}I-labeled AII binding substantiated the presence of AII receptors in regions of the mammalian brain which reliably produced quantifiable, dose-dependent physiological responses to AII (12–16b). It was the introduction of the nonpeptidergic AII antagonists, however, which excited the scientific community with substantiation of the existence of AII receptor subtypes.

The notion of angiotensin receptor heterogeneity had been introduced several years prior (17, 18), but experimental evidence was lacking.

The crux of this chapter begins with the results and methodological detail of autoradiographic experiments demonstrating the presence of AII receptors in the brain. However, a brief discussion of accepted nomenclature is first warranted. As is too often case, failure to adopt a universally accepted nomenclature for receptor classification leads to confusion and erroneous conclusions. We will follow the nomenclature suggested in the January 1992 *TIPS Receptor Nomenclature Supplement* (19). Therefore the AII type 1 receptor (previously referred to as the AII_1, or AII type B) is henceforth written as AT_1. Likewise, the AII type 2 receptor will be written as AT_2.

Autoradiographic Localization of Angiotensin II Type 1 and 2 Receptors in Brain

Methods

Tissue Preparation

Male Long-Evans rats (Simonson Laboratories, Gilroy, CA) are euthanized by lethal intraperitoneal injection (100 mg/kg) of sodium pentobarbital (Nembutal). When the animals no longer respond to painful stimuli (scrotal and tail pinch) they are intracardially perfused with cold (4°C) isotonic saline. The brains are rapidly removed over ice and slowly lowered into cold isopentane (approximately −45 to −50°C). [*Technical note:* If the isopentane is too cold (below −50°C), the brains may split along the corpus callosum, making sectioning and localization of labeled midline structures extremely difficult. It is also important to lower the brains into the isopentane very slowly and to complete the freezing process by allowing the brains to remain on dry ice, loosely covered in heavy-duty aluminum foil, for approximately 30 min.]

The brains are cut in a cryostat into 10-μm-thick sections and thaw-mounted onto chrome alum-subbed slides. Because some of the nuclei associated with AII activity are extremely small (e.g., subfornical organ), it is advisable to take as thin a section as practical and to use immediately adjacent sections to define nonspecific binding.

Autoradiography

Tissue sections are preincubated for 30 min at 22°C (room temperature) in 30 mM sodium phosphate buffer (pH 7.2) containing 0.4% (w/v) bovine serum albumin, 10 mM $MgCl_2$, 150 mM, NaCl, and 5 mM ethylene glycol bis

(β-aminoethyl ether) N,N,N',N-tetraacetic acid (EGTA). Unless otherwise stated, all reagents are purchased from Sigma (St. Louis, MO). Previous experiments proved this buffer to be ideal for optimizing specific binding of ^{125}I-labeled AII to rat brain (16a,b). The sections are then incubated for 60 min at room temperature in the same buffer containing 0.5 nM ^{125}I-labeled [Tyr^4]-AII (2200 Ci/mmol, Du Pont–NEN, Boston, MA). Nonspecific binding is defined by the addition of 3 μM [Val^5] AII. To discriminate between AT$_1$ and AT$_2$ receptors, additional sections are incubated with the radioligand plus 10 μM Losartan or 3 μM PD12377. The concentrations of the two competitors have been selected from the work of Chiu *et al.* (20) to maximize displacement of one or the other receptor sites. The incubation is terminated by vigorously dipping then rinsing the slide-mounted sections twice for 5 min each time in the specified buffer at 4°C (ice bath temperature). The rinses are followed by once again vigorously dipping the sections in two separate cold distilled water baths. [*Technical note:* It is always advisable to precede drying of autoradiographic slides with a cold water dip to remove salts which may adhere to the tissues, but the viscosity of this buffer warrants particular attention to this last step.]

The tissues are dried by gently blowing cold, desiccated air over them. The slides are stored desiccated at 4°C (refrigerated) for at least 24 hr. Autoradiograms are generated by exposing radiation-sensitive film (Hyper-film, Amersham, Arlington Heights, IL) to the dried sections in a X-ray cassette. Commercially prepared standards (Amersham microscales) are included in the cassettes to facilitate quantitation of radiolabeling. Exposure time is approximately 72 hr. Under safelight conditions [Kodak (Rochester, NY) GBX-2 filter], the Hyperfilm is developed by immersing in each of the following: Kodak D-19 developer (5 min), stop bath (2% acetic acid) (30 sec), and Kodak Rapidfix (5 min). The film is then washed in running water for 15 min followed by a brief rinse (about 1 min) in Kodak Photoflo.

Results and Discussion

Densitometric analyses are performed using a microcomputer imaging device system (MCID) from Imaging Research Inc. (St. Catherines, ON). Examples of total binding and binding in the presence of [Val^5]AII (nonspecific), Losartan, or PD123177 are presented in Fig. 1. The highest levels of specific binding (total binding − nonspecific binding) were found in regions previously shown to have high concentrations of AII receptors [e.g., subfornical organ (SFO), median eminence (ME), and nucleus of the solitary tract (NTS)] (15, 16a,b). Specific binding in these regions represented approximately 92% of total binding. Losartan inhibited 99, 97, and 97% of the specific binding in

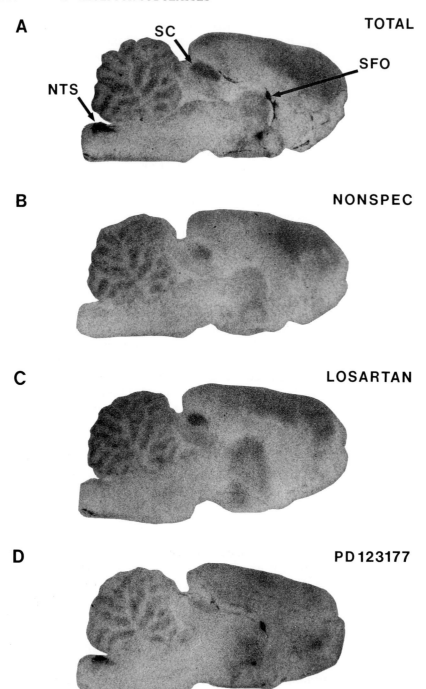

TABLE I   Analysis of Angiotensin II
Binding in Rat Brain[a]

Region[b]	Specific binding (fmol/mg tissue)	%Specific binding displaced	
		Losartan	PD123177
SFO	0.678 ± 0.090	99	6
NTS	0.874 ± 0.052	97	37
PV	0.165 ± 0.033	87	13
ME	0.500 ± 0.131	97	13
SC	0.103 ± 0.016	50	89

[a] Densitometric quantitation of ^{125}I-labeled AII binding in rat brain nuclei expressed as fmol/mg tissue bound and percent of binding displaced by 10 μM Losartan or 1 μM PD123177. Each value represents the mean ± S.E.M. from three animals.

[b] SFO, Subfornical organ; NTS, nucleus of the solitary tract; PV, paraventricular nucleus; ME, median eminence; SC, superior colliculus.

the SFO, ME, and NTS, respectively. In contrast, PD123177 inhibited only 6, 13, and 33% of the specific binding in these regions.

A summary of the data is presented in Table I. These data suggest that both AT_1 (Losartan sensitive) and AT_2 (PD123177 sensitive) receptors exist and are differentially distributed (though there is some overlap) in the CNS. Additional confirmation of these findings can be found in the work of Tsutsumi and Saavedra (21), Rowe et al. (22), and Gehlert et al. (23).

Chiu and co-workers (24) provided evidence of the functional specificity of these receptor subtypes by showing that the AT_1 antagonist Losartan was effective in decreasing blood pressure when administered in vivo to rats. The AT_2 antagonist PD123177 was not in the paradigm. A comprehensive study by Dudley et al. (25), supports these conclusions. Thus, there is compelling evidence that functionally specific AII receptor subtypes exist in discrete nuclei in the CNS and that the AT_1 subtype alone is associated with hypertension.

FIG. 1   Photomacrographs of 10-μm sagittal sections through the brain of a Long-Evans rat indicating binding in the subfornical organ (SFO), nucleus of the solitary tract (NTS), and superior colliculus (SC) following incubation in 0.5 nM ^{125}I-labeled AII alone (A) or in the presence of 3 μM [val[5]]AII (B), 1 μM Losartan (C), or 1 μM PD123177 (D). Bar, 2 mm.

Inhibition Experiments, Autoradiography

Because of the nature of ^{125}I-labeled AII binding in the CNS (i.e., to very small, discrete nuclei), autoradiography was deemed the most suitable approach to analyze binding of the competitive inhibitors, Losartan and PD123177. This method, however, is not without its problems. For example, it is important that the optical density (OD) range of the standards included with each film exposure extend beyond the limits of the OD of the tissue samples; that is, the OD range of the standards should bracket the OD range of the samples. This seems a simple criterion to meet, but one must keep in mind that, in an inhibition experiment, the concentration of the competitive inhibitor is increased over a broad range (usually about $\pm 1000 \times$ the IC_{50}). Consequently, the amount of radioactivity available to expose the film will decrease substantially as the concentration of inhibitor increases. In addition, radiation-sensitive film does not respond equally at particularly high or low OD values. Consequently, it may be necessary to do multiple exposures of the tissues in order to ensure a representative sample and an accurate assessment of bound values. For a thorough discussion of autoradiography and its principles, the reader is referred to Yamamura *et al.* (26).

Methods

Tissue Preparation
Tissue preparation is conducted as previously described with the exception that more animals are needed to ensure that representative samples of each nucleus are present at each concentration of the inhibitor.

Autoradiography
Experimental conditions (e.g., buffer, pH, and temperature) are precisely as described above with the exception of the concentrations of Losartan and PD123177. The concentrations of each compound range from 10^{-9} (1 nM) to 10^{-3} (1 mM). Nonspecific binding is defined by the addition of 3 μM [Val^5] AII. The tissues are apposed to Hyperfilm for approximately 72 hr. The ^{125}I standards (Amersham) are of a sufficient range so as to make multiple exposures unnecessary for this experiment.

Results and Discussion

The results of the competition experiment in the SFO, NTS, and superior colliculus (SC) for Losartan and PD123177 are presented in Fig. 2. Nonlinear regression analyses (InPlot, Graphpad Software, San Diego, CA) revealed that Losartan inhibited ^{125}I-labeled AII binding in the SFO and NTS in accordance with mass action principles (i.e., Hill coefficients did not differ significantly from 1). Curve fitting to two sites did not differ from fitting to one. The derived IC_{50} values were similar in both areas: 1.61×10^{-7} (SFO) and 2.83×10^{-7} (NTS). In contrast, nonlinear regression analysis of PD123177 competitive inhibition curves revealed a significant deviation from mass action in competition for ^{125}I-labeled AII binding in both the SFO and NTS. Once again, however, fitting to two sites did not differ significantly from fitting to one. Hill coefficients were -0.14 ($p < 0.05$) in the SFO and -0.41 ($p < 0.05$) in the NTS. Derived IC_{50} values were 1.92×10^{-3} and 1.13×10^{-5}, respectively. Competition curves in the SC indicated binding to a single site for both Losartan and PD123177, but PD123177 had a 10-fold greater affinity for the site involved. This suggested that AII receptors in this region are of the AT_2 type.

The data support conclusions reached from previous findings indicating heterogeneity of AII receptors, but they also raise questions concerning the nature of these sites. McQueen and Semple (27), for example, suggested that the apparent heterogeneity of the AII receptor may be due to its capacity to exist in two different forms. Timmermans et al. (2) proposed an elegant coupling model in the vascular AII receptor to account for the differences seen in concentration–contractile (dose–response) curves for AII when blocked by EXP3892 (a nonpeptide AII antagonist) or by Losartan. These authors found that EXP3892 caused a nonparallel rightward shift in dose-related AII-induced contractions in rabbit aorta, whereas Losartan caused a parallel shift. Furthermore, Losartan antagonized the EXP3892-induced suppression or the response to AII. These authors propose explanations of the underlying mechanisms which imply a complexity of the AII receptor system which has yet to be elaborated. Likewise, the Hill coefficients listed above for PD123177 as a competitive inhibitor of AII insinuate some type of negative cooperatively at the site. Further experimentation is required before an accurate model of the AII receptor system in the CNS can be proposed.

Autoradiographic Localization of Angiotensin II Receptor Subtypes in Spontaneously Hypertensive versus Normotensive Rats

The spontaneously hypertensive rat (SHR) and its normotensive counterpart, the Wistar Kyoto (WKY) rat, have served as animal models for human essential hypertension since 1966. Gehlert et al. (16b) described the distribu-

A

Subfornical Organ

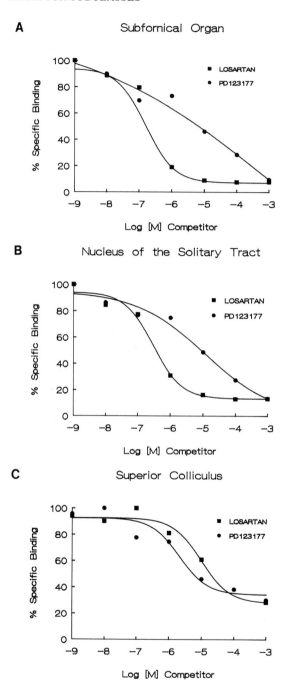

B

Nucleus of the Solitary Tract

C

Superior Colliculus

tion of AII receptors in the brains of SHR and WKY rats. They found that the concentration of AII receptors in nuclei relevant to hypertension, such as the SFO in SHR rats, was as much as 368% greater than in WKY controls. The availability of AT_1-selective antagonists prompted another look into this distribution.

Methods

Tissue Preparation

SHR and WKY rats are obtained from Harlan Sprague Dawley, Inc. (Indianapolis, IN). The animals are euthanized and the tissue prepared and sectioned as previously described.

Autoradiography

Labeling of the brain sections is performed as described above. Adjacent sections are incubated in ^{25}I-labeled AII or in ^{125}I-labeled AII plus 1 μM Losartan, 1 μM PD123177, or 1 μM [Val^5] AII.

Results and Discussion

Binding in hypertensive relevant nuclei such as the SFO and paraventricular nuclei (PV) of the SHR animals was generally higher than in the WKY animals but not to the degree reported previously (16b). Comparative binding of selected nuclei are presented in Table II. In addition, the percentage of specific binding displaced by either 1 μM Losartan or 1 μM PD123177 was equivalent for SHR and WKY animals, implying that the distribution and proportion of AT_1 and AT_2 receptors do not differ in these groups. Thus these data lend support to the hypothesis that it is the AT_1 receptors which are implicated in hypertension. Table III summarizes these data.

FIG. 2 Competitive inhibition curves in the subfornical organ (SFO) (A), nucleus of the solitary tract (NTS) (B), and superior colliculus (SC) (C) for Losartan or PD123177 versus 0.5 nM ^{125}I-labeled AII. Curves for each brain area and competitor were generated using nonlinear regression analysis. Derived IC_{50} values are as follows: in the SFO, Losartan = 1.61×10^{-7}, PD123177 = 1.92×10^{-3}; in the NTS, Losartan = 2.83×10^{-7}, PD123177 = 1.13×10^{-5}; in the SC, Losartan = 1×10^{-5}, PD123177 = 2×10^{-6}.

TABLE II   Analysis of Angiotensin II Binding in
Hypertensive and Normotensive Rats[a]

Region[b]	Specific binding (fmol/mg tissue)		WKY (%)
	SHR	WKY	
SFO	19.562 ± 2.11	13.68 ± 1.06	143
PV	9.152 ± 1.14	3.94 ± 0.76	232
MPO	16.45 ± 2.02	7.25 ± 1.01	227
PA	4.71 ± 1.10	2.63 ± 0.93	179
PCTX	6.41 ± 2.23	4.35 ± 0.82	147
NTS	7.68 ± 1.91	2.52 ± 0.66	305
SC	2.44 ± 1.00	1.20 ± 0.46	203

[a] Densitometric quantitation of ^{125}I-labeled AII binding in brain nuclei of spontaneously hypertensive (SHR) and Wistar Kyoto (WKY) rats. Values represent the means ± S.E.M. for three animals of each strain.
[b] SFO, Subfornical organ; PV, paraventricular nucleus; MPO, median preoptic region; PA, posterior amygdaloid nucleus; PCTX, pyriform cortex; NTS, nucleus of the solitary tract; SC, superior colliculus.

Conclusion

The availability of compounds such as Losartan and PD123177 and the enthusiasm with which scientists have greeted their discovery has catapulted research on the AII system into a new era. The data presented or cited in this chapter are irrefutable evidence of the existence of multiple AII receptor subtypes in the CNS. Though there seems to be some differences in findings with respect to density and distribution, the majority of research is in close accord, particularly with respect to the functional specificity of the AT_1 receptor. One cannot overlook the heuristic value of this research either. The confirmation of the existence of multiple AII receptors has raised as many questions as it has answered. The nature of these subtypes remains to be elucidated. Are there two distinct subtypes, perhaps allosterically coupled, or is there an AII–receptor complex such as the model suggested by De Chaffoy de Courcelles and elaborated on by Timmermans et al. (2)? Whatever the final outcome, it is clear that pharmacotherapeutic management of hypertension and related cardiovascular disorders will become more precise, and with continued research these diseases may even disappear.

In addition to the etiologic role of AII in the CNS in essential hypertension, a recent report by Barnes et al. (28) implicates the CNS AII system in cognitive function as well. These researchers found that a low dose (10 ng/kg) of Losartan enhanced the performance of albino male mice in an active

TABLE III   Percent Specific Binding Displaced[a]

Region	SHR		WKY	
	Losartan	PD123177	Losartan	PD123177
SFO	92	21	85	12
PV	84	34	81	38
MPO	87	60	89	67
PA	45	17	75	7
PCTX	80	5	83	25
NTX	79	47	69	34
SC	17	37	22	34

[a] Densitometric quantitation of percent ^{125}I-labeled AII binding displaceable by 1 μM Losartan or 1 μM PD123177 in SHR and WKY rats (n = 3/strain). For abbreviations, see Table II.

avoidance task. Their findings concur with earlier work demonstrating cognitive enhancing capabilities of ACE inhibitors (29, 30).

Furthermore, Gehlert *et al.* (23) found substantial amounts of Losartan-sensitive ^{125}I-labeled [Sar^1-Ile^8]AII binding in regions of the brain known to be involved in learning and memory in mammals, such as medial septum, the CA1 region of the hippocampus, pre- and parasubiculum, and basal forebrain nuclei. There are of course, far-reaching implications of these data, and while it is true that a concise theory to explain such empirical evidence has yet to emerge, the importance to pursuing this aspect of the AII system is evident.

References

1. J. C. Garrison and M. J. Peach, *in* "The Pharmacological Basis of Therapeutics" (A. G. Gilman, T. W. Rall, A. S. Niles, and P. Taylor, eds.), p. 749. Pergamon, New York, 1990.
2. P. B. M. W. M. Timmermans, P. C. Wong, A. T. Chiu, and W. F. Herblin, *Trends Pharmacol. Sci.* **12,** 55 (1991).
3. Y. Furukawa, S. Kishimoto, and K. Nishikawa U.S. Patent No. 4,340,598 (1982).
4. Y. Furukawa, S. Kishimoto, and K. Nishikawa, U.S. Patent No. 4,355,040 (1982).
5a. A. T. Chiu, D. J. Carini, A. L. Johnson, D. E. McCall, W. A. Price, M. J. M. C. Thoolan, P. C. Wong, R. I. Taber, and P. B. M. W. M. Timmermans, *Eur. J. Pharmacol.* **157,** 13 (1988).
5b. A. T. Chiu, D. E. McCall, W. A. Price, P. C. Wong, D. J. Carini, J. V. Duncia, R. R. Wexler, S. E. Yoo, A. L. Johnson, and P. B. M. W. M. Timmermans, *J. Pharmacol. Exp. Ther.* **252,** 711 (1990).
5c. A. T. Chiu, D. E. McCall, P. E. Aldrich, and P. B. M. W. M. Timmermans, *Biochem. Biophys. Res. Commun.* **172,** 1195 (1990).
6a. A. T. Chiu, W. A. Price, M. J. M. C. Thoolen, D. J. Carini, A. L. Johnson,

R. I. Taber, and P. B. M. W. M. Timmermans, *J. Pharmacol. Exp. Ther.* **247,** 1 (1988).

6b. P. C. Wong, W. A. Price, A. T. Chiu, J. V. Duncia, D. J. Carini, P. R. Wexler, A. L. Johnson, and P. B. M. W. M. Timmermans, *J. Pharmacol. Exp. Ther.* **252,** 719 (1990).

7. B. K. Bickerton and J. P. Buckley, *Proc. Soc. Exp. Biol. Med.* **106,** 834 (1961).

8. J. P. Bennett and S. H. Snyder, *J. Biol. Chem.* **251,** 7423 (1976).

9. F. E. Cole, E. D. Frohlich, and A. A. MacPhee, *Brain Res.* **154,** 178 (1978).

10. J. Ciriello and F. R. Calaresu, *Am. J. Physiol.* **139,** R137 (1980).

11. J. W. Harding, L. P. Stone, and J. W. Wright, *Brain Res.* **205,** 265 (1981).

12. M. L. Mangiapane and J. B. Simpson, *Am. J. Physiol.* **239,** R382 (1980).

13. J. F. E. Mann, P. W. Schiller, E. L. Schiffrin, R. Boucher, and J. Genest, *Am. J. Physiol.* **241,** R124 (1991).

14. J. W. Maran and F. E. Yates, *Am. J. Physiol.* **223,** E273 (1977).

15. F. A. O. Mendelsohn, R. Quirion, J. M. Saaveda, G. Aguilera, and K. J. Catt, *Proc. Natl. Acad. Sci. U.S.A.* **81,** 1575 (1984).

16a. D. R. Gehlert, R. C. Speth, and J. K. Wamsley, *Neuroscience (Oxford)* **18,** 837 (1986).

16b. D. R. Gehlert, R. C. Speth, and J. K. Wamsley, *Peptides* **7,** 1021 (1986).

17. M. J. Peach, F. M. Bumpus, and P. A. Khairallah, *J. Pharmacol. Exp. Ther.* **167,** 291 (1969).

18. M. J. Peach and N. R. Levens, *Adv. Exp. Med.* **130,** 171 (1980).

19. S. Watson and A. Abbott, *Trends Pharmacol. Sci.* (Receptor Nomenclature Suppl., 3rd Ed.), 7 (1992).

20. A. T. Chiu, W. F. Herblin, D. E. McCall, R. J. Ardecky, D. J. Carini, J. V. Duncia, L. J. Pease, P. C. Wong, R. R. Wexler, A. L. Johnson, and P. B. M. W. M. Timmermans, *Biochem. Biophys. Res. Commun.* **165,** 196 (1989).

21. K. Tsutsumi and J. M. Saavedra, *J. Neurochem.* 56(1), 348 (1991).

22. P. B. Rowe, K. L. Grove, D. L. Saylor, and R. C. Speth, *Regul. Pept.* **33,** 45 (1991).

23. D. R. Gehlert, S. L. Gackenheimer, and D. A. Shober, *Neuroscience (Oxford)* **44**(2), 510 (1991).

24. P. C. Wong, W. A. Price, A. T. Chiu, J. V. Duncia, D. J. Carini, R. R. Wexler, A. L. Johnson, and P. B. M. W. M. Timmermans, *J. Pharmacol. Exp. Ther.* **252,** 726 (1990).

25. D. T. Dudley, R. L. Panek, T. C. Major, G. H. Lu, R. F. Bruns, B. A. Klinketus, J. C. Hodges, and R. E. Weissar, *Mol. Pharmacol.* **38,** 370 (1990).

26. H. I. Yamamura, S. J. Enna, and M. J. Kuhar, *in* ''Methods in Neurotransmitter Receptor Analysis'' Raven, New York, 1990.

27. J. McQueen and P. F. Semple, *Mol. Pharmacol.* **35,** 809 (1989).

28. N. M. Barnes, S. Champaneria, B. Costall, M. E. Kelly, D. A. Murphy, and R. J. Naylor, *Neuroreport* **1,** 239 (1990).

29. J. M. Barnes, N. M. Barnes, and J. Caughlan, *in* ''Current Advances in ACE Inhibition'' (G. A. MacGregor and P. S. Sever, eds.), p. 159. Churchill Livingston, London, 1988.

30. B. Costall, J. Caughlan, and Z. P. Horovitz, *Pharmacol. Biochem. Behav.* **33,** 573 (1989).

[14] Radioligand Binding Studies of the Three Major Classes of Neurokinin Receptors

Rémi Quirion and Than-Vinh Dam

Introduction

The existence of three major classes of substance P (SP)/neurokinin (NK) receptors has been clearly demonstrated by their recent cloning and molecular characterization (1–4). The NK-1 receptor is preferentially recognized by SP, whereas the NK-2 and NK-3 subtypes possess preferential affinity for neurokinin A (NKA) and neurokinin (NKB), respectively (2–4). However, none of the endogenous NKs are fully selective for a single receptor class, and they can activate all three subtypes under certain conditions (2, 5). Consequently, it is most important to investigate a given NK receptor subtype using ligands and conditions which ensure high specificity and selectivity. Characteristics of assays used for the three major classes of NK receptors are described here for both membrane binding and receptor autoradiography.

General Considerations

Purity and Stability of Peptide Radioligands

The great majority of radiolabeled probes which have been used thus far to study NK receptors are of a peptidergic nature. Thus, they are all relatively unstable and sensitive to radiolysis. Consequently, it is highly recommended that one obtain only the quantity of probes required for a given series of experiments. It is not appropriate to order larger batches of ligands and to store them. In the long run, no savings will be made, experiments having to be repeated because of poor results related to the use of nonoptimal radioligands. If the purchase (or preparation by iodination) of large quantities of ligands cannot be avoided, the proper handling and storage of probes must be assured as well as their repurification [by high-performance liquid chromatography (HPLC)] before use. Usually, this can be achieved by dividing a freshly received sample into small individual aliquots (5–10 μl) which

Methods in Neurosciences, Volume 12
Copyright © 1993 by Academic Press, Inc. All rights of reproduction in any form reserved.

are then flushed with nitrogen or argon to prevent oxidation before being kept at −20°C (or −80°C)until use (6). This applies chiefly to ligands having a long half-life (tritiated), but the purity of iodinated probes must also be checked just before use.

Results obtained from the use of inadequate probes can be misleading and can suggest, for example, the existence of multiple receptor affinity states (multiphasic saturation or competition curves) or the detection of a new class of receptor sites. It should also be kept in mind that even with the development of nonpeptide NK receptor radioligands such as CP-96345 and others (7–9), it will still be of importance to evaluate the purity of each batch of radiolabeled probes before use. This ensures a better reproducibility of the data and minimizes the study of artifactual information.

Membrane Preparations

A variety of crude membrane homogenates and synaptosomal preparations has been used for the study of NK receptor subtypes. Whole cell preparations have also been used with success. Globally, receptor binding kinetics obtained using various assays are relatively similar, the critical variables being the use of divalent cations (Mg^{2+} or Mn^{2+}) to ensure adequate receptor binding affinity and the inclusion of peptidase inhibitors to prevent the degradation of radiolabeled probes during the incubation period (10–14).

A typical membrane preparation is described here for the NK-1 or NK-2 receptors using the selective radioligand [^{3}H][Sar^9,Met(O$_2$)^{11}]substance P and ^{125}I-labeled NKA, respectively (15–18). Tissues (mostly brains, minus cerebellum) are homogenized in 20 volumes of a buffer containing 50 mM Tris-HCl (pH 7.4 at 4°C), 120 mM NaCl, and 5 mM KCl, using a Brinkmann (Westbury, NY) Polytron at setting 6 for 20–25 sec. The preparation is then centrifuged at 49,000 g for 10 min, the supernatants discarded, and the pellets resuspended in a buffer containing 50 mM Tris-HCl (pH 7.4 at 4°C), 300 mM KCl, and 10 mM disodium EDTA and incubated on ice for 30 min. At the end of this period, the preparations are centrifuged at 49,000 g for 10 min, the supernatants discarded, the pellets washed at least 3 times with 50 mM Tris-HCl (pH 7.4 at 4°C) before a final resuspension in 60 volumes of this buffer. Aliquots are taken for protein determination, or the membrane preparation is frozen at −80°C for use at a later date. It is generally possible to keep frozen homogenates for up to few months without apparent loss of binding affinity or capacity.

For the study of the NK-3 receptor subtype using the selective ligand [^{3}H]senktide, optimal results are obtained by preparing brain membrane homogenates according to the following protocol (19). Brains are dissected on ice, with only the cerebral cortices being kept and placed in ice-cold 10% sucrose (w/v). Tissues are then homogenized in 10 volumes of 10% sucrose using a Teflon–glass homogenizer (10–12 strokes, 800 rpm). Homogenates are next centrifuged at 20,000 g for 10 min. The crude synaptosomal pellet is lysed by resuspension using a Polytron at setting 7 for 5 sec in 10 volumes of 5 mM HEPES buffer at pH 7.5. This suspension is centrifuged at 30,000 g for 30 min. Supernatants are discarded and pellets resuspended in the same buffer and recentrifuged at 30,000 g for an additional 30 min. The final pellet is resuspended in 20 mM Krebs–HEPES buffer $\times$ 19 (for buffer salt composition) (pH 7.4) to give a protein concentration of 2–3 mg/ml. Membranes are used immediately or stored at $-80°C$. Using this protocol, binding data were found to be relatively less variable than those obtained using the first protocol described above.

Receptor Autoradiography

Tissue sections (mostly brains) are prepared as described in detail (20). When using this method, it is of utmost importance to ensure the preparation of the highest quality sections. Consequently, it is critical to ensure that sections strongly adhere to the slides (plastic or glass) by adequately cleaning and coating them. We use the following protocol, which is rather elaborate but gives good results even under low isotonic incubation conditions.

1. Place slides in stainless steel racks.
2. Fill a sink or large container with hot water and a detergent (Sparkleen-type). Soak racks in this solution for 2 hr.
3. Remove soapy water and rinse *extensively* for up to 12 hr to remove all detergent.
4. Soak racks (for up to 1 hr) in room temperature deionized water (repeat twice).
5. Replace the water with absolute alcohol and let soak overnight.
6. Remove the alcohol and replace with deionized water. Let soak for a few hours.
7. In the meantime, prepare a *fresh* subbing solution by slowly adding to 1 liter of stirring deionized water at 50°C, 5 g of gelatin powder (Sigma Chemicals, St. Louis, MO) (275 Bloom). Cool to 30°C before adding 500 mg of chromium potassium sulfate. Filter the mixture.
8. Remove racks from water, drain, and slowly lower them in and out of the subbing solution. Drain off excess coating solution.
9. Dry the slides in an oven at 37°C for at least 2 hr.
10. Store the gelatin-coated slides in clean, dust-free slide boxes. Under these conditions, coated slides can be kept for years.

For the preparation of sections, tissues are snap-frozen in 2-methylbutane at −40°C. For an adult rat brain, 15–20 sec usually suffices for adequate freezing before placing brains on a bed of dry ice until ready to be mounted on cryostat holders using embedding matrix (Lipshaw, etc.). Following an equilibrium period of 2–3 hr in the cryostat at −15 to −20°C, 8 to 20-μm-thick sections are carefully thaw-mounted on the precleaned gelatin-coated slides. At this step, freezing and thawing of sections should be avoided by maintaining them at constant subzero temperatures. Section-mounted slides are then dried overnight under reduced pressure in a desiccator at 4°C before storage at −80°C until use. Sections carefully prepared according to this protocol should have a translucid, glassy appearance (not whitish) and can be kept for up to 1 year for a variety of receptor binding sites.

Neurokinin Subtype 1 Receptors

Radioligands

To date, the great majority of NK-1/substance P receptor binding studies have been performed using [^{3}H]substance P and [^{125}I]Bolton–Hunter-substance P as radioligands (see Refs. 2–5). However, it is well known that endogenous ligands lack appropriate selectivity for a given receptor subtype. Thus, it was of importance to develop more selective and metabolically stable probes for this receptor class. Recently, we reported on the characterization of [^{3}H][Sar^9,Met(O$_2$)^{11}]substance P as a highly selective NK-1 radioligand (5, 17, 18). Others have used this same analog in its iodinated form (21, 22) and [4,5-^{3}H-Leu^{10}]substance P has recently been reported as another peptidergic NK-1 receptor probe (23). Moreover, the development of the nonpeptide *antagonist* CP-96345 in its radiolabeled form (8) is generating great interest, as the use of such highly stable probes can have major advantages over peptide radioligands (longer shelf-life, resistance to enzymatic degradation, limited adhesion to plastic and glass, etc.). However, the development of nonpeptide receptor *agonists* is still in its infancy, suggesting that peptide agonist radioligands will still be used in the foreseeable future.

NK-1 Receptor Membrane Binding Assays

In a series of experiments, we observed that optimal incubation conditions for rat brain membrane preparations are as follows: 200 μl of the homogenate obtained as described above is incubated for 90 min at 25°C in 400 μl of a buffer containing 50 mM Tris-HCl (pH 7.4 at 4°C), 3 mM MnCl$_2$, 0.02%

bovine serum albumin to minimize losses of ligand to nonreceptor surfaces, 40 μg/ml bacitracin (Sigma, St. Louis, MO), 2 μg/ml chymostatin (Sigma), and 4 μg/ml leupeptin (Sigma) to ensure the stability of the radiolabeled ligand during the incubation period, and various concentrations of [^{3}H][Sar^9,Met(O$_2$)^{11}]substance P for saturation (0.01 to 50 nM; 32.5 Ci/mmol, NET-1025, Du Pont, Boston, MA) or competition (usually 1–3 nM) experiments. Nonspecific binding is determined in the presence of 1 μM unlabeled ligand or substance P. We observed that, under such assay conditions, the radiolabeled probe was highly stable during incubation.

Incubations are terminated by rapid filtration (Millipore, Beford, MA) or Brandel (Gaithersburg, MD) filtration units] through Whatman (Clifton, NJ) GF/C filters presoaked in 0.1% polyethyleneimine (PEI) for at least 3 hr prior to filtration. The use of PEI-treated GF/C filters minimizes binding of ligand to filters. Filters are then washed 3 times with 3–4 ml of cold 50 mM Tris-HCl buffer (pH 7.4). After drying the filters overnight, binding of ligand is quantitated by counting filters in 5 ml Ecolite (+) (ICN Radiochemicals, Costa Mesa, CA) scintillation cocktail using a Beckmann (Fullerton, CA) counter with 40–50% efficiency. Binding data are finally analyzed using computerized curve fitting programs (Bio-Soft, Lundon, LIGAND). Under such assay conditions, [^{3}H][Sar^9,Met(O$_2$)^{11}]substance P apparently bound to a single class of high-affinity (K_d 1.4 $\pm$ 0.5 nM), low-capacity (B_{max} 160 $\pm$ 3.0 fmol/mg protein) sites in rat brain membrane homogenates. At concentrations approximating K_d values, specific binding reached rapid equilibrium (<30 min) at 25°C and represented between 70 and 75% of total binding. The ligand selectivity profile clearly revealed that [^{3}H][Sar^9,Met(O$_2^{1-}$^{1}]substance P specifically and selectively labeled the NK-1/SP receptor subtype since only NK-1 competitors such as substance P and unlabeled [Sar^9,Met(O$_2$)^{11}]substance P behave as potent competitors (K_i in low nanomolar range) in this assay whereas selective NK-2 ([Nle^{10}]NKA$_{4-10}$) and NK-3 (senktide) analog are virtually inactive (17).

NK-1 Receptor Autoradiography

On the day of the experiments, sections are slowly warmed to the freezing point by placing them on a bed of wet ice for up to 60 min. They are then placed in Cytomailers (5 slides per mailer) and incubated for 90 min at room temperature in the incubation buffer (10–12 ml per mailer) which includes the mixture of enzyme inhibitors used for membrane binding assays and 3.0 nM [^{3}H][Sar^9,Met(O$_2^{11}$]substance P. Nonspecific binding is determined in the presence of 1 μM substance P or [Sar^9,Met(O$_2$)^{11}]substance P. At the end of the incubation, slides are placed in stainless steel racks and transferred sequentially through four washes (1 min each) in 50 mM Tris-HCl buffer (pH 7.4) at 4°C, followed by a rapid dip in cold, deionized water to remove salts. Incubated slides are next rapidly dried under a stream of cold air using a cold air dryer, placed in X-ray cassettes alongside standards (tritium or iodine, Amersham, Arlington Heights, IL), tightly juxtaposed against tritium-sensitive films [LKB (Piscataway, NJ) Ultrofilm or Amersham Hyperfilm], and stored at room temperature for up to 30 days. We observed no clear-cut advantages to keeping cassettes at 4°C (or below), the exposure time not being significantly shortened while film backgrounds seem somewhat higher. After appropriate exposure, films are developed in the darkroom using a Kodak (Rochester, NY) D-19 developer (at 16–18°C; 4 min). Development is stopped using an acid bath (30 sec) followed by a hardening step (Rapid Fix, Kodak, 4 min) and a final rinsing period in cold, running tap water (30 min). Films are then analyzed and quantified using a computerized image analyzer (e.g., MCID System Imaging Research Inc., St. Catherines, ON). Pictures are taken directly from films (not TV monitor) using a Tech-Pan film (Kodak) to ensure better quality prints. Incubated sections can then be either processed for high resolution wet emulsion autoradiography (20) for more detailed anatomical studies or stained (cresyl violet, acetylcholinesterase, etc.) to facilitate anatomical identification.

NK-1 sites visualized using this selective ligand are widely but discretely distributed in the rat brain (Figs. 1 and 2). High densities of sites are especially concentrated in the dentate gyrus of the hippocampus, the superior colliculus, and the amygdalo-hippocampal area. Detailed descriptions of the distribution of NK-1 sites using [^{3}H][Sar^9,Met(O$_2^{11}$)]substance P have been reported elsewhere (17, 18). As shown in Fig. 1 and 2, nonspecific labeling is negligible in most brain regions.

Neurokinin Subtype 2 Receptors

Radioligands

Selective NK-2 receptor radioligands have yet to be developed. Thus far, labeling of the selective agonist [β-Ala^8]NKA$_{4-10}$ (24) failed to generate a useful probe having adequate affinity and stability for radioreceptor assays (T.-V. Dam, Y. Dumont, and R. Quirion, unpublished results, 1990). Consequently, the NK-2 preferential endogenous ligand, NKA, has been used either in its tritiated ([^{3}H]NKA) (25, 26) or iodinated {(2-[^{125}I]iodohistidyl^1 NKA [^{125}I]NKA), 2200 Ci/mmol; Amersham Co.} forms (5, 15, 16, 27). It now appears that only the iodinated probe possesses adequate affinity, stability, and specific activity to permit the detection of genuine NK-2 sites (5, 15, 16, 27). Extended NKA-related endogenous ligands such as [γ-^{125}I] pre-

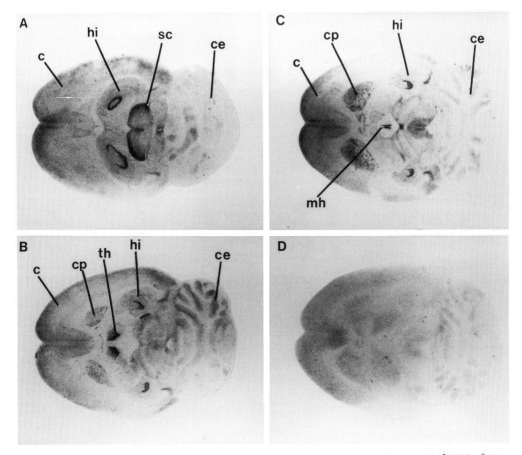

FIG. 1 Photomicrographs of the autoradiographic distribution of [^{3}H][Sar^9,Met(O$_2^{11}$]substance P/neurokinin-1 binding sites in horizontal sections of rat brain. Specific labeling (A–C) is discretely distributed, being most abundant in the striatum, dentate gyrus of the hippocampus, and superior colliculus. Nonspecific labeling observed in presence of 1.0 μM substance P is very low; contrasts needed to be increased in order to visualize it (D). c, Cortex; ce, cerebellum; cp, caudate putamen; hi, hippocampus; mh, medial habenula; sc, superior colliculus; th, thalamus.

protachykinin (72–92) peptide amide (γ-PPT$_{72-92}$; homemade) have also recently been found to be useful NK-2 receptor probes (5, 16, 28). However, using either [^{125}I]NKA or its extended forms, it is most important to clearly demonstrate the labeling of NK-2 receptor binding sites by adequate ligand selectivity profile studies since these radiolabeled probes have strong cross-reactivities for the other two NK receptor classes (2, 5). This can be achieved by evaluating the respective potency of a series of highly selective NK-1,

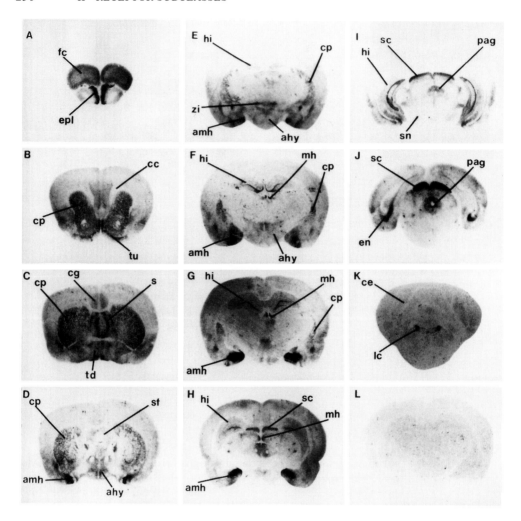

FIG. 2   Photomicrographs of the autoradiographic distribution of [Sar^9,Met(O$_2$)^{11}]-substance P/neurokinin-1 binding sites in coronal sections of rat brain. Note that these sites are broadly, but discretely, distributed throughout the rat brain (see Ref. 17 for details of the anatomical distribution). (L) Section incubated in the presence of 1.0 μM substance P to determine nonspecific labeling. ahy, Anterior hypothalamus; amh, amygdalo-hippocampal area; cc, corpus callosum; ce, cerebellum; cg, cingulate cortex; cp, caudate putamen; en, entorhinal cortex; epl, external plexiform layer of the olfactory bulb; fc, frontal cortex; hi, hippocampus; lc, locus coeruleus; mh, medial habenula; pag, periaqueductal gray matter; s, septum, sc, superior colliculus; sf, septofimbrial nucleus; sn, substantia nigra; td, diagonal band of broca; tu, olfactory tubercule; zi, zona incerta.

NK-2, and NK-3 receptor agonists and antagonists for radiolabeled NKA (or its extended homologs) binding sites. NK-2 analogs should be active in the low nanomolar range, whereas NK-1 and NK-3 ligands should be inactive or only weakly active in competing for putative NK-2 labeled sites (5, 16). We have been able to use both [^{125}I]NKA and [γ-^{125}I]PPT$_{72-92}$ under such conditions (5, 15, 16). It is now expected that a newly characterized, highly potent NK-2 receptor antagonist could be developed as a useful radioligand in the near future (D. Regoli, personal communication, 1992).

NK-2 Membrane Binding Assays

Homogenate preparations and incubation assay conditions used thus far for the characterization of NK-2 receptors are basically idential to those developed for the NK-1 receptor class. The only exceptions are the use of subcortical areas of the adult rat brain, picomolar concentrations of radioligands for competition experiments, and $1 \mu M$ unlabeled NKA or γ-PPT$_{72-92}$ for the determination of nonspecific labeling (16). Under these conditions, [γ-^{125}I]PPT$_{72-92}$, for example, apparently labeled a single class of high-affinity (K_d 0.46 nM), low-capacity (B_{max} 4.12 fmol/mg protein) sites in rat subcorticol membrane homogenates. At 25 pM, [γ-^{125}I]PPT$_{72-92}$ binding reached equilibrium between 45 and 50 min at 25°C and represented approximately 45–50% of total binding. The ligand selectivity profile revealed that unlabeled γ-PPT$_{72-92}$, NKA, and the amphibian tachykinin eledoisin competed for those sites, with K_i values in the low nanomolar range, whereas selective NK-1 {[Sar^9,Met(O$_2$)^{11}]substance P} and NK-3 ([MePhe^7]neurokinin B) agonists were basically inactive at up to micromolar concentrations (16). This clearly reveals the NK-2 receptor profile of the sites labeled using this probe.

NK-2 Receptor Autoradiography

The protocol developed for NK-1 sites is also used for NK-2 receptor autoradiography in the presence of either 25pM [γ-^{125}I]PPT$_{72-92}$ or 50 pM [^{125}I]NKA. Nonspecific labeling is determined as described for membrane preparations. Film exposure ranges between 10 and 30 days depending on the concentration of the radioligand used (15, 16).

As shown in Fig. 3, putative NK-2 sites labeled with [^{125}I]NKA are discretely distributed in the rat brain, with the highest concentration of sites seen in various layers of the hippocampal formation and the amygdalo-hippocampal area. A detailed description of the localization of NK-2 sites in the rat brain was reported elsewhere (15, 16, 28).

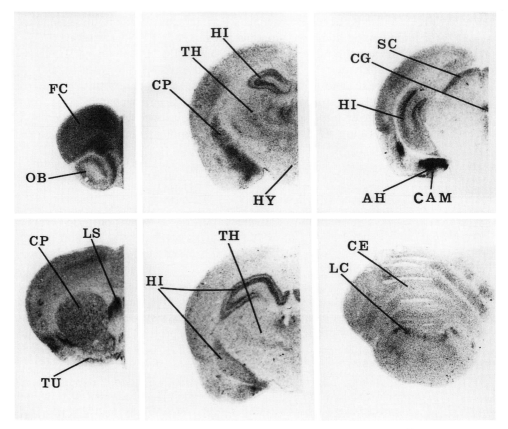

FIG. 3   Photomicrographs of the autoradiographic distribution of ^{125}I-labeled neuroki-
nin A/putative neurokinin-2 (NK-2) binding sites in coronal sections of rat brain.
These sites are especially abundant in the lateral septum (LS), hippocampal formation
(HI), and amygdalo-hippocampal area (AH). AH, Amygdalo-hippocampal area;
CAM, amygdaloid body; CE, cerebellum; CG, central gray area; CP, caudate puta-
men; FC frontal cortex; HI, hippocampus; HY, hypothalamus; LC, locus coeruleus;
LS, lateral septum; OB, olfactory bulb; SC, superior colliculus; TH, thalamus; TU,
olfactory tubercule.

Neurokinin Subtype 3

Radioligands

[3H]Senktide (55 Ci/mmol, NET-997, Du Pont) has recently been developed
as a highly selective NK-3 receptor ligand (19, 29). It is currently the *only*
selective probe to charaterize this receptor subtype, as others including [125]I-
labeled eledoisin (4, 5) and ^{125}I-labeled scyliorhinin II (30) lack adequate

selectivity for this receptor class (5). This is especially relevant for radiolabeled eledoisin, which should no longer be used as an NK-3 ligand since it also possesses very high affinity for both NK-1 and NK-2 receptors. It would be helpful to have access to an iodinated NK-3 receptor probe, but attempts in that regard have thus far failed with senktide. With regard to [³H]senktide, only freshly repurified batches should be used in assays since it is a fairly unstable radioligand which cannot be stored for extended periods of time without repurification and analytical characterization.

NK-3 Receptor Membrane Binding Assays

[³H]Senktide binding assays have been performed in the rat cortex according to the protocol described above for the NK-1 receptor using various concentrations (0.1–125 nM) of [³H]senktide for saturation experiments or 3.0 nM for the evaluation of the competition profile. Eledoisin at 1 μM is used to determine nonspecific labeling. For this assay, GF/C filters are washed 3 times with 3–4 ml of cold 50 mM Tris-HCl buffer containing 0.01% sodium dodecyl sulfate (Bio-Rad, Richmond, CA) to improve the signal-to-noise ratio (29). Under these conditions [³H]senktide is stable and apparently binds to a single class of high-affinity (K_d 2.8 ± 1.0 nM), low-capacity (B_{max} 31.2 ± 3.0 fmol/mg protein) sites in the rat cortex. As expected, only potent NK-3 receptor agonists and antagonists are able to compete for [³H]senktide binding in this preparation (29).

Other assay conditions have also been used to evaluate NK-3 sites with [³H]senktide (19). Following the preparation of membrane homogenates as described above for the NK-3 assay, approximately 250 μg of membrane protein is incubated for 60 min at room temperature in a final volume of 200 μl of buffer (pH 7.4) containing 20 mM Krebs–HEPES, 0.1% bovine serum albumin, 2 mM MnCl$_2$, 2 μg/ml chymostatin, 4 μg/ml leupeptin, 40 μg/ml bacitracin, and various concentrations of [³H]senktide. The reaction is terminated by adding 2 ml of ice-cold buffer (20 mM Krebs–HEPES, 2 mM MnCl$_2$, 0.05% bovine serum albumin, and 0.01% sodium dodecyl sulfate) before filtration through GF/C filters which are presoaked for 1–2 hr in a solution of 0.3% PEI/0.5% Triton X-100 to reduce binding of ligand to filters and using a Brandel Cell Harvester followed by 3 washes of ice-cold incubation buffer. Nonspecific binding is determined using 1 μM senktide. Under these conditions, specific [³H]senktide binding represented between 70 and 75% of total binding at low nanomolar concentrations (19). This is slightly better than the percentage (60–70%) obtained under the first assay conditions described above (29).

Specific [³H]senktide binding reached equilibrium in approximately 30 min at room temperature and remained stable for at least 90 min. K_d and B_{max}

values are relatively higher under the second assay conditions (K_d 8.5 $\pm$ 0.4 nM; B_{max} 76.3 $\pm$ 1.6 fmol/mg protein) (19). The reasons for this apparent discrepancy is not evident but may relate to the respective purity of the various batches of ligand used, with impurities possibly altering kinetic data. However, under both assay conditions, only selective NK-3 ligands compete with [^{3}H]senktide binding. Selective NK-1 and NK-2 analogs are basically inactive (19, 29).

NK-3 Receptor Autoradiography

The visualization of NK-3 receptors by quantitative autoradiography is performed exactly as described above for the other two classes of NK receptors. Film exposures are usually up to 4 weeks.

As shown in Fig. 4, [^{3}H]senktide/NK-3 sites are mostly located in midcortical layers in the rat brain. High amounts of specific labeling are also seen in the zona incerta and interpeduncular nucleus (Fig. 4). A detailed description of the distribution of NK-3 receptors in the rat brain was recently reported elsewhere (29). Nonspecific labeling observed in the presence of 1 μM eledoisin varied according to brain regions, being relatively high in the pineal gland and certain hypothalamic nuclei (Fig. 4).

Concluding Comments

Specific membrane binding assay conditions and quantitative receptor autoradiography protocols have been developed to characterize the three major classes of NK receptors. However, the characterization of highly selective, specific, and stable peptidergic and nonpeptidergic radioligands has lagged behind. Such tools are extremely useful to further investigate, in detail, the precise characteristics of a given NK receptor class. The recent development of [^{3}H][Sar^9,Met(O$_2$)^{11}]substance P and [^{3}H]senktide as selective NK-1 and NK-3 radioligands is a first step in that regard. Nonpeptide, highly selective and potent radiolabeled antagonists should be available soon as radioligands. This should allow for the comparative investigation of the properties of agonist and antagonist receptor binding kinetics under a variety of experimental conditions. Moreover, the availability of highly selective radioligands for each NK receptor class should permit more detailed investigations of their respective localization at the cellular and ultrastructural level. Such information should be extremely useful to further our understanding of the respective role of each NK receptor in the organism. Also, it could provide additional

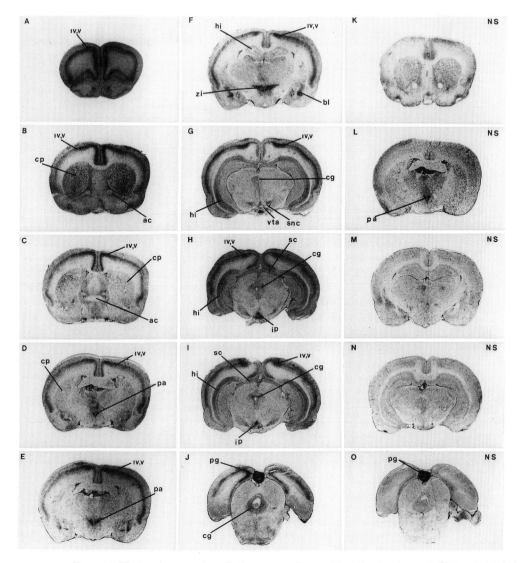

Fig. 4 Photomicrographs of the autoradiographic distribution of [³H]senktide/ neurokinin-3 binding sites in coronal sections of rat brain. This receptor class is most abundant in cortical areas, the zona incerta (zi), and the interpeduncular nucleus (ip). Nonspecific labeling detected in presence of 1.0 μM eledoisin is depicted in sections K to O (see Ref. 29 for details of the anatomical distribution). ac, Anterior commissura; bl, basolateral amygdaloid nucleus; cg, central gray; cp, caudate putamen; hi, hippocampus; ip, interpeduncular nucleus; pa, paraventricular nucleus of the hypothalamus; pg, pineal gland; sc, superior colliculus; snc, substantia nigra pars compacta; vta, ventral tegmental area; zi, zona incerta; IV, V, laminae IV and V of the cerebral cortex.

evidence for the existence of further NK receptor subtypes, at least in some species (31–34).

Acknowledgments

The expert secretarial assistance of Mrs. Joan Currie is acknowledged. Research was supported by grants for the Medical Research Council of Canada and the Funds de la Recherche en Santé de Quebec.

References

1. H. Ohkubo and S. Nakanishi, *Ann. N.Y. Acad. Sci.* **632,** 53 (1991).
2. R. Quirion and T.-V. Dam, *Regulat. Pept.* **22,** 18 (1988).
3. S. Guard and S. P. Watson, *Neurochem. Int.* **17,** 149 (1990).
4. E. Burcher, C. J. Mussap, D. P. Geragthy, J. McClure-Sharp, and D. T. Watkins, *Ann. N.Y. Acad. Sci.* **632,** 123 (1991).
5. R. Quirion, T.-V. Dam, and S. Guard, *Ann. N.Y. Acad. Sci.* **632,** 137 (1991).
6. R. Quirion and P. Gaudreau, *Neurosci. Biobehav. Rev.* **9,** 413 (1985).
7. R. M. Snider, J. W. Constantine, J. A. Lowe III, K. P. Longo, W. S. Label, H. A. Woody, S. E. Drozda, M. C. Desai, F. J. Vinick, R. W. Spencer, and H. J. Hess, *Science* **251,** 435 (1991).
8. S. McClean, A. H. Ganong, T. F. Seeger, D. K. Bryce, K. G. Pratt, L. S. Reynolds, C. J. Siok, J. A. Lowe, III, and J. Heym, *Science* **251,** 437 (1991).
9. C. Garret, A. Carruette, V. Fardin, S. Moussaoui, J. F. Peyronel, J. C. Blanchard, and P. Laduron, *Proc. Natl. Acad. Sci. U.S.A.* **88,** 10208 (1991).
10. C.-M. Lee, J. A. Javitch, and S. H. Snyder, *Mol. Pharmacol.* **23,** 563 (1983).
11. C. H. Park, V. J. Massari, R. Quirion, Y. Tizabi, C. W. Shults, and T. L. O'Donohue, *Peptides* **5,** 833 (1984).
12. J.-C. Beaujouan, Y. Torrens, A. Herbert, M. C. Daguet, J. Glowinski, and A. Prochiantz, *Mol. Pharmacol.* **22,** 48 (1982).
13. Y. Torrens, J.-C. Beaujouan, A. Viger, and J. Glowinski, *Naunyn-Schmiedeberg's Arch. Pharmacol.* **324,** 134 (1983).
14. M. A. Cascieri and T. Liang, *J. Biol. Chem.* **258,** 5158 (1983).
15. T.-V. Dam, E. Escher, and R. Quirion, *Brain Res.* **453,** 372 (1988).
16. T.-V. Dam, Y. Takeda, J. E. Krause, E. Escher, and R. Quirion, *Proc. Natl. Acad. Sci. U.S.A.* **87,** 246 (1990).
17. T.-V. Dam, B. Martinelli, and R. Quirion, *Brain Res.* **531,** 333 (1990).
18. S. Guard, T.-V. Dam, S. P. Watson, B. Martinelli, K. J. Watling, and R. Quirion, *Biotechnol. Update* **6,** (1991).
19. S. Guard, S. P. Watson, J. E. Maggio, H. P. Too, and K. J. Watling, *Br. J. Pharmacol.* **99,** 767 (1990).
20. M. Herkenham and C. B. Pert, *J. Neurosci.* **2,** 1129 (1982).
21. D. Regoli, F. Nantel, C. Tousignant, D. Jukic, N. Rouissi, N. E. Rhaleb, S.

Télémaque, G. Drapeau, and P. d'Orléans-Juste, *Ann. N.Y. Acad. Sci.* **632,** 170 (1991).

22. R. Lew, D. P. Gerathy, G. Drapeau, D. Regoli, and E. Burcher, *Eur. J. Pharmacol.* **184,** 97 (1990).
23. D. Aharony, C. A. Catanese, and D. P. Woodhouse, *J. Pharmacol. Exp. Ther.* **259,** 146 (1991).
24. G. Drapeau, P. d'Orléans-Juste, S. Dion, N. E. Rhaleb, and D. Regoli, *Neuropeptides* **10,** 43 (1987).
25. L. Bergstrom, J.-C. Beaujouan, Y. Torrens, M. Saffroy, J. Glowinski, S. Lavielle, G. Chassaing, A. Marquet, P. d'Orléans-Juste, S. Dion, and D. Regoli, *Mol. Pharmacol.* **32,** 227 (1988).
26. M. Saffroy, J.-C. Beaujouan, Y. Torrens, J. Besseyre, L. Bergstrom, and J. Glowinski, *Peptides* **9,** 227 (1988).
27. K. Yasphal, T.-V. Dam, and R. Quirion, *Brain Res.* **506,** 259 (1990).
28. T.-V. Dam, Y. Takada, J. E. Krause, and R. Quirion, *Ann. N.Y. Acad. Sci.* **632,** 137 (1991).
29. T.-V. Dam, E. Escher, and R. Quirion, *Brain Res.* **506,** 175 (1990).
30. C. J. Mussap and E. Burcher, *Peptides* **11,** 827 (1990).
31. B. D. Gitter, D. C. Waters, R. F. Burns, N. R. Mason, J. A. Nixon, and J. J. Howbert, *Eur. J. Pharmacol.* **197,** 237 (1991).
32. T.-V. Dam and R. Quirion, *Peptides* **7,** 855 (1986).
33. C. A. Maggi, R. Patacchini, M. Astolfi, P. Rovero, S. Guiliani, and A. Giachetti, *Ann. N.Y. Acad. Sci.* **632,** 184 (1991).
34. S. H. Buck, B. O. Fanger, and P. L. M. van Giersbergen, *Ann. N.Y. Acad. Sci.* **632,** 112 (1991).

[15] Brain Serotonin Receptor Subtypes: Radioligand Binding Assays, Second Messengers, Ligand Autoradiography, and *in Situ* Hybridization Histochemistry

José M. Palacios, Guadalupe Mengod, and Daniel Hoyer

Introduction

The introduction of radioligand binding, the development of compounds acting selectively on receptor subpopulations, and the use of anatomical techniques for identifying regions enriched in receptor subtypes have resulted in a dramatic increase of our understanding of serotonin (5-hydroxytrypta-mine, 5-HT) receptors. The more recent application of molecular biology into this field has provided detailed insights into the identity, protein structure, and cellular localization of some 5-HT receptor subtypes. Classically, the tools used to establish serotoninergic pathways at the presynaptic level have been histofluorescence, autoradiography, and immunohistochemistry. However, autoradiography was the only technique allowing the visualization of pre- and postsynaptic receptor sites for 5-HT until recently. $[^3H]LSD$ and $[^3H]5$-HT were the first serotoninergic ligands introduced. Following the discovery of the serotoninergic component of several neuroleptics, $[^3H]$spip-erone was also found to be a ligand for 5-HT receptors. After the seminal publications of Leysen *et al.* (1) identifying 5-HT ($5-HT_2$) recognition sites and that of Peroutka and Snyder (2) subclassifying $5-HT_1$ and $5-HT_2$ sites, the field literally exploded due to the availability of a variety of ligands selective for one or the other of the 5-HT receptors. Although a heterogeneity of $5-HT_2$ binding sites has never been convincingly established, the existence of different subtypes of $5-HT_1$ binding sites ($5-HT_{1A}$ and $5-HT_{1B}$) was demonstrated in rat brain (3). Another type of $[^3H]5$-HT binding site was identified by autoradiography and characterized in choroid plexus by homogenate binding and autoradiographic studies (4, 5). This site was named $5-HT_{1C}$. More recently, a novel subtype of $5-HT_1$ site was defined by autoradiography in human brain, subsequently characterized by radioligand binding in homogenates (6), and named $5-HT_{1D}$. These sites appear to represent a heteroge-

Methods in Neurosciences, Volume 12

neous population, as suggested by the competition profile of 5-carboxamido-tryptamine (5-CT) and sumatriptan for these sites (7, 8).

Additional 5-HT$_1$ binding sites have been reported, namely, 5-HT$_{1E}$ (8) and 5-HT$_{1P}$ (9), but their functional significance is not well established. 5-HT$_3$ receptors, characterized for about 30 years in functional studies, were ascribed a corresponding binding site only recently (10). Since then, numerous autoradiographical studies have been reported using a variety of ligands (see Table I). The latest class of 5-HT receptor was first described in primary cultures from mouse colliculi and named 5-HT$_4$ (11). To date no successful labeling of these sites has been reported. The radioligands routinely used in the autoradiographic study of 5-HT receptor sites are summarized in Table I.

Molecular cloning studies have now confirmed the existence of these pharmacologically defined receptor subtypes and identified new gene products. Thus, five distinct G-protein-coupled mammalian serotonin receptors [5HT$_{1A}$ (12–14), 5-HT$_{1B}$ (15–17), 5-HT$_{1C}$ (18, 19), 5-HT$_{1D}$ (20, 21), and 5-HT$_2$ (19, 22) and one ligand-gated ion channel serotonin receptor [5-HT$_3$ (23)] have been cloned and sequenced. Their characteristics are summarized in Table I.

Species differences have been reported for all classes of 5-HT receptors. For example, 5-HT$_{1B}$ receptors have only been identified pharmacologically in some rodents (rat, mouse, hamster) and in the opossum, whereas 5-HT$_{1D}$ receptors were found in most mammals, but also in some rodents (guinea pig), rabbits, and birds (24). The 5-HT$_{1R}$ site described in rabbit caudate nucleus (25) represents a species variant of the 5-HT$_{1D}$ site. The situation may be more complicated, since in both rat and human brain at least two genes coding for 5-HT$_{1D}$-like receptors appear to coexist (see above). Similarly, species differences in pharmacological characteristics have been reported for 5-HT$_2$ and 5-HT$_3$ receptors.

In this chapter we review the methodology we have used to study serotonin receptor subtypes at the pharmacological (binding studies and second messenger assays) and anatomical (autoradiography and *in situ* hybridization) levels.

Binding Studies

5-HT$_{1A}$ Receptors

Although many radioligands have been reported to label 5-HT$_{1A}$ sites (Table I), the first choice is still [^{3}H]8-OH-DPAT(8-hydroxy-dipropylaminotetraline) (26), which shows high affinity and selectivity for these sites. Other ligands

TABLE I Serotonin Receptor Subtypes: Characteristics and Radioligands

Receptor	Selective agonists	Selective antagonists	Second messenger	Gene	Protein structure			Radioligands
					Number of Amino acids 7 transmembrane domains (TM)	Third cytoplasmic loop length	C terminus length	
5-HT$_{1A}$	8-OH-DPAT	—	Adenylate cyclase	5-ht1a Human Rat 7 TM	421 422	128 90	18 18	[3H]8-OH-DPAT, [3H]5-HT, [3H]ipsapirone, [3H]LSD, [125I]BH-8-MeO-N-PAT, [3H]WB4101, [3H]PAPP, [3H]spiroxatrine, [3H]lisuride, [125I]BH-DPAT, [125I]I-PAPP, [3H]buspirone, [3H]NAN 190, [3H]BMY 7378
5-HT$_{1B}$	CP 93129	Cyanopindolol	Adenylate cyclase	5-ht1b Human Rat Mouse 7 TM	390 386 386	84 82 87	17 17 21	[125I]-CYP, [3H]5-HT, [125I]-GTI, [3H]dihydroergotamine, [3H]CP 93129
5-HT$_{1C}$	α-Methyl-5-HT DOI	Ritanserin, Pizotifen, Mesulergine	Phosphatidylinositide turnover	5-ht1c Human Rat 7 TM	458 460	76 78	86 86	[3H]Mesulergine, [3H]5-HT, [3H]LSD, [125I]-methyl-LSD, [125I]-MIL, [3H]DOB, [125I]-DOI, [125I]-SCH 23982
5-HT$_{1D}$	Sumatriptan	—	Adenylate cyclase	5-ht1d Human Rat Canine 7 TM	377 — 377	72 — 84	19 — 17	[125I]-GTI, [3H]5-HT, [3H]5-CT
5-HT$_2$	α-Methyl-5-HT, DOI	Ritanserin, Pizotifen, LY 53857	Phosphatidylinositide turnover	5-ht2 Human Rat 7 TM	471 471	70 70	86 86	[3H]Ketanserin, [3H]mesulergine, [3H]spiperone, [125I]-LSD, [125I]-methyl-LSD, [3H]LSD, [125I]-MIL, [125I]-SCH 23982, [3H]SCH 23390, [3H]mianserin, [3H]DOB, [125I]-DOI
5-HT$_3$	2-Methyl-5-HT, 1-phenylbiguanide	ICS 205930, Ondansetron, MDL72222	Ion channel	5-ht3 Mouse 4 TM	487	—	—	[3H]ICS 205930, [3H]zacopride, [3H]GR 65630, [3H]GR 67330, [3H]BRL 43694, [3H]quipazine, [3H]QICS 205930, [3H]LY 278584, [125I]-zacopride

are either less selective or have not become generally available. Using [^{3}H]8-OH-DPAT, 5-HT$_{1A}$ sites have been described in cortex and hippocampus, the two most accessible brain regions, in a variety of species including rat, pig, calf, and man. The striatum should be avoided, as the site labeled by this radioligand seems to represent a 5-HT transporter.

5-HT$_{1B}$ Receptors

Under appropriate conditions, 5-HT$_{1B}$ sites can be labeled with [^{3}H]5-HT in the presence of adequate concentrations of drugs blocking 5-HT$_{1A}$ and 5-HT$_{1C}$ sites (e.g., 100 nM 8-OH-DPAT and mesulergine); however, additional sites may be bound (e.g., 5-HT$_{1E}$, see below). Thus, we recommend the use of ^{125}I-labeled cynanopindolol (CYP) (27, 28) in the presence of 30 μM isoprenaline (to block β-adrenoceptor binding). 5-HT$_{1B}$ binding can be performed in rat, mouse, or hamster cortex or hippocampus. Recently, ^{125}I-labeled GTI (serotonin-5-O-carboxymethylglycyl-^{125}I-tyrosinamide) was also reported to label 5-HT$_{1B}$ sites (29). Other ligands have been described, but they are either nonselective or not commercially available (see Table II).

5-HT$_{1C}$ Receptors

5-HT$_{1C}$ sites were initially labeled with [^{3}H]5-HT (thus the name 5-HT1) and [^{3}H]mesulergine (4). Other ligands are available, such as ^{125}I-labeled lysergic acid diethylamide LSD (5) or ^{125}I-labeled 2-(4-iodo-2,5-dimethoxyphenyl)-1-methylethylamine DOI (30), but most of these will label 5-HT$_2$ receptors as well (see Ref. 31). There is presently no selective 5-HT$_{1C}$ ligand; however, fortunately enough, the choroid plexus has a very high concentration of these sites and thus represents the tissue of choice. 5-HT$_{1C}$ sites have been demonstrated in choroid plexus of a variety of species, including rat, mouse, pig, and man.

5-HT$_{1D}$ Sites

5-HT$_{1D}$ sites can be labeled using [^{3}H]5-HT in the presence of drugs which block binding to 5-HT$_{1A}$ and 5-HT$_{1C}$ receptors (6). 5-HT$_{1D}$ binding should be performed in striatum of calf or species known to express 5-HT$_{1D}$ sites (32, 33). However, under these conditions binding may not be homogeneous (32) (see below). More recently, ^{125}I-labeled GTI was described as labeling a homogeneous population of 5-HT$_{1D}$ sites in human substantia nigra (34); however, this ligand also labels 5-HT$_{1B}$ sites (see above), and it is not established whether the sites labeled are homogeneous in all cases.

5-HT$_{1E}$ Sites

Leonhardt *et al.* (8) described the 5-HT$_{1E}$ binding site in human and rat brain using [^{3}H]5-HT under conditions where 5-HT$_{1A}$, 5-HT$_{1B}$, 5-HT$_{1C}$, and 5-HT$_{1D}$

TABLE II Radioligands and Binding Conditions for Studying Serotonin Receptors in Membrane Preparations

Binding site	Membrane source, protein conc.	Radioligand, concentration	Nonspecific binding conc.	Incubation conditions			Reaction mixture
				Volume (ml)	Temperature (°C)	Time (min)	
5-HT1A[a]	Rat hippocampus, 120 μg/ml	[^{3}H]8-OH-DPAT, 0.5 nM	5-HT, 10 μM	1	25	30	50 mM Tris-HCl, 4 mM CaCl$_2$, 0.1% ascorbate, 10 μM pargyline, pH 7.4 (at 25°C)
5-HT1B[b]	Rat cortex, 80 μg/ml	^{125}I-Cyanopindolol, 0.15 nM	5-HT, 10 μM	0.25	37	90	10 mM Tris-HCl, 154 mM NaCl, 10 mM pargyline, 30 μM isoprenaline. pH 7.7 (at 25°C)
5-HT1C[c]	Pig choroid plexus, 60 μg/ml	[^{3}H]Mesulergine, 1 nM	5-HT, 1 μM	1	37	30	50 mM Tris-HCl, 4 mM CaCl$_2$, 0.1% ascorbate, 10 μM pargyline, pH 7.7 (at 25°C)
5-HT1D[d]	Calf caudate, 180 μg/ml	[^{3}H]5-HT, 2.5 nM	5-HT, 10 μM	1	37	30	50 mM Tris-HCl, 4 mM CaCl$_2$, 0.1% ascorbate, pH 7.7 (at 25°C), plus 100 nM 8-OH-DPAT, 100 nM mesulergine, 10 μM pargyline
5-HT1D[e]	Calf substantia nigra, 400 μg/ml	^{125}I-GTI, 0.1 nM	5-HT, 20–10 μM	0.25	37	30	50 mM Tris-HCl, 4 mM CaCl$_2$, pH 7.7 (at 25°C)
5-HT$_2$[f]	Pig frontal cortex, 280 μg/ml	[^{3}H]Ketanserin, 2 nM	Mianserin, 1 μM	1	37	30	50 mM Tris-HCl, 4 mM CaCl$_2$, 0.1% ascorbate, 10 μM pargyline, pH 7.7 (at 25°C)
5-HT$_3$[g]	Mouse neuroblastoma–rat glioma hybrid cells (NG 108-15), 1.2 × 10^6 cells/ml	[^{3}H]GR 65630, 1 nM	(S)-Zacopride, 1 μM	0.25	25	45	10 mM Tris-HCl, 154 mM NaCl, 5.4 mM KCl, 1.2 mM KH$_2$PO$_4$, 2.5 mM CaCl$_2$, 1.0 mM MgCl$_2$, 11 mM glucose, pH 7.4

[a] H. Gozlan, S. El Mestikawy, L. Pichat, J. Glowinski, and M. Hamon Nature (London) **305**, 140 (1983).

[b] D. Hoyer, G. Engel, and H. O. Kalkman. Eur. J. Pharmacol. **118**, 1 (1985).

[c] A. Pazos, D. Hoyer, and J. M. Palacios. Eur. J. Pharmacol. **106**, 539 (1984).

[d] C. Waeber, P. Schoeffter, J. M. Palacios, and D. Hoyer, Naunyn-Schmiedeberg's Arch. Pharmacol. **337**, 595 (1988).

[e] A. T. Bruinvels, B. Landwehrmeyer, C. Waeber, J. M. Palacios, and D. Hoyer, Eur. J. Pharmacol. **202**, 89 (1991).

[f] A. Pazos, D. Hoyer, and J. M. Palacios, Eur. J. Pharmacol. **106**, 531 (1984).

[g] N. A. Sharif, E. H. F. Wang, D. N. Lowry, E. Stefanich, A. D. Michel, R. M. Eglen, and R. L. Whiting, Br. J. Pharmacol. **102**, 919 (1991).

binding is blocked (in the presence of 100 nM 5-CT and mesulergine). The sites have a 5-HT$_1$-like pharmacology but low affinity for 5-CT. The functional relevance of this site remains to be established.

5-HT$_{1P}$ Sites

5-HT$_{1P}$ sites are labeled with [^{3}H]indalpine and [^{3}H]5-HT. They have been found in brain and especially in the gastrointestinal tract (9).

5-HT$_2$ Sites

Leysen *et al.* (1) reported [^{3}H]spiperone binding to a population of 5-HT sites which corresponds to the 5-HT$_2$ class; however, this ligand binds with even higher affinity to dopamine (D$_2$, D$_3$, D$_4$) receptors. It is preferable to use [^{3}H]ketanserin (35), which is the most commonly employed radioligand for 5-HT$_2$ receptors; cortex, platelets, and some smooth muscle cell lines can be used as a source. Many other ligands label 5-HT$_2$ sites such as ^{125}I-labeled LSD (36), ^{125}I-labeled methyl-iodo-LSD(MIL,) and agonists such as [^{3}H] 2-(4-bromo-2,5-dimethoxyphenyl)-1-methylethylamine (DOB) (37) or ^{125}I-labeled DOI (30). It is now obvious that 5-HT$_2$ sites are not only labeled with antagonists, in contrast to initial claims (2). Two subtypes of 5-HT$_2$ receptors have been proposed (38), labeled by [^{3}H]DOB (5-HT$_{2A}$) and [^{3}H]ketanserin (5-HT$_{2A}$ and 5-HT$_{2B}$), but it appears that agonists label a subpopulation of 5-HT$_2$ receptors in a high-affinity state, as demonstrated more recently (39). It is remarkable that many of the radioligands used, including the agonists but not ketanserin and spiperone, label both 5-HT$_{1C}$ and 5-HT$_2$ receptors (31).

5-HT$_3$ Sites

5-HT$_3$ receptors were initially labeled using [^{3}H]GR 65630 (10) and [^{3}H]ICS 205930 (40), but there are now a variety of radioligands available for these receptors (41–43). 5-HT$_3$ binding studies can be performed in brain (e.g., entorhinal cortex), but the density of the sites is in general fairly low. An alternative is to use neuroblastoma hybrid cell lines (e.g., NG 108-15, N1E-115, or NCB 20; see Refs. 40 and 44), which possess a homogeneous population of receptors. There are few ligands that are commercially available. The most useful appear to be [^{3}H]GR 67330 (45), [^{3}H]BRL 43694 (46), and [^{3}H]zacopride (41); [^{3}H]quipazine (42) should be avoided as its degradation products may label 5-HT uptake sites. There is at present little evidence for receptor subtypes from binding, but it should only be a matter of time until there is.

5-HT$_4$ Sites

We are not aware of any report on 5-HT$_4$ receptor binding, since no high-affinity agonist or antagonist has been commonly available.

Second Messenger Studies

$5\text{-}HT_{1A}$, $5\text{-}HT_{1B}$, $5\text{-}HT_{1D}$, and $5\text{-}HT_4$ receptors are negatively or positively coupled to adenylate cyclase. $5\text{-}HT_{1C}$ and $5\text{-}HT_2$ receptors are linked to phospholipase C/protein kinase C and calcium mobilization. $5\text{-}HT_3$ receptors are ligand-gated ion channels, and there is no evidence that these receptors are coupled to G-protein-mediated events. Some reports have suggested activation of guanylate cyclase, phospholipase A2, and K^+ channels; however, we restrict this presentation to the most commonly used and accepted models.

5-HT Receptor-Mediated Modulation of Adenylate Cyclase Activity

5-HT₁ₐ Receptors

5-HT-stimulated adenylate cyclase has been studied thoroughly in guinea pig and rat hippocampus (47). To the best of our knowledge, the pharmacological profile of 5-HT-stimulated adenylate cyclase activity is mediated by $5\text{-}HT_{1A}$ receptors (48). However, $5\text{-}HT_{1A}$ receptor-mediated inhibition of forskolin-stimulated adenylate cyclase was subsequently reported in rat, mouse, and guinea pig hippocampus (e.g., Ref. 49), which is now generally regarded as a standard $5\text{-}HT_{1A}$ model. In calf hippocampus, the rank order of potency of agonists and antagonists to inhibit forskolin-stimulated adenylate cyclase correlated highly with $5\text{-}HT_{1A}$ binding (50).

5-HT₁ᵦ Receptors

$5\text{-}HT_{1B}$ receptors are negatively coupled to adenylate cyclase (51) in homogenates of rat substantia nigra, which possess a high density of almost exclusively $5\text{-}HT_{1B}$ sites. In this preparation, the rank order of potency of both agonists and antagonists correlates best with affinity values for $5\text{-}HT_{1B}$ binding sites (52). Similar findings have been reported in an opossum kidney and hamster cell lines (e.g., Ref. 53).

5-HT₁ᴅ Receptors

Activation of $5\text{-}HT_{1D}$ receptors leads to inhibition of forskolin-stimulated adenylate cyclase activity in calf and guinea pig substantia nigra (33, 54), which contain a high proportion of $5\text{-}HT_{1D}$ sites. It has been clearly established that the pharmacological profile of the inhibition of adenylate cyclase in substantia nigra is of the $5\text{-}HT_{1B}$ type in rat, whereas it is of the 5-

HT$_{1D}$ type in guinea pig and calf, in agreement with radioligand binding and autoradiographic studies performed in these species.

5-HT$_4$ Receptors

5-HT-stimulated adenylate cyclase activity has been reported in the adult hippocampus and especially in the colliculus; however, the response is usually more robust in fetal brain cells. Recently, Bockaert and collaborators in a series of elegant papers characterized this activity in cultures of neonatal mouse colliculi neurons and hippocampal preparations (11). 5-HT$_4$ activity has also been found in hippocampus homogenates (55), but this response is fairly small and may be hidden by the 5-HT$_{1A}$ receptor-mediated stimulation of cyclase activity. Kaumann *et al.* (56) have recently reported the presence of a 5-HT$_4$ receptor in human heart which activates the adenylate cyclase/ protein kinase A (PKA) cascade.

5-HT Receptor-Mediated Modulation of Phospholipase C, Protein Kinase C, and Calcium Mobilization

5-HT$_{1A}$ Receptors

5-HT$_{1A}$ receptor-mediated inhibition of carbachol-stimulated accumulation of inositol phosphates has been reported in neonatal but not in adult rat hippocampus (57); nevertheless, inhibition of forskolin-stimulated adenylate cyclase activity should be the method of choice.

5-HT$_{1C}$ Receptors

5-HT$_{1C}$ receptor activation in rat, mouse, and pig choroid plexus leads to the activation of phospholipase C and accumulation of inositol phosphates (58), and this systems has been widely used (59). In adult rat hippocampus 5-HT$_{1C}$ receptors mediate stimulation of phospholipase C activity; however, the choroid plexus is much more robust and represents the best model.

5-HT$_2$ Receptors

5-HT-stimulated hydrolysis of inositol lipids has been analyzed in various brain regions (60), in blood platelets (61), in vascular smooth muscle (62), and in a variety of cultured cells (63, 64). Depending on the tissue, one can also measure Ca^{2+} mobilization and/or protein kinase C activity. Depending on the equipment and/or the interest, one or the other of these methods and tissues can be used.

Ligand-Gated Ion Channels

5-HT₃ Receptors

Binding studies suggest that 5-HT$_3$ receptors are not linked to G proteins. Evidence from electrophysiological studies show that 5-HT$_3$ receptors activate fast ion channels (65), as clearly demonstrated by Derkach *et al.* (66) using patch clamping with the guinea pig ileum. The vagus nerve and the nodose and superior cervical ganglia represent other possible tissue sources. Electrophysiology can also be applied in neuroblastoma cells such as the NG 108-15, N1E-115, or NCB 20 lines (65, 67).

Receptor Ligand Autoradiography

The procedures for *in vitro* autoradiographic visualization of receptors have been extensively described. Incubation conditions are initially taken from those used in membrane assays and tested by measuring the radioactivity bound to tissue sections. Conditions providing the best total to nonspecific binding ratio and the highest level of selectivity (see ref 68). Owing to the poor selectivity of the majority of these compounds, selective blockers have to be included in the incubation medium to avoid labeling of other receptor subtypes.

Ten-micron-thick frozen sections are brought to room temperature and preincubated in Tris-HCl buffer to eliminate endogenous 5-HT. Then they are incubated in buffer containing enzyme inhibitors (in particular the monoamine oxidase inhibitor pargylin), ascorbic acid to prevent chemical oxidation, and the radioactive ligand at a concentration ranging from 0.5 to 5 nM for tritiated ligands and 0.01 to 0.05 nM for iodinated ligands. Displacing compounds are added in parallel conditions to assess the pharmacological characteristics of the binding. Finally, the sections are washed in ice-cold buffer, briefly immersed in cold distilled water to remove salts (which could generate autoradiographic artifacts), and rapidly dried under a stream of cold air. Exposure time to radiation-sensitive films is of the order of a few days for iodinated ligands and of a few weeks or months for tritiated compounds. After developing, the optical densities of the films are measured by means of a computer-aided image analysis system.

One of the main limitations of light microscopic *in vitro* receptor autoradiography is its lack of cellular resolution. Although the procedure allows the localization of receptor sites to specific brain nuclei or cell layers, it is impossible to conclude, in most cases, which cells actually bear the receptors. A complementary, although indirect, approach is to combine autoradiography with more or less specific mechanical or chemical lesions of cell populations or pathways. This approach has been used in attempts to assign brain 5-HT receptors to different neuronal populations and pathways in rat brain (69). The detailed conditions used for different ligands are summarized in Table III. Relevant data are discussed below for each receptor subtype.

5-HT$_{1A}$ Receptors

5-HT$_{1A}$ receptors have been most frequently labeled using [^{3}H]8-OH-DPAT. This agonist is one of the rare specific compounds used to label 5-HT receptors. Moreover, although tritiated, it is available at a high specific activity (>200 Ci/mmol), allowing exposure times as short as 2 weeks. Of the various other ligands used (Table I), [^{125}I]BH-8-MeO-N-PAT is the only one presenting some advantages to [^{3}H]8-OH-DPAT, owing to its high specific activity (1000 Ci/mmol) (70). The shorter exposure times needed (3–4 days) are, however, counteracted by the short half-life of the radioligand and its limited availability. The different incubation conditions for the autoradiographic localization of 5-HT$_{1A}$ binding sites are depicted in Table III.

The highest densities of binding sites for [^{3}H]8-OH-DPAT (71, 72) are found in the hippocampus, followed by the nucleus raphe dorsalis, the lateral septum, and the cerebral cortex. Some amygdaloid nuclei and the olfactory bulb also contain significant densities of these sites. Other regions rich in 5-HT$_{1A}$ receptors are the hypothalamus and substantia gelatinosa of spinal cord. Basal ganglia (caudate, putamen, pallidum) and associated regions (substantia nigra) are virtually devoid of [^{3}H]8-OH-DPAT binding sites.

5-HT$_{1B}$ and 5-HT$_{1D}$ Receptors

Available data show that 5-HT$_{1B}$ and 5-HT$_{1D}$ receptors present a very similar regional and cellular distribution, similar signal transduction mechanism, and similar autoreceptor function in the cortex (24). Differences between 5-HT$_{1B}$ and 5-HT$_{1D}$ subtypes are found in their drug binding profile. Rat, at variance with human (17), 5-HT$_{1B}$ receptors characteristically display a high affinity for β-adrenoceptor blocking agents (73); rodent 5-HT$_{1B}$ sites are thus conveniently labeled using ^{125}I-labeled CYP in the presence of isoproterenol, which permits overnight exposures. Both 5-HT$_{1B}$ and 5-HT$_{1D}$ receptors can also be labeled using [^{3}H]5-HT in the presence of blockers of the other 5-HT$_1$ receptor subtypes (typically 100 nM 8-OH-DPAT and mesulergine) (24), but the exact nature of the sites labeled under these conditions remains to be established.

5-HT$_{1B}$ and 5-HT$_{1D}$ receptors are both enriched in basal ganglia and associated structures (24, 71, 72). They are similarly concentrated in substantia nigra pars reticulata and in globus pallidus as well as in the superficial gray

TABLE III Autoradiographic Localization of Serotonin Receptors, Ligands and Incubation Conditions

Receptor subtype	Ligand	[Ligand] (nM)	Preincubation protocol	Buffer	Incubation temperature	Time (min)	Washing protocol	Blank generation	Exposure time
5-HT$_{1A}$	[^{3}H]5-HT	2	30 min at room temperature in incubation buffer	0.17 M Tris-HCl, 4 mM CaCl$_2$, 0.01% ascorbic acid (pH 7.6)	room temperature	60	1 × 5 min (4°C)	10 nM 8-OH-DPAT	60 days
	[^{3}H]8-OH-DPAT	2	30 min at room temperature in incubation buffer	0.17 M Tris-HCl, 4 mM CaCl$_2$, 0.01% ascorbic acid (pH 7.6) 1 μM paragyline and 1 μM fluoxetine	room temperature	60	2 × 5 min (4°C)	1 μM 5-HT	60 days
	[^{3}H]TVX-Q 7821	9	30 min at 25°C in 0.17 M Tris-HCl (pH 7.7)	0.17 M Tris-HCl, 4 mM CaCl$_2$, 0.01% ascorbic acid (pH 7.7) plus 1 μM paragyline and 1 μM fluoxetine	room temperature	60	4 × 15 sec (4°C)	10 μM 5-HT or 10 μM TVX-Q 7821	35 days
	[^{3}H]LSD	5	5 min at room temperature in incubation buffer (−pargyline)	0.17 M Tris-HCl, 4 mM CaCl$_2$, 0.01% ascorbic acid, 0.15 M NaCl, 10 μM pargyline (pH 7.6)	room temperature	60	2 × 10 min (4°C)	100 nM 5-HT	100 days

Receptor	Radioligand	K_d (nM)	Preincubation	Incubation buffer	Temperature	Time (min)	Wash	Nonspecific	Stability
5-HT$_{1B}$	[^{3}H]5-HT	2	30 min at room temperature in incubation buffer	0.17 M Tris-HCl, 4 mM CaCl$_2$, 0.01% ascorbic acid plus 100 nM PAT and mesulergine	room temperature	60	1 × 5 min (4°C)	10 nM (−)21009	60 days
	^{125}I-CYP	0.012	10 min at room temperature in 170 mM Tris-HCl (pH 7.4) plus 150 mM NaCl	Same as in preincubation plus 3 μM isoproterenol and 100 nM PAT	room temperature	120	2 × 15 min (4°C)	100 nM 5-HT	3 days
	^{125}I-GTI	0.020	2 × 15 min in 170 mM Tris-HCl (pH 7.5), 4 mM CaCl$_2$	Same as in preincubation plus 10 μM pargyline, 0.01% ascorbic acid	room temperature	60	2 × 5 min (4°C)	10 μM 5-HT	7 days
5-HT$_{1C}$	[^{3}H]5-HT	2	30 min at room temperature in incubation buffer	0.17 M Tris-HCl, 4 mM CaCl$_2$, 0.01% ascorbic acid (pH 7.6) plus 100 nM spiperone	room temperature	60	1 × 5 min (4°C)	100 nM 5-HT	60 days
	[^{3}H]LSD	2	30 min at room temperature in incubation buffer	0.17 M Tris-HCl, 4 mM CaCl$_2$, 0.01% ascorbic acid (pH 7.6) plus 100 nM spiperone	room temperature	60	1 × 5 min (4°C)	100 nM 5-HT	60 days

(continued)

TABLE III (*continued*)

Receptor subtype	Ligand	[Ligand] (nM)	Preincubation protocol	Buffer	Incubation temperature	Time (min)	Washing protocol	Blank generation	Exposure time
	^{125}I-LSD	0.05–1.5	30 min at room temperature in 0.17 mM Tris-HCl plus 1% BSA (pH 7.4)	50 mM Tris-HCl (pH 7.6), or same as in preincubation plus 150 mM NaCl	room temperature	120 or 60	2 × 15 min (4°C) or 4 × 20 min (4°C)	100 nM 5-HT	12–24 hr
	[^{3}H]Mesulergine	5	15 min at room temperature in incubation buffer	0.17 M Tris-HCl (pH 7.7) plus 100 nM spiperone	room temperature	120	2 × 10 min (4°C)	100 nM 5-HT	30 days
5-HT$_{1D}$	[^{3}H]5-HT	0.012	10 min at room temperature in 170 mM Tris-HCl (pH 7.4) plus 150 mM NaCl	Same as in preincubation plus 100 nM PAT 8 OH DPAT and mesulergine	room temperature	120	2 × 15 min (4°C)	100 nM 5-HT	3 days
	^{125}I-GTI	0.020	2 × 15 min in 170 mM Tris-HCl (pH 7.5), 4 mM CaCl$_2$	Same as preincubation plus 10 μM pargyline, 0.01% ascorbic acid	room temperature	60	2 × 5 min (4°C)	10 μM 5-HT	7 days
5-HT$_2$	^{125}I-DOI	0.2	15 sec at room temperature in incubation buffer	50 mM Tris-HCl, pH 7.4 4 mM CaCl$_2$, 0.1% BSA, 0.1% ascorbic acid, 10 nM mesulergine	room temperature	90	2 × 10 min (4°C)	100 nM 5-HT	4 days
	[^{3}H]Ketanserin	2	15 min at room temperature in incubation buffer	0.17 M Tris-HCl (pH 7.7)	room temperature	120	2 × 10 min (4°C)	1 μM methylsergide	30 days

250

	Concentration (nM)	Preincubation	Incubation buffer	Temperature	Time (min)	Wash	Nonspecific	Half-life
[³H]Mesulergine	5	15 min at room temperature in incubation buffer	0.17 M Tris-HCl (pH 7.7)	room temperature	120	2 × 10 min (4°C)	300 nM ketanserin	30 days
[³H]Spiperone	0.4	None	0.17 M Tris-HCl, 1 mM MgCl₂, 2 mM CaCl₂, 5 mM KCl, 120 mM NaCl (pH 7.7)	room temperature	30	2 × 5 min (4°C)	300 nM ketanserin	90 days
[³H]LSD	5	5 min at room temperature in incubation buffer (−pargyline)	0.17 M Tris-HCl, 4 mM CaCl₂, 0.15 M NaCl, 0.01% ascorbic acid, 10 μM pargyline (pH 7.6)	room temperature	60	2 × 10 min (4°C)	1 μM cinanserin	100 days
¹²⁵I-LSD	0.05–1.5	0.17 mM Tris-HCl plus 1% BSA (pH 7.4)	50 mM Tris-HCl (pH 7.6), or same as in preincubation plus 150 mM NaCl	room temperature	120 or 60	2 × 5 min (4°C) or 4 × 20 min (4°C)	0.1 or 1 μM ketanserin	12–24 hr
5-HT₃ [³H]ICS 205930	1	30 min at room temperature in incubation buffer	50 mM Tris-HCl, 157 mM NaCl, 4 mM CaCl₂, 0.01% ascorbic acid (pH 7.5)	room temperature	60	1 × 5 min (4°C)	50 μM 5-HT	60 days
[³H]LY 278584	2	30 min at room temperature in 50 mM Tris-HCl pH 7.4, 150 mM NaCl	Same as in preincubation plus 4 mM CaCl₂, 0.01% ascorbic acid, 2 mM pargyline	room temperature	30	2 × 10 min (4°C)	10 μM 5-HT	120–180 days

(continued)

TABLE III (continued)

Receptor subtype	Ligand	[Ligand] (nM)	Preincubation protocol	Buffer	Incubation temperature	Time (min)	Washing protocol	Blank generation	Exposure time
	[³H](S)-Zacopride	2	30 min at room temperature in incubation buffer	2.5 mM Tris-HCl, 157 mM NaCl, 0.01% ascorbic acid (pH 7.5)	room temperature	60	2 × 5 min (4°C)	1 μM ICS 205930	60 days
	[³H]GR 65630	0.2	30 min at room temperature in incubation buffer	50 mM HEPES, 0.01% ascorbic acid (pH 7.4)	room temperature	60	2 × 3 min (4°C)	1 μM ICS 205930	90 days
	[³H]BRL 43694	1	None	50 mM HEPES (pH 7.4)	room temperature	45	2 × 10 min (4°C)	100 μM metoclopramide	60 days

layer of superior colliculus, central gray, and caudate/putamen. All cortical areas also contain intermediate densities of 5-HT$_{1B}$/5-HT$_{1D}$ binding sites.

^{125}I-labeled GTI binds selectively to 5-HT$_{1B}$ (29) or 5-HT$_{1D}$ sites (74). Considering the very low concentration of this ligand in the incubation medium (about 100 pM), it is a useful tool to investigate the multiple affinity states that have been proposed for 5-HT$_{1B}$/5-HT$_{1D}$ sites (75). Recent studies have shown that ^{125}I-labeled GTI labels only a fraction of the sites labeled by [^{3}H]5-HT (under 5-HT$_{1D}$ conditions) and is displaced by 5-CT in a monophasic manner, in contrast with the typical profile observed using [^{3}H]5-HT (74). The chemical nature of this ligand allows its cross-linking with the receptor and the visualization of 5-HT$_{1B}$/5-HT$_{1D}$ sites at higher resolution (29, 75a).

5-HT$_{1C}$ Receptors

The conditions for selective visualization of 5-HT$_{1C}$ receptor binding sites are summarized in Table III. In the rat brain as in all mammalian species examined, the highest density of 5-HT$_{1C}$ receptor binding sites (selectively though not specifically labeled by [^{3}H]mesulergine and ^{125}I-labeled DOI) is localized in the choroid plexus of all the ventricles (71, 72). Lower, but significant, densities are observed heterogeneously distributed throughout the brain (76); the nucleus presenting the highest level of [^{3}H]mesulergine labeling is the anterior olfactory nucleus. Intermediate densities are observed in nucleus accumbens, primary olfactory cortex, and nucleus caudate/putamen, particularly in its ventral part. The paratenial, periventricular, and paraventricular nuclei of the thalamus are also labeled. In the midbrain, the substantia nigra pars compacta and reticulata display important levels of 5-HT$_{1C}$ binding sites, as well as the central gray and the deepest part of the superior colliculus. Binding is also observed in parts of the medial geniculate nucleus. The gray matter of the pons, medulla, and spinal cord exhibits a low homogeneous binding.

5-HT$_2$ Receptors

The ligands and optimal incubation conditions to visualize 5-HT$_2$ receptors by autoradiography are shown in Table III. 5-HT$_2$ receptors have been found in most cortical areas of all mammalian species so far investigated (71, 77). In rat the highest concentrations are localized in the claustrum, olfactory tubercle, and layer IV of the neocortex (77). Very low densities of binding sites are observed over the septal nuclei, as well as in the bed nucleus of the stria terminalis. In hippocampal formation, the concentrations of 5-HT$_2$ receptors are low or very low in all the areas, except in the ventral dentate gyrus, which presents an intermediate density of binding sites. In basal ganglia, intermediate levels of binding sites are found in the nucleus accumbens and in the body and tail of the caudate/putamen.

5-HT$_3$ Receptors

5-HT$_3$ receptors were first visualized in mouse brain using [^{3}H]ICS 205930 as a ligand (78) and shortly afterward in human brain (79). [^{3}H]ICS 205930 and several other 5-HT$_3$ radioligands developed later (Tables I and III) revealed binding sites in regions where they were present at very high concentrations, such as the dorsal nucleus of the vagus nerve. These areas are, however, very small and hard to dissect for homogenate binding studies, explaining why 5-HT$_3$ binding sites were difficult to detect in brain.

The (S) isomer of [^{3}H]zacopride and some newly developed 5-HT$_3$ radioligands (e.g., LY 278584) are useful to visualize receptors in forebrain (80–82). These regions contain low densities of 5-HT$_3$ sites (with the exception of the entorhinal cortex), which were only faintly labeled by ligands such as [^{3}H]ICS 205930. Thus, in rat brain, 5-HT$_3$ receptors have been observed in various subdivisions of the cortical amygdaloid nucleus, in the tenia tecta, external plexiform layer of the olfactory bulb, pyramidal cell layer, and stratum lacunosum moleculare of dorsal CA1, CA2, and CA3 fields, as well as in the granule cell layer of the dentate gyrus (C. Waeber and J. M. Palacios, unpublished). In human brain, [^{3}H](S)-zacopride binding sites are also found in striatum, particularly in its ventral part, probably in the matrix compartment (C. Waeber and J. M. Palacios, unpublished).

In Situ Hybridization Histochemistry

In situ hybridization histochemistry (ISHH) allows the autoradiographic visualization of the cells containing the mRNA under study by using labeled probes (either single-strand DNA or RNA). Oligonucleotide probes are single-stranded, synthetic DNA fragments which are designed to hybridize with the mRNA of interest. One of the advantages of using oligonucleotide probes is the possibility of obtaining probes highly specific for the mRNA of interest. This is especially important when multiple subtypes of receptors exist for a given neurotransmitter, since oligonucleotides can be designed to hybridize with regions of the mRNA which share very little or no identity among the different subtypes. In this way, probes which will recognize only one of the mRNA subtypes can be obtained and used to map the expression of the various subtypes (see Refs. 83 and 84 for reviews).

Methodology

The oligonucleotides used to hybridize to the different serotonin receptor subtype mRNAs are summarized in Table IV. In general, three different oligonucleotides are used as independent hybridization probes per subtype receptor from regions which share no homology among them, namely, the amino terminus, third cytoplasmic loop, and carboxy terminus of the coding region. The oligonucleotides are labeled at their 3' end with terminal deoxy-nucleotidyltransferase and $[\alpha^{32}P]dATP$. They are used as hybridization probes with 20-μm-thick sections that have been fixed with paraformaldehyde and treated with pronase (85). The hybridization buffer contains 40 or 50% formamide, 10% dextran sulfate, 0.6 M NaCl, 10 mM Tris-HCl, pH 7.5, 1 mM EDTA, 1× Denhardt's solution [0.02% Ficoll, 0.02% bovine serum albumin (BSA), 0.02% polyvinylpyrrolidone], and 0.5 mg/ml yeast tRNA. Labeled oligonucleotide probe is added at a final concentration of 0.4–0.8 pmol/ml, and hybridizations are carried out overnight at 42°C. The sections are then washed with four 1-hr rinses in 0.6 M NaCl, 10 mM Tris-HCl, pH 7.5, 1 mM EDTA at the appropriate temperature. Autoradiograms are generated by apposition of the hybridized tissue sections to β_{max} films (Amersham International, Amersham, UK) or by dipping the sections into Kodak (Rochester, NY) NTB-3 nuclear track emulsion.

To assess the specificity of the hybridization signal obtained, several experiments are done (84). For Northern blot analysis with RNA extracted from different regions of the brain, the RNA detected should be of a size corresponding to that originally described and with a regional distribution in agreement with the pattern observed in ISHH experiments. For each mRNA analyzed, at least two oligonucleotides complementary to different regions of the mRNA are synthesized and used separately as hybridization probes in consecutive tissue sections. The hybridization pattern obtained must be the same for both probes. Specific hybridization signals will disappear when an excess of unlabeled oligonucleotide is included in the hybridization solution, the remaining signal giving an idea of the background levels. Different patterns of hybridization in serial sections with probes for related but different mRNAs (i.e., different subtypes of the same family) further supports the specificity of the signal. The thermal stability of the hybrids should be determined. Table IV summarizes the probes and hybridization conditions we have used to study 5-HT receptor subtype mRNAs.

5-HT$_{1A}$ mRNA

The highest levels of hybridization to 5-HT$_{1A}$ mRNA are observed in the septum, hippocampus, entorhinal cortex, interpeduncular nucleus, and dor-

TABLE IV Probes and Conditions for Visualization of Serotonin Receptor mRNAs by *in Situ* Hybridization Histochemistry

Receptor	Oligonucleotide	Sequence limits	Hybridization conditions	Washing conditions	Anatomical localization
5-HT$_{1A}$	Oligo N	aa 1–16	40% formamide, 42°C	4 × 1 hr, 60°C	Hippocampus, raphe, septum, cortex
5-HT$_{1B}$	1b/N	aa 1–16	50% formamide, 42°C	4 × 1 hr, 60°C	Striatum, hippocampus, cortex, raphe, cerebellum
5-HT$_{1C}$	1c/2	aa 442–457	50% formamide, 42°C	4 × 1 hr, 60°C	Choroid plexus, olfactory nucleus, cortex, amygdala, lateral habenula, hippocampus, nucleus subthalamicus, substantia nigra
5-HT$_{1D}$	1d/N	aa 1–16	50% formamide, 42°C	4 × 1 hr, 60°C	Striatum
5-HT$_2$	s2/N	bp 669–716	40% formamide, 42°C	4 × 1 hr, 60°C	Neocortex, claustrum, olfactory bulb, pontine nuclei, motor trigeminal nerve, nucleus facialis, nucleus hypoglossus

sal raphe nucleus (85). Positive hybridization signals are also present in other areas, such as the olfactory bulb, cerebral cortex, some thalamic and hypothalamic nuclei, several nuclei of the brain stem, including all the remaining raphe nuclei, nucleus of the solitary tract and nucleus of the spinal tract of the trigeminus, and the dorsal horn of the spinal cord. The distribution and abundance of 5-HT$_{1A}$ receptor mRNA in different rat brain areas generally correlated with those of the binding sites, suggesting that 5-HT$_{1A}$ receptors are predominantly somatodendritic receptors.

5-HT$_{1B}$ mRNA

In rat brain, the highest levels of 5-HT$_{1B}$ mRNA are found in the CA1 region of hippocampus (15, 17). A strong signal is seen over neurons of the dorsal and median raphe nuclei, as well as in the central gray, red nuclei, and dorsal interpeduncular nucleus. It is also present in the accessory olfactory nucleus, olfactory tubercle, striatum, layer IV of neocortex, entorhinal cortex, and several thalamic nuclei. In the cerebellum, specific labeling is confined to Purkinje cells. A very similar distribution is found for the mouse 5-HT$_{1B}$ mRNA (16). In human brain (17), 5-HT$_{1B}$ mRNA is found in the caudate and putamen nuclei, lamina V of the neocortex, and in the Purkinje cell layer of the cerebellar cortex. The same distribution is observed in monkey brain (17).

5-HT$_{1C}$ mRNA

Detailed studies have been carried out for the anatomical localization of 5-HT$_{1C}$ mRNA (76, 86, 87). As expected, the highest densities of hybridization signals were found in the choroid plexus. However, and rather unexpectedly because of the low density of receptors, high levels of hybridization were also seen in many areas of the brain. These include the anterior olfactory nucleus, pyriform cortex, amygdala, some thalamic nuclei, especially the lateral habenula, the CA3 area of the hippocampal formation, the cingulate cortex, some components of the basal ganglia and associated areas, particularly the subthalamic nucleus, and the substantia nigra. In general, the distribution of 5-HT$_{1C}$ mRNA corresponded well to that of the 5-HT$_{1C}$ receptors visualized with [^{3}H]mesulergine (76). Nevertheless, exceptions were seen, particularly in the lateral habenula and in the subthalamic nucleus. In these two brain nuclei the density of 5-HT$_{1C}$ binding sites was much lower than that expected from the high levels of mRNA.

5-HT$_{1D}$ mRNA

Cells containing 5-HT$_{1D}$ mRNA have been detected in basal ganglia (87a), particularly rich in caudate and putamen nuclei, but none were detectable in the globus pallidus and substantia nigra. Hybridization signals were also observed in the nucleus raphe dorsalis. These results indicate that 5-HT_{1D} receptors exist not only as autoreceptors on serotoninergic terminals but also as heteroreceptors on neurons which do not use serotonin as neurotransmitter. In all species examined to date, 5-HT_{1D} mRNA is less abundant than 5-HT_{1B} mRNA.

5-HT_2 mRNA

The highest levels of hybridization to 5-HT_2 mRNA are observed in the frontal cortex (88). The rest of neocortex, the pyriform, and the entorhinal cortex were also rich in 5-HT_2 receptor transcripts, as well as some subcortical areas, such as the claustrum and endopiriform nucleus. Positive signals were also present in the olfactory bulb, anterior olfactory nucleus, and several brain stem nuclei including the pontine nuclei and motor trigeminal, facial, and hypoglossal nuclei. Intermediate densities of hybridization signals were seen in the lateral and dorsal part of caudate nucleus, in the nucleus accumbens, as well as in the olfactory tubercle. In contrast, the cerebellum and several brain stem and thalamic areas did not show any specific signal (88). This distribution compares well with that of 5-HT_2 binding sites labeled with ligands such as ($\pm$) $[^{125}\text{I}]\text{DOI}$, $[^3\text{H}]$mesulergine, or $[^3\text{H}]$ketanserin.

Note Added in Proof: Several new 5HT receptor subtypes have been cloned and sequenced. A clone S31 with the characteristics of a serotonin receptor was reported by Levy *et al.* (89) and shortly afterwards identified as the 5HT_{1E} receptor (90). The 5HT_5 receptor isolated by Hen and co-workers (91, 92) appears to be the mouse equivalent of this receptor. The 5HT_6 receptor, also characterized by this group is another member of the "5HT_{1D}-like" family, although presenting a lower degree of homology with the previously isolated 5HT_{1D} receptor genes. A new member of the 5HT_2 receptor family, the rat stomach fundus 5HT receptor, has been isolated by two independent groups (93, 94) and named 5HT_{2F}. Also, a radioligand for the 5HT_4 receptor, the compound $[^3\text{H}]$-GR 113808, has been reported by Grossman *et al.* (95).

References

1. J. E. Leysen, C. J. E. Niemegeers, J. P. Tollenaere, and P. M. Laduron, *Nature(London)* **272,** 168 (1978).
2. S. J. Peroutka and S. H. Snyder, *Mol. Pharmacol.* **16,** 687 (1979).
3. N. W. Pedigo, H. I. Yamamura, and D. L. Nelson, *J. Neurochem.* **36,** 220 (1981).
4. A. Pazos, D. Hoyer, and J. M. Palacios, *Eur. J. Pharmacol.* **106,** 539 (1984).
5. K. A. Yagaloff and P. R. Hartig, *J. Neurosci.* **5,** 3178 (1985).

6. R. E. Heuring and S. J. Peroutka, *J. Neurosci.* **7,** 894 (1987).

7. M. J. Sumner and P. P. A. Humphrey, *Br. J. Pharmacol.* **98,** 29 (1989).

8. S. Leonhardt, K. Herrick-Davis, and M. Titeler, *Neurochemistry* **53,** 465 (1989).

9. T. A. Branchek, G. M. Mawe, and M. D. Gershon, *J. Neurosci.* **8,** 2582 (1988).

10. G. J. Kilpatrick, B. J. Jones, and M. B. Tyers, *Nature (London)* **330,** 746 (1987).

11. A. Dumuis, R. Bouhelal, M. Sebben, R. Cory, and J. Bockaert, *Mol. Pharmacol.* **34,** 880 (1988).

12. B. K. Kobilka, T. Frielle, S. Collins, T. Yang-Feng, T. S. Kobilka, U. Francke, R. J. Lefkowitz, and M. G. Caron, *Nature (London)* **329,** 75 (1987).

13. A. Fargin, J. R. Raymond, M. J. Lohse, B. K. Kobilka, M. G. Caron, and R. J. Lefkowitz, *Nature (London)* **335,** 358 (1988).

14. P. Albert, O. Y. Zhou, H. Van Tol, J. Bunzow, and O. Civelli, *J. Biol. Chem.* **265,** 5825 (1990).

15. M. M. Voigt, D. J. Laurie, P. H. Seeburg, and A. Bach, *EMBO J.* **10,** 4017 (1991).

16. L. Maroteaux, F. Saudou, N. Amlaiky, U. Boschert, J. L. Plassat, and R. Hen, *Proc. Natl. Acad. Sci. U.S.A.* **89,** 3020 (1992).

17. H. Jin, D. Oksenberg, A. Askenazi, S. J. Peroutka, R. Rozmahel, Y. Yang, G. Mengod, J. M. Palacios, and B. F. O'Dowd, *J. Biol. Chem.* **267,** 5735 (1992).

18. D. Julius, A. B. MacDermott, R. Axel, and T. Jessel, *Science* **241,** 558 (1988).

19. A. G. Saltzman, B. Morse, M. M. Whitman, Y. Ivanshchenko, M. Jaye, and S. Felder, *Biochem. Biophys. Res. Commun.* **181,** 1469 (1991).

20. M. W. Hamblin and M. A. Metcalf, *Mol. Pharmacol.* **40,** 143 (1991).

21. N. Adham, P. Romanienko, P. Hartig, R. L. Weinshank, and T. Branchek, *Mol. Pharmacol.* **41,** 1 (1992).

22. D. B. Pritchett, A. W. J. Bach, M. Wozny, O. Taleb, R. Dal Toso, J. C. Shih, and P. H. Seeburg, *EMBO J.* **7,** 4135 (1988).

23. A. V. Marick, A. S. Peterson, A. J. Brake, R. M. Myers, and D. Julius, *Science* **254,** 432 (1992).

24. C. Waeber, P. Schoeffter, D. Hoyer, and J. M. Palacios, *Neurochem. Res.* **15,** 567 (1990).

25. W.-C. Xiong and D. L. Nelson, *Life Sci.* **45,** 1433 (1989).

26. H. Gozlan, S. El Mestikawy, L. Pichat, J. Glowinski, and M. Hamon, *Nature (London)* **305,** 140 (1983).

27. D. Hoyer, G. Engel, and H. O. Kalkman, *Eur. J. Pharmacol.* **118,** 1 (1985).

28. D. Hoyer, G. Engel, and H. O. Kalkman, *Eur. J. Pharmacol.* **118,** 13 (1985).

29. P. Boulenguez, J. Chauveau, L. Segu, A. Morel, J. Lanoir, and M. Delaage, *Eur. J. Pharmacol.* **194,** 91 (1991).

30. R. A. Glennon, M. R. Seggel, W. H. Soine, K. Herrick-Davis, R. A. Lyon, and M. Titeler, *J. Med. Chem.* **31,** 5 (1988).

31. D. Hoyer *Trends Pharmacol. Sci.* **9,** 89 (1988).

32. C. Waeber, P. Schoeffter, J. M. Palacios, and D. Hoyer, *Naunyn-Schmiedeberg's Arch. Pharmacol.* **337,** 595 (1988).

33. C. Waeber, P. Schoeffter, J. M. Palacios, and D. Hoyer, *Naunyn-Schmeideberg's Arch. Pharmacol.* **340,** 479 (1989).

34. A. T. Bruinvels, B. Landwehrmeyer, C. Waeber, J. M. Palacios, and D. Hoyer, *Eur. J. Pharmacol.* **202,** 89 (1991).

35. J. E. Leysen, C. J. E. Niemegeers, J. M. Van Nueten, and P. M. Laduron, *Mol. Pharmacol.* **21,** 301 (1982).

36. G. Engel, E. Müller-Schweinitzer, and J. M. Palacios, *Naunyn-Schmiedeberg's Arch. Pharmacol.* **325,** 328 (1984).

37. M. Titeler, K. Herrick, R. A. Lyon, J. D. McKenney, and R. A. Glennon, *Eur. J. Pharmacol.* **117,** 145 (1985).

38. S. J. Peroutka, A. Hamik, M. A. Harrington, A. J. Hoffman, C. A. Mathis, P. A. Pierce, and S. S.-H. Wang, *Mol. Pharmacol.* **34,** 537 (1988).

39. M. Teitler, S. Leonhardt, E. Weisberg, and B. J. Hoffman, *Mol. Pharmacol.* **38,** 594 (1990).

40. D. Hoyer and H. C. Neijt, *Eur. J. Pharmacol.* **143,** 191 (1987).

41. N. M. Barnes, B. Costall, and R. J. Naylor, *Br. J. Pharmacol.* **94,** 319P (1988).

42. S. J. Peroutka and A. Hamik, *Eur. J. Pharmacol.* **148,** 297 (1988).

43. N. A. Sharif, E. H. F. Wang, D. N. Lowry, E. Stefanich, A. D. Michel, R. M. Eglen, and R. L. Whiting, *Br. J. Pharmacol.* **102,** 919 (1991).

44. D. Hoyer and H. C. Neijt, *Mol. Pharmacol.* **33,** 303 (1988).

45. G. J. Kilpatrick, A. Butler, R. M. Hagan, B. J. Jones, and M. B. Tyers, *Naunyn-Schmeideberg's Arch. Pharmacol.* **342,** 22 (1990).

46. D. R. Nelson and D. R. Thomas, *Biochem. Pharmacol.* **38,** 1693 (1989).

47. A. Shenker, S. Maayani, H. Weinstein, and J. P. Green, *Life Sci.* **32,** 2335 (1983).

48. R. Markstein, D. Hoyer, and G. Engel, *Naunyn-Schmiedeberg's Arch. Pharmacol.* **333,** 335 (1986).

49. M. De Vivo and S. Maayani, *J. Pharmacol. Exp. Ther.* **238,** 248 (1986).

50. P. Schoeffter and D. Hoyer, *Br. J. Pharmacol.* **95,** 975 (1988).

51. R. Bouhelal, L. Smounya, and J. Bockaert, *Eur. J. Pharmacol.* **151,** 189 (1988).

52. P. Schoeffter and D. Hoyer, *Naunyn-Schmeideberg's Arch. Pharmacol.* **340,** 285 (1989).

53. T. J. Murphy and D. B. Bylund, *J. Pharmacol. Exp. Ther.* **249,** 535 (1989).

54. P. Schoeffter, C. Waeber, J. M. Palacios, and D. Hoyer, *Naunyn-Schmiedeberg's Arch. Pharmacol.* **337,** 602 (1988).

55. A. Shenker, S. Maayani, H. Weinstein, and J. P. Green, *Mol. Pharmacol.* **31,** 357 (1987).

56. A. J. Kaumann, L. Sanders, A. M. Brown, K. J. Murray, and M. J. Brown, *Br. J. Pharmacol.* **100,** 879 (1990).

57. Y. Claustre, J. Benavides, and B. Scatton, *Eur. J. Pharmacol.* **149,** 149 (1988).

58. P. J. Conn and E. Sanders-Bush, *J. Neurochem.* **47,** 1754 (1986).

59. D. Hoyer, P. Schoeffter, C. Waeber, J. M. Palacios, and A. Dravid, *Naunyn-Schmiedeberg's Arch. Pharmacol.* **339,** 252 (1989).

60. P. J. Conn and E. Sanders-Blush, *J. Pharmacol. Exp. Ther.* **234,** 195 (1985).

61. D. De Chaffoy de Courcelles, J. E. Leysen, F. De Clerck, H. Van Belle, and P. A. J. Janssen, *J. Biol. Chem.* **260,** 7603 (1985).

62. R. N. Cory, P. Berta, J. Haiech, and J. Bockaert, *Eur. J. Pharmacol.* **131,** 153 (1986).

63. V. M. Doyle, J. A. Creba, U. T. Rüegg, and D. Hoyer, *Naunyn-Schmiedeberg's Arch. Pharmacol.* **333,** 98 (1986).
64. R. N. Cory, B. Rouot, G. Guillon, F. Sladeczek, M.-N. Balestre, and J. Bockaert, *J. Pharmacol. Exp. Ther.* **241,** 258 (1986).
65. H. C. Neijt, I. J. te Duits, and H. P. M. Vijverberg, *Neuropharmacology* **27,** 301 (1988).
66. V. Derkach, A. Surprenant, and R. A. North, *Nature (London)* **339,** 706 (1989).
67. J. A. Peters and J. L. Lambert, *Trends Pharmacol. Sci.* **10,** 172 (1989).
68. J. M. Palacios, R. Cortés, and M. M. Dietl, *in* "Molecular Neuroanatomy" (F. W. Van Leeuwen, R. M. Buijs, C. W. Pool, and O. Pach, eds.), p. 95. Elsevier, Amsterdam, 1988.
69. J. M. Palacios and M. M. Dietl, *in* "The Serotonin Receptors" (E. Sanders-Bush, ed.), p. 89. Humana Press, Clifton, New Jersey, 1988.
70. H. Gozlan, M. Ponchant, G. Daval, D. Vergé, F. Ménard, A. Vanhove, J. P. Beaucourt, and M. Hamon, *J. Pharmacol. Exp. Ther.* **244,** 751 (1988).
71. J. M. Palacios, C. Waeber, D. Hoyer, and G. Mengod, *Ann. N.Y. Acad. Sci.* **600,** 36 (1990).
72. A. Pazos and J. M. Palacios, *Brain Res.* **346,** 205 (1985).
73. A. Pazos, G. Engel, and J. M. Palacios, *Brain Res.* **343,** 403 (1985).
74. J. M. Palacios, C. Waeber, A. Bruinvels, and D. Hoyer, *Mol. Brain Res.* **13,** 175 (1992).
75. C. Waeber and J. M. Palacios, *in* "Serotonin: Molecular Biology, Receptors and Functional Effects" (J. R. Fozard and P. R. Saxena, eds.), p. 107. Birkhäuser, Basel, Switzerland, 1991.
75a. L. Segu, J. Chauveau, P. Boulenguez, A. Christolomme, and J. Lanoir. In Abstracts of Serotonin, pp 36, Birmingham UK (1991).
76. G. Mengod, H. Nguyen, H. Le, C. Waeber, H. Lübbert, and J. M. Palacios, *Neuroscience (Oxford)* **35,** 577 (1990).
77. A. Pazos, R. Cortés, and J. M. Palacios, *Brain Res.* **346,** 231 (1985).
78. C. Waeber, K. Dixon, D. Hoyer, and J. M. Palacios, *Eur. J. Pharmacol.* **151,** 351 (1988).
79. C. Waeber, D. Hoyer, and J. M. Palacios, *Neuroscience (Oxford)* **31,** 393 (1989).
80. J. M. Barnes, N. M. Barnes, S. Champaneria, B. Costall, and R. J. Naylor, *Neuropharmacology* **29,** 1037 (1990).
81. C. Waeber, L. M. Pinkus, and J. M. Palacios, *Eur. J. Pharmacol.* **181,** 283 (1990).
82. D. R. Gehlert, S. L. Gackenheimer, D. T. Wong, and D. W. Robertson, *Brain Res.* **553,** 149 (1991).
83. M. T. Vilaró, M. I. Martinez-Mir, M. Sarasa, M. Pompeiano, J. M. Palacios, and G. Mengod, *Curr. Aspects Neurosci.* **3,** 1 (1991).
84. J. M. Palacios, G. Mengod, M. Sarasa, M. T. Vilaró, M. Pompeiano, and M. I. Martinez-Mir, *J. Rec. Res.* **11,** 459 (1991).
85. M. Pompeiano, J. M. Palacios, and G. Mengod, *J. Neurosci.* **12,** 440 (1992).
86. B. J. Hoffman and E. Mezey, *FEBS Lett.* **247,** 453 (1989).
87. S. M. Molineaux, T. M. Jessell, R. Axel, and D. Julius, *Proc. Natl. Acad. Sci. U.S.A.* **86,** 6793 (1989).

87a. R. Cortés, G. Mengod, B. O'Dowd, F. Artigas, and J. M. Palacios, *Soc. Neurosci. Abstr.* **18,** 463 (1992).

88. G. Mengod, M. Pompeiano, M. I. Martinez-Mir, and J. M. Palacios, *Brain Res.* **524,** 139 (1990).

89. F. O. Levy, T. Guderman, M. Birnbaumer, A. J. Kaumann, and L. Birnbaumer *FEBS Lett.* **296,** 201 (1992).

90. G. McAllister, A. Charlesworth, C. Snodin, M. S. Beer, A. J. Noble, D. N. Middlemas, L. L. Iverson, and P. Whiting *Proc. Natl. Acad. Sci. U.S.A.* **89,** 5517–5521 (1992).

91. R. Hen, U. Boschert, F. Saudau, N. Amlaiky, J. L. Plassat, D. Ait Amara, J. Lanoir, and L. Segu, *Soc. Neurosci. Abstr.* **18,** 211 (1992).

92. N. Amlaiky, J. L. Plassat, S. Ramboz, U. Boschert, and R. Hen, *Soc. Neurosci. Abstr.* **18,** 212 (1992).

93. M. Foguet, H. Nguyen, H. Le, and H. Lübert *NeuroReport* **3,** 345 (1992).

94. M. Baez, D. Kursar, D. B. Wainscott, M. L. Cohen, and D. L. Nelson "2nd International Symposium on Serotonin" (abstr.) p. 50 (1992).

95. C. J. Grossman, G. J. Kilpatrick, K. T. Bunce, A. W. Oxford, J. D. Gale, J. F. Whitehead, and P. P. A. Humphrey "2nd International Symposium on Serotonin" (abstr.) p.31 (1992).

[16]   Methyllycaconitine: A New Probe That Discriminates between Nicotinic Acetylcholine Receptor Subclasses

S. Wonnacott, E. X. Albuquerque, and D. Bertrand

Introduction

Species of *Delphinium* (larkspurs) have long been recognized for their toxicity, both to cattle and to insects (1). Professor Michael Benn and colleagues at the University of Calgary isolated and identified methyllycaconitine (MLA) as the principal toxic agent in *Delphinium brownii* (1). MLA is the 2-methylsuccinylanthranilic acid ester of the norditerpenoid alkaloid lycoctonine (Fig. 1). Following the observation that poisoned cattle exhibited muscle paralysis, MLA was shown to reversibly block the rat phrenic nerve–diaphragm preparation, with an ED_{50} of 2.3 μM (1). In vertebrates, neuromuscular transmission is mediated by an α-bungarotoxin (αBgt)-sensitive nicotinic acetylcholine receptor (nAChR), whereas in insects αBgt-sensitive nAChR constitute the major excitatory receptor in the central nervous system (CNS). Binding assays carried out on fly head homogenates indicated that MLA inhibited [^{3}H]αBgt binding to this preparation with a K_i of $2.5 \times 10^{-10}M$ (2), that is, 4 orders of magnitude greater potency than the blockade of mammalian neuromuscular transmission. These disparate observations provided the first hint of the extraordinary discrimination of nAChR subtypes by MLA.

Binding Assays

Conventional competition binding assays, in which serial dilutions of MLA have been assessed for their ability to inhibit specific radioligand binding (Fig. 2a,b), have defined the potency and selectivity of MLA for neuronal αBgt binding sites of both vertebrate and invertebrate nervous tissues (3, 4) (Table I). Thus, αBgt binding sites in vertebrate brain and ganglia and in insect nervous systems are blocked by MLA concentrations in the nanomolar range. In striking contrast, αBgt-sensitive nAChR in vertebrate muscle and αBgt-insensitive nAChR on neurons (identified by high-affinity [^{3}H]nicotine binding in vertebrate brain and by ^{125}I-labeled neuronal bungarotoxin in autonomic neurons) require micromolar concentrations of MLA for inhibition

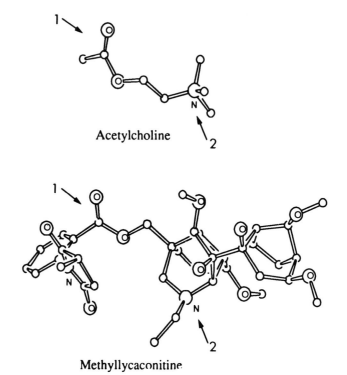

Acetylcholine

Methyllycaconitine

FIG. 1   Putative structures of ACh and MLA. The conformations of ACh and MLA were constructed using the X-ray structures of cytisine and aconitine. Hydrogen atoms are not shown, and carbon atoms are unlabeled. Arrows indicate the carbonyl function and nitrogen atom in ACh (and the corresponding positions in MLA) implicated in the pharmacophore model of nicotinic cholinergic ligand binding (22). (Reproduced from Ref. 4, with permission).

of radioligand binding. Similar assays have also demonstrated that MLA has no muscarinic activity (4).

To further explore the mechanism of interaction of MLA with αBgt-sensitive nAChR, saturation binding assays of ^{125}I-labeled αBgt binding to hippocampal membranes were carried out in the absence and presence of 10 nM MLA (5). MLA produced a shift in affinity for the radioligand (K_d), with no change in the total number of binding sites (B_{max}; Fig. 2c). This analysis is consistent with a competitive mode of inhibition.

MLA has also been utilized to characterize the binding of ^{125}I-labeled αBgt to oocytes expressing the candidate gene (α7) for the neuronal αBgt-sensitive

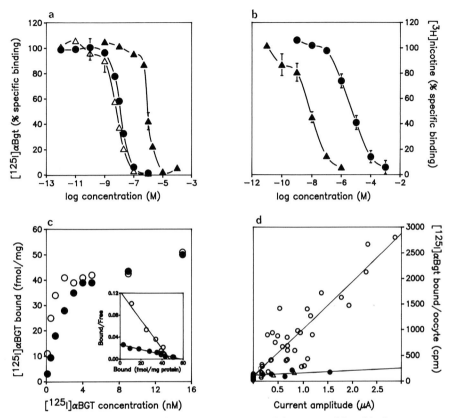

FIG. 2—Nicotinic potency of MLA demonstrated in competition binding assays. (a) Inhibition curves for MLA (●), α-cobratoxin (△), and (+)-anatoxin-a (▲) versus ^{125}I-labeled αBgt binding to rat hippocampal membranes. IC$_{50}$ values were 12 nM, 6.5 nM, and 0.96 μM, respectively. (Reproduced from Ref. 5, with permission). (b) Inhibition curves for MLA (●) and anatoxin-a (▲) versus [^{3}H]nicotine binding to rat forebrain membranes. IC$_{50}$ values were 7.5 μM and 7.0 nM, respectively. (c) Effect of MLA on the saturable binding of ^{125}I-labeled αBgt to hippocampal membranes. Saturation binding assays were carried out in the absence (○) and presence (●) of 10 nM MLA. Inset: Scatchard analysis of saturable binding, yielding the following kinetic constants: Kd 0.54 ± 0.18 nM (control), 1.89 ± 0.22 nM (+MLA) B$_{max}$ 51.1 ± 4.7 fmol/mg protein (control), 55.3 ± 1.2 fmol/mg protein (+MLA). (Reproduced from Ref. 5, with permission.) (d) Inhibition of ^{125}I-labeled αBgt binding to oocytes expressing the α7 nicotinic subunit. A positive correlation was found between the magnitude of acetylcholine-evoked current and amount of ^{125}I-labeled αBgt bound by individual oocytes previously injected with the α7 cDNA(○) (10). The binding of ^{125}I-labeled αBgt (10 nM) to oocytes was totally abolished by coincubation with 1 μM unlabeled αBgt (△) or 1 μM MLA (●).

TABLE I   Comparison of Binding Data for Methyllycaconitine in Different Preparations

Species	Tissue	Preparation	Radioligand	K_i (M) for MLA	Ref.
Vertebrates					
Torpedo	Electric organ	Purified nAChR	^{125}I-αBgt	$1.1 \pm 0.6 \times 10^{-6}$	3
Frog	Muscle	Detergent extract	^{125}I-αBgt	$1.0 \pm 0.2 \times 10^{-5}$	4
Human	Muscle	Detergent extract	^{125}I-αBgt	$7.8 \pm 2.0 \times 10^{-6}$	4
	Cell line (muscle)	TE671 cells	^{125}I-αBgt	$6.3 \pm 1.4 \times 10^{-5}$	4
Rat	Brain (forebrain)	P2 membranes	[^{3}H]Nicotine	$3.7 \pm 0.7 \times 10^{-6}$	3
	(striatum)		[^{3}H]Nicotine	4.4×10^{-6}	11
	(forebrain)		^{125}I-αBgt	$1.4 \pm 0.2 \times 10^{-9}$	4
	(hippocampus)		^{125}I-αBgt	4.3×10^{-9}	5
	(forebrain)		[^{3}H]QNB	$>10^{-4}$	4
Chick	Ciliary ganglion	Neurons	^{125}I-αBgt	$2.8 \pm 0.2 \times 10^{-9}$	14
			^{125}I-nBgt	$1.4 \pm 0.2 \times 10^{-7}$	14
Invertebrates					
Fly	Head	Homogenate	[^{3}H]αBgt	$2.5 \pm 0.5 \times 10^{-10}$	2
Locust	Supra-esophogeal ganglion	Membranes	^{125}I-αBgt	1.8×10^{-8}	3
Cockroach	Abdominal and thoracic ganglia	Nerve cord homogenate	^{125}I-αBgt	1.4×10^{-9}	15

nAChR (6) (Fig. 2d). Individual oocytes were shown to bind ^{125}I-labeled αBgt in proportion to the magnitude of the acetylcholine-evoked current exhibited by the oocyte, indicating that the number of ^{125}I-labeled αBgt binding sites is a reflection of the number of functional nAChR expressed. The binding of ^{125}I-labeled αBgt to oocytes was totally abolished by coincubation with 1 μM MLA (Fig. 2d), in accord with its potency for neuronal αBgt-sensitive nAChR.

Functional Assays

Earlier electrophysiological experiments (1) have suggested that, at the neuromuscular junction, MLA acts as a competitive inhibitor. However, it was not known if MLA would have any action on nerve to nerve transmission in the peripheral or central nervous system. To address this, MLA was studied in several preparations. The toxin has been found to block all the known neuronal nAChR on which it was tested, but the MLA concentration necessary for blockade differed widely between nAChR subtypes. As a good correlation between binding affinity and functional potency emerges, MLA can be exploited either as a pharmacological agent or as a diagnostic tool for distinguishing neuronal nAChR subtypes.

Nicotinic Acetylcholine Receptor Subtypes Expressed in Xenopus oocytes

Nicotinic acetylcholine receptors are encoded by a gene family that is, in turn, part of the larger superfamily of ligand-gated ion channels that includes receptors for γ-aminobutyric acid (GABA) and glycine (7). Numerous putative nAChR subunits have been identified by gene cloning, and, although the subunit composition of nAChR *in situ* is still uncertain, many of the isolated subunits have been found to reconstitute functional nAChR when expressed in pairwise combinations in *Xenopus* oocytes (Fig. 3). Moreover, the physiological and pharmacological properties of reconstituted nAChR resemble those of native receptors (8).

The predominant agonist-binding (α) subunit in the vertebrate brain in α4, which, in combination with the structural (β) subunit β2, appears to correspond to the nicotinic binding site labeled with high affinity by tritiated agonists (9). A different α subunit, α3, predominates in autonomic ganglia, where it may be associated with either β2 or β4 subunits, or both. Receptors reconstituted with these subunits are insensitive to αBgt. Recently, however, two genes coding for a neuronal αBgt-binding subunit, and designated α7 and α8, have been cloned from chicken brain. These subunits differ from the other known α subunits and, when expressed in *Xenopus* oocytes, can form homooligomeric receptors in absence of any structural subunit (10). However, it remains to be established if such homooligomeric channels occur *in situ*.

The effects of MLA have been characterized on *Xenopus* oocytes expressing the chick α4:β2, α3β2, and α7 nAChR subtypes (6, 11). In the initial protocol (Fig. 3, Protocol 1), oocytes were exposed to a static incubation with MLA (in order to conserve the toxin) and then transferred to the recording chamber where responses to acetylcholine (ACh) were recorded under perfusion. The α4β2 nAChR was not affected by a relatively high concentration of MLA (1 μM), whereas the α7 subtype was completely blocked. On continuous washing, α7 responsiveness was slowly recovered over a 15-min period. Thus MLA produces a long-lasting blockade of α7 nAChR. Attempts to determine the dose–response relationship for antagonism of α7 nAChR have been foiled by a variability in the responsiveness of different batches of oocytes, implicating a synergism between MLA and some other variable, presumably related to the different conditions of the oocytes. Thus, although a precise IC_{50} cannot yet be given for MLA on the reconstituted α7 nAChR, MLA always produces a potent, reversible blockade of α7 responses, such that these nAChR are at least as sensitive to MLA as the αBgt binding sites monitored in rat brain membranes (Table I).

To disclose MLA antagonism on α3β2 or α4β2, the earlier experimental protocol was modified and the compound applied directly in perfusion (Fig.

PROTOCOL 1

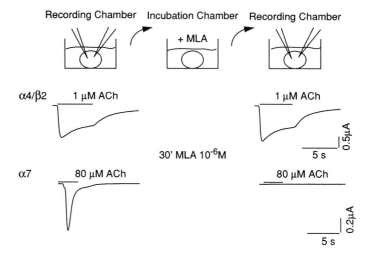

PROTOCOL 2: Perfusion

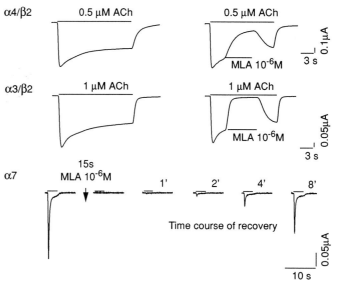

Fig. 3—Inhibition by MLA of three neuronal nAChR reconstituted in *Xenopus* oocytes. Protocol 1 was used to test the action of MLA on α4β2 and α7 receptors. ACh-evoked currents are first measured in the perfusion chamber, and the oocytes are then placed in a small volume (50–100 μl) of solution containing MLA. After 30

3, Protocol 2). Perfusion application of MLA revealed that these nAChR are reversibly blocked by MLA, but, in contrast to the case of $\alpha 7$, recovery of the responses to agonist is very fast upon removal of the toxin. Dose–response curves yielded IC_{50} values of 0.08 and 0.65 μM for $\alpha 3\beta 2$ and $\alpha 4\beta 2$ subunit combinations, respectively (11). Further examination of the effects of MLA on the $\alpha 4\beta 2$ subtype established that the toxin inhibits the receptor in a voltage-independent manner and that MLA is a competitive antagonist.

Functional Nicotinic Acetylcholine Receptors in Rat Brain

Presynaptic Receptors on Dopaminergic Striatal Nerve Terminals

The nicotinic modulation of neurotransmitter release via nAChR localized on presynaptic terminals appears to be a major role of nAChR in the CNS. Such nAChR can be studied in a perfused synaptosome preparation by monitoring the release of prelabeled transmitter in response to nicotinic drugs. The nicotine-evoked release of [³H]dopamine from striatal nerve terminals has been particularly well characterized. This presynaptic nAChR is insensitive to αBgt and has been tentatively correlated with the $\alpha 4\beta 2$ subtype (12). MLA (10 μM) was found to inhibit nicotine-evoked [³H]dopamine release by more than 80%, whereas 1 μM MLA had little or no effect (11). Thus, the sensitivity to blockade by MLA is comparable to its K_i for inhibition of [³H]nicotine binding to the same preparation (Table I). MLA had no effect on basal or K^+-evoked release.

αBungarotoxin-Sensitive Receptors on Primary Cultures of Rat Hippocampus

Electrophysiological recordings from cultured hippocampal neurons have disclosed ACh-evoked currents blocked by snake α-toxins (13). This is in contrast to the majority of neuronal nAChR which seem to be insensitive to

---

min of incubation the oocytes are tested again in the perfusion chamber. Typical recordings, obtained with this protocol on oocytes expressing $\alpha 4\beta 2$ and $\alpha 7$ receptors, are presented. Application of 1 μM MLA had no effect on the $\alpha 4\beta 2$ receptor but abolished the ACh-evoked current of the $\alpha 7$ receptor. Protocol 2 employed bath application of MLA to inhibit reconstituted receptors with $\alpha 4\beta 2$, $\alpha 3\beta 2$, and $\alpha 7$ subunits. A fast decrease of the ACh-evoked current is observed when MLA is applied simultaneously with ACh by perfusion. The $\alpha 4\beta 2$ and $\alpha 3\beta 2$ receptors display a fast recovery from MLA inhibition, whereas the $\alpha 7$ receptor recovers very slowly. Cells were clamped at -100 mV and challenged with ACh concentrations near the EC_{50}.

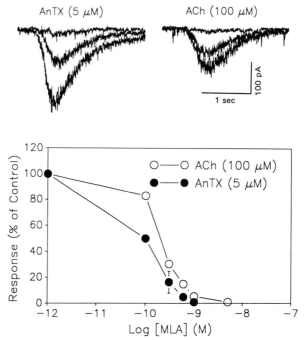

FIG. 4—Inhibition by MLA of nicotinic currents recorded from cultured fetal rat hippocampal neurons. Whole cell currents were evoked by (+)-anatoxin-a (AnTX, 5 μM) and acetylcholine (ACh, 100 μM) at −50 mV holding potential. Traces represent (from bottom to top) 0 (control), 0.1 and 0.6 nM MLA, respectively. The concentration-dependent inhibition by MLA of agonist-evoked whole cell currents from several experiments is illustrated at bottom. Neurons grown for 15–30 days in culture were used.

αBgt. Using this preparation of cultured neurons, a rather detailed assessment of MLA on this α7-like nAChR has been undertaken (5) (Figs. 4 and 5). Perfusion with 0.1–5 nM MLA elicited a dose dependent inhibition of inward currents evoked by ACh or by the nicotinic agonist anatoxin-a, with IC$_{50}$ values of 0.2 and 0.1 nM, respectively. The inhibition was reversible but, as observed with the α7 nAChR expressed in *Xenopus* oocytes (see above), the onset of blockade (about 2 min) was faster than the offset (8–15 min).

The antagonism of ACh-evoked whole cell currents by MLA was voltage independent, and MLA could protect against the pseudo-irreversible blockade produced by αBgt. These observations support a competitive mode of

AnTX (1 μM), control

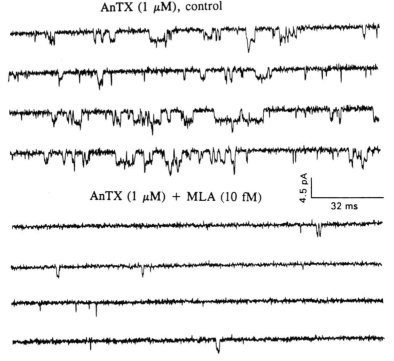

AnTX (1 μM) + MLA (10 fM)

4.5 pA

32 ms

FIG. 5—Inhibition by MLA of nicotinic single channel currents recorded from cultured fetal rat hippocampal neurons. Single channel currents were evoked by (+)-anatoxin-a (AnTX, 1 μM) from an outside-out patch at −120 mV holding potential. A 15-day-old neuron in culture was used. At bottom is shown the decrease in channel opening frequency produced by 10 fM MLA.

antagonism, arising from an interaction at or near the agonist binding site, which is implicit in the ligand binding data (Table I). The action of MLA was specific: no antagonism of whole cell currents evoked by N-methyl-D-aspartate (NMDA), quisqualate, kainate, or GABA was observed (5).

Single channel recordings from outside-out patches obtained from cultured hippocampal neurons revealed an exquisite sensitivity of nAChR to MLA (Fig. 5). The frequency of channel openings activated by 1 μM anatoxin-a was decreased by MLA concentrations as low as 1 fM and completely abolished by 1 pM MLA. MLA had no effect on mean channel open time of the predominant conductance state, which suggests that MLA inhibits nAChR in the closed conformation and again supports a competitive mode of action.

Autonomic Ganglia

Rat Superior Cervical Ganglion

Micromolar concentrations of MLA are required to produce a reversible blockade of nicotinic responses in rat superior cervical ganglion (11), where the $\alpha3$ agonist-binding subunit is held to be responsible for nicotinic transmission. This low sensitivity to MLA has been demonstrated by extracellular recording from isolated ganglia and by intracellular recording of whole cell currents from cultured neurons (dissociated from ganglia from fetal (15–19 day) rats) (11). For example, whole cell currents activated by 30 μM 1,1-dimethyl-4-phenylpiperazinium iodide (DMPP) were inhibited by 47 and 88% by 1 and 10 μM MLA, respectively. In this preparation, the antagonism by MLA was also voltage independent.

α-Bungarotoxin-sensitive Receptors in Chick Ciliary Ganglion

Autonomic ganglia possess membrane components that bind αBgt. In the chick ciliary ganglion, these sites appear to be localized outside the synapse. The neurons express the $\alpha7$ gene, and αBgt-sensitive sites on these neurons can be equated with the $\alpha7$ gene product (although the possibility of other nicotinic subunits contributing to the receptor protein has not been ruled out). Recently, αBgt-sensitive nAChR on chick ciliary ganglion neurons have been shown to function in raising intracellular free Ca^{2+} concentrations in these neurons (14). Using a fluorescence assay that involved loading the neurons with a Ca^{2+}-sensitive fluorescent dye, fluo-3, it was demonstrated that 0.1–1.0 μM nicotine increased the fluorescence signal in a manner blockable by d-tubocurarine, αBgt and MLA. The EC_{50} values for αBgt and MLA (1–2 nM) agreed with the K_1 values of 0.7 and 2.8 nM, respectively, for inhibition of ^{125}I-labeled αBgt binding to the same neurons. At the concentrations investigated, MLA was specific for this subtype of nAChR. Thus, MLA has contributed to the definitive allocation of a function to this enigmatic αBgt-sensitive nAChR in autonomic neurons.

Invertebrate Preparations

The effects of MLA have been examined in some detail (15) by electrophysiological recording from the cell body membrane of the fast coxal depressor motorneuron (Df), that can be identified visually in the desheathed metathoracic ganglion of the cockroach. Bath application of 1 μM MLA totally abolished the response to iontophoretically applied ACh; the IC_{50} for this antagonism was about 0.1 μM. The blockade was not readily reversible but was voltage independent.

The sensitivity of this functional response to MLA was 2 orders of magnitude lower than its ability to inhibit ^{125}I-labeled αBgt binding to the same tissue (Table I). Factors such as accessibility to nAChR in the *in situ* preparation are invoked as an explanation of this discrepancy (15). Using somewhat different techniques to record from the sixth abdominal ganglion of the cockroach, MLA produced a slow, concentration-dependent blockade of electrically applied stimuli (16). In this preparation, the lowest concentrations of αBgt and MLA to fully block synaptic responses within 1 hr were reported to be 10 and 100 nM, respectively.

In contrast, using patch clamp techniques to study ACh-induced currents sensitive to αBgt in cockroach embryonic cultures, MLA was exceedingly potent and inhibited whole cell currents with an IC$_{50}$ of 3×10^{-11} M (17). The inhibition could be partially reversed by washing for 20 minutes.

The only report to date of the effects of MLA in a noninsect invertebrate preparation is a preliminary examination of MLA on the muscle cells of the nematode *Ascaris*. These cells have a nAChR likened to the vertebrate ganglionic nAChR. Acetylcholine-evoked conductances were reversibly blocked by MLA, with an IC$_{50}$ of 0.23 μM (18). However, in this preparation the response to acetylcholine was also "weakly" inhibited by αBgt, so it is unclear how the nematode nAChR compares with other nAChR subtypes.

Discussion

The common finding in all of the functional studies documented here is that MLA inhibited nicotinic responses in every preparation examined, and, where characterized, its action was consistently found to be reversible and competitive. Thus MLA is a welcome addition to the nicotinic pharmacopoeia, where competitive antagonists are very few in number: the majority of nonpolypeptide nicotinic antagonists act noncompetitively at the level of the nicotinic channel. Competitive antagonists offer two advantages over channel-blocking drugs: they tend to be more specific (noncompetitive antagonists are often found to interact with other receptor channels) and also provide an excellent correlation between binding data (K_i) and functional potency (IC$_{50}$) for a given preparation. This is very evident for MLA, at least in the vertebrate studies.

It is also clear, however, that the potency of MLA varies among nAChR subtypes. αBgt-insensitive nAChR in brain and ganglia, notably $\alpha 3\beta 2$ and $\alpha 4\beta 2$ reconstituted nAChR, and muscle nAChR require micromolar concentrations of MLA for inhibition, making this toxin a useful but unremarkable antagonist. In contrast, the exquisite sensitivity to MLA of αBgt-sensitive neuronal nAChR, and of the putative corresponding gene product, $\alpha 7$, reconstituted in *Xenopus* oocytes, defines MLA as a unique probe for discriminating this enigmatic nAChR. In this respect, MLA promises to be particularly useful, since αBgt-sensitive nAChR in brain and ganglia have, until recently (5, 13, 14, 19), eluded functional demonstration. Hitherto, functional characterization of this receptor has relied on αBgt for its discrimination. The poor accessibility and slow onset of inhibition of this large polypeptide has constrained this research. MLA is as potent as αBgt yet avoids these concerns. In many experiments its reversibility will also be an advantage over αBgt, and MLA is also a more selective agent, having more than a 1000-fold preference for the neuronal over the muscle nAChR.

It is surprising to find that MLA has a similarly high potency toward insect αBgt-sensitive nAChR, shown convincingly in binding assays (Table I) and whole cell recordings (17). However, the considerable homology between the vertebrate $\alpha 7$ gene product (10) and a locust αBgt-sensitive nAChR subunit sequence (20) may underlie the common strong recognition of MLA. This creates another role for MLA as a pharmacological probe of evolutionary relationships between nAChRs.

MLA is a new ligand with considerable potential in nicotinic receptor research. Moreover, its small size affords the opportunity for chemical modification of its structure, to generate derivatives to elucidate the structural basis of its nicotinic potency. Such studies are already underway (21). Thus we can anticipate that MLA will make a major contribution to nAChR research, not least in unveiling the physiological role of αBgt-sensitive nAChR in the mammalian CNS.

Acknowledgements

We are grateful to the following colleagues for contributions to the studies described here: Alison Drasdo and Victor Cockcroft at the University of Bath; Malcolm Caulfield at University College, London; Manickavasagon Alkondon, Edna Pereira, and Mabel Zelle at the University of Maryland School of Medicine, Marc Ballivet at Sciences II, Geneva and Sonia Bertrand at CMU, Geneva. We are indebted to Professor Michael Benn for providing the MLA used in our studies.

References

1. V. Nambi-Aiyar, M. H. Benn, T. Hanna, J. Jacyno, S. H. Roth, and J. L. Wilkens, *Experientia* **35**, 1367 (1979).
2. K. R. Jennings, D. G. Brown, and D. P. Wright, *Experientia* **42**, 611 (1986).
3. D. R. E. MacAllan, G. G. Lunt, S. Wonnacott, K. L. Swanson, H. Rapoport, and E. X. Albuquerque, *FEBS Lett.* **226**, 357 (1988).

4. J. M. Ward, V. B. Cockcroft, G. G. Lunt, F. S. Smillie, and S. Wonnacott, *FEBS Lett.* **270**, 45 (1990).

5. M. Alkondon, E. F. R. Pereira, S. Wonnacott, and E. X. Albuquerque, *Mol. Pharmacol.* **41**, 802-808 (1992).

6. D. Bertrand, S. Bertrand, M. Ballivet, and S. Wonnacott, *Soc. Neurosci. Abstr.* **17**, 384.11 (1991).

7. V. B. Cockcroft, D. J. Osguthorpe, E. A. Barnard, and G. G. Lunt, *Mol. Neurobiol.* **4**, 129–169 (1990).

8. S. Couturier, L. Erkmans, S. Valera, D. Rungger, S. Bertrand, J. Boulter, M. Ballivet, and D. Bertrand, *J. Biol. Chem.* **265**, 17560 (1990).

9. P. Whiting, R. Schoepfer, J. Lindstrom, and T. Priestley, *Mol. Pharmacol.* **40**, 463 (1991).

10. S. Couturier, D. Bertrand, J.-M. Matter, M.-C. Hernandes, S. Bertrand, N. Millar, S. Valera, T. Barkas, and M. A. Ballivet, *Neuron* **5**, 847 (1990).

11. A. L. Drasdo, M. P. Caulfield, D. Bertrand, S. Bertrand, and S. Wonnacott, *Mol. Cell. Neurosci.* **3**, 237–243 (1992).

12. C. Rapier, G. G. Lunt, and S. Wonnacott, *J. Neurochem.* **54**, 937 (1990).

13. M. Alkondon and E. X. Albuquerque, *Eur. J. Pharmacol.* **191**, 505 (1990).

14. S. Vijayarghaven, P. C. Pugh, Z.-W. Zhang, M. M. Rathouz, and D. K. Berg, *Neuron* **8**, 353 (1992).

15. D. B. Sattelle, R. D. Pinnock, and S. C. R. Lummis, J. Exp. Biol. **142**, 215 (1989).

16. K. R. Jennings, D. G. Brown, D. P. Wright, and A. E. Chalmers, *in* "Sites of Action of Neurotoxic Pesticides" (R. M. Hollingworth and M. D. Green, eds.), p. 274. American Chemical Society, Washington, D. C., 1987.

17. H. Cheung, B. S. Clarke, and D. J. Beadle, *Pestic. Sci.* (in press).

18. L. Colquhoun, R. Gallagher, L. Holden-dye, and R. J. Walker, *Neurosci. Lett.* **S38**, S113 (1990).

19. M. Listerud, A. B. Brussaard, P. Devay, D. R. Colman, and L. W. Role, *Science* **254**, 1518 (1991).

20. J. Marshall, S. D. Buckingham, R. Shingai, G. G. Lunt, M. W. Goosey, M. G. Darlison, D. B. Satelle, and E. A. Barnard, *EMBO J.* **9**, 4391 (1990).

21. P. A. Coates, I. S. Blagbrough, B. V. L. Potter, M. G. Rowan, N. F. Thomas, T. Lewis, and S. Wonnacott, FIDIA Res. Found. Symp. Ser., p. 54 (1992).

22. R. P. Sheridan, R. Nilakantan, J. S. Dixon, and R. Venkataraghavan, *J. Med. Chem.* **29**, 899 (1986).

Section III

Receptor Localization

[17] Localization and Quantification of Brain Somatostatin Receptors by Light Microscopic Autoradiography

Jacques Epelbaum and Jérôme Bertherat

Introduction

Somatostatin is a tetradecapeptide (SRIF$_{14}$, somatotropin release inhibiting factor) originally isolated as a neurohormone from sheep hypothalamus (1) and subsequently demonstrated by immunohistochemistry to be widely distributed throughout the neuroaxis in most vertebrates (see Ref. 2 for review). Biochemical studies demonstrated that SRIF immuno-reactivity corresponded in part to an N-terminally extended peptide, SRIF$_{28}$ (3). Both SRIF$_{14}$ and SRIF$_{28}$ are alternately generated from a single gene in mammals (5). Except for a SRIF$_{28}$-selective neuronal system recently discovered from the nucleus of the solitary tract to magnocellular hypothalamic nuclei (6), all SRIF-immunoreactive pathways appear to contain both SRIF$_{14}$ and SRIF$_{28}$.

Characterization of Somatostatin Binding Sites

In accordance with its wide distribution, multiple biochemical and behavioral effects of SRIF have been reported, ranging from stimulatory or inhibitory actions on the release and metabolism of other neuropeptides and neurotransmitters to increases in learning and memory abilities (see Ref. 7 for review). SRIF being a hydrophobic molecule, its multiple physiological effects are likely to derive from its binding to an outer membrane receptor. Such receptors with nanomolar affinities for either of the two endogenous peptides have been characterized by conventional radioligand binding techniques on membrane or homogenate preparations (8, 9). By these techniques, the guanine nucleotide and ionic dependency of SRIF binding was evidenced (10) and the existence of two subtypes postulated on the basis of the differential displacement of radiolabeled SRIF ligands by synthetic analogs (11–13) or the differential effects of SRIF$_{14}$ and SRIF$_{28}$ on cerebral neurons in culture (14) or supraoptic neurons *in vivo* (15). The plurality of SRIF binding sites has recently been confirmed by the cloning and sequencing of cDNAs encoding two SRIF binding sites of 391 and 369 amino acids which are members of the seven-transmembrane domain superfamily of G protein-coupled receptors (16). Both molecules display equivalent affinities for $SRIF_{14}$ and $SRIF_{28}$, but the shortest one is more highly expressed in brain tissues.

Localization of Somatostatin Receptors

The distribution of SRIF binding sites in mammalian brain has been reported by film radioautography of cryostat-cut sections after incubation with various iodinated or tritiated ligands (17–25). There is a general regional match between the endogenous peptide concentration and the localization of the binding sites. Briefly, SRIF binding sites are highly concentrated in the deeper cortical layers (V and VI) throughout the rostrocaudal axis, most regions of the limbic system such as the hippocampus (molecular and granular layers of the dentate gyrus, all but the pyramidal layer of CA1), the amygdala (mainly basolateral), and the septum and medial habenula. In the hypothalamus, the first studies reported only moderate quantities of binding sites, but preincubations with desaturating concentrations of guanine nucleotides revealed an important proportion of SRIF binding sites previously occupied by the endogenous peptide (26). Regions involved in sensory functions are also enriched in SRIF binding sites, for example, the olfactory tubercles and anterior olfactory nuclei, inferior and superior colliculi, and the retina. Moderate levels are present in the caudate putamen and thalamic nuclei. Cerebellum is totally devoid of binding in adult rats, but binding sites are transiently expressed in the external granular cell layers in the first 3 weeks after birth (27).

Why Light Microscopic Autoradiography of Somatostatin Receptors?

While a mere film autoradiographic technique (''dry'' autoradiography) is quite adequate for revealing the distribution of a given binding site at the regional and subregional level, its limited power of resolution does not allow for cellular localization. Such a level of resolution can only be achieved by directly coating the radiolabeled sections with liquid nuclear emulsion, provided the bound radioactive ligand molecules are tightly cross-linked to their membrane binding sites with divalent aldehydes (28, 29). The purpose of this chapter is to describe the application of such ''wet'' autoradiographic techniques to the distribution and quantification of SRIF binding sites in brain using ^{125}I-labeled $[Try^0,DTrp_8]SRIF_{14}$. This radiolabeled agonist is a degradation-resistant ligand, previously shown to label with high sensitivity and specificity both subtypes of central SRIF binding sites (19, 30). Furthermore the presence of two lysine residues, in addition to the N-terminal free amino group, allows for secure cross-linking to tissue proteins with divalent aldehydes.

Material and Methods

Tissue Preparation

Rats are sacrificed by decapitation and their brains rapidly frozen by immersion in liquid isopentane at $-40°C$ for 20 sec. The brains are then deposited on dry ice for 10 min and stored at $-80°C$ in air-tight containers until used. With these conditions, we have kept brains for up to 10 months without any loss in binding. Sections (from 5 to 20 μm thick) are cut on a cryostat at $-18°C$ and thaw-mounted onto slides coated with 2% gelatin (Sigma, St. Louis, MO, G 2625, 20 g/liter; chrome alum 0.5 g/liter). To dissolve the gelatin, double-distilled water is added at 70°C before chrome alum is added. The hot gelatin solution is filtered and the slides are dipped for 3 sec and blotted on paper. Sections are then dried overnight at 50°C and kept at 4°C until use.

Iodination of [Tyr⁰,DTrp⁸]SRIF₁₄

Iodination of $[Tyr^0,DTrp^8]SRIF_{14}$

[Tyr⁰,DTrp⁸]SRIF₁₄ (Peninsula, Palo Alto, CA) is iodinated by the chloramine-T method. Five micrograms of peptide is incubated under agitation for 40–45 sec with 2 mCi Na¹²⁵I (IMS-30, Amersham International, Amersham, UK) and 4 μg chloramine-T in 70 μl of a 50 mM sodium phosphate buffer at pH 7.4. The reaction is stopped by addition of 12 μg of $Na_2S_2O_5$ and 100 μl of 10% bovine serum albumin (BSA) in the same buffer solution. Monoiodinated ¹²⁵I-labeled [Tyr⁰,DTrp⁸]SRIF₁₄ is purified on a 10 × 1 cm column of carboxylmethyl cellulose (CM-52, Whatman, Maidstone, Kent, UK), preequilibrated with 2 mM ammonium acetate (pH 4.6), by stepwise elution from 2 to 200 mM ammonium acetate at the same pH. The specific activity of the tracer, calculated from the elution profile, ranges between 750 and 800 Ci/mmol. The tracer is kept at $-80°C$ up to 3 weeks.

¹²⁵I-labeled [Tyr⁰,DTrp⁸]SRIF₁₄ Autoradiography

For receptor labeling, sections are brought to room temperature and preincubated for 15 min in 50 mM TrisHCl buffer, pH 7.6 containing 0.25 M sucrose

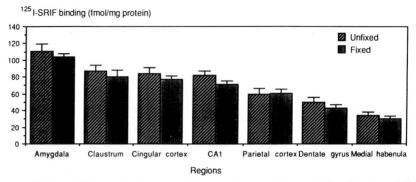

FIG. 1 Effect of histological processing on the retention and distribution of ^{125}I-labeled [Tyr^0,DTrp^8]SRIF$_{14}$. Two series of 20-μm-thick adjacent coronal sections from rat forebrain taken at the level of the basolateral amygdala (between parts C and D of Fig. 3) were incubated with ^{125}I-labeled [Tyr^0,DTrp^8]SRIF$_{14}$ (0.3 nM) and washed in parallel. They were then either air-dried (unfixed) or postfixed in 4% glutaraldehyde (fixed) as indicated in the text. Both series were autoradiographed by apposition to tritum-sensitive film (^{3}H-Ultrofilm, LKB, Uppsala, Sweden), exposed in light-proof cassettes for 2–5 days, developed for 4 min in Kodak LX-24, rinsed in water, and fixed with Kodak rapid fixer. The regional distribution of the label was measured by computer-assisted densitometry (Biocom RAG200 program). Data are expressed as fmol/mg protein and represent the means ± s.e.m. of seven to nine measurements in three animals. The mean percentage of retention in the seven regions tested is 91 ± 2. The coefficient of correlation between fixed and unfixed series is 0.991.

and 0.2% BSA. Sections are incubated for 45 min at room temperature in the same medium to which ^{125}I-labeled [Tyr^0,DTrp^8]SRIF$_{14}$ in various amounts, MgCl$_2$ (5 mM), and bacitracin (50 μM) are added. Sections adjacent to those used for total binding are incubated in the presence of 1 μM nonradioactive SRIF$_{14}$ to determine nonspecific binding. After incubation, the sections are rinsed in two consecutive ice-cold baths of supplemented Tris buffer (5 min/bath) and immediately fixed by immersion in 4% glutaraldehyde in 50 mM phosphate buffer, pH 7.5, for 30 min at 4°C. This procedure ensures irreversible cross-linking of more than 90% of bound ^{125}I-labeled [Tyr^0,DTrp^8]SRIF$_{14}$ molecules to tissue proteins independently of the region measured (Fig. 1). After fixation, sections are dehydrated in graded concentrations (70, 95, and 100%) of ethanol (2 times, 5 min each), defatted in xylene (2 times, 15 min), rehydrated (ethanol at 100, 95, and 70 then water, 2 times, 5 min each), and then dried at 37°C overnight. Sections are autoradiographed by dipping in Kodak (Rochester, NY) NTB2 or Ilford (Cheshire, UK) K5 emulsions diluted 1:1 with distilled water following the respective technical information sheets. After 4 to 6 weeks of exposure at 4°C, the autoradiographs are developed in Kodak Dektol diluted 1:2 with distilled water, fixed in Kodak Ektaflow, counterstained with cresyl violet, and coverslipped.

Image Analysis and Quantification

The sections are examined with a CCD video camera fixed on a photomicroscope and coupled to a computerized image analysis system using the RAG200 program (Biocom, les Ulis, France). After observation under bright-field illumination for cellular localization, dark-field illumination is used for grain quantification. Conversion of optical density to radioactivity units is performed by references to standards prepared from brain paste. Briefly, half a rat brain is homogenized with a Polytron in 1 ml of Tris buffer, and 100 μl of brain paste is mixed with 100 μl of buffer containing increasing amounts of ^{125}I-labeled [Tyr^0,DTrp^8]SRIF$_{14}$. Three microliters of each solution is deposited in duplicate in diamond-engraved circles (6 to 8 mm in diameter) on gelatin-coated slides. The standard slides are then dried, fixed, dehydrated, defatted, and rehydrated in parallel with experimental brain sections. A comparison of standard curves obtained from film and liquid emulsion autoradiograms is given in Fig. 2 (top), and the correlation between values obtained for triplicate measurements by liquid radioautography is also shown (Fig. 2, bottom). Although the total densitometric range of the standard curve is lower with the liquid emulsion technique, the relationship to the standard is maintained with both techniques, as is the reproducibility of the experimental measurements.

Results

Localization of ^{125}I-Labeled [Tyr^0,DTrp^8]SRIF$_{14}$ Binding Sites by Light Microscopic Autoradiography

At low magnification of liquid emulsion-processed sections (Fig. 3), the regional distribution of specifically ^{125}I-labeled [Tyr^0,DTrp^8]SRIF$_{14}$ binding sites is very similar to that previously reported using film autoradiography. In conformity with these previous reports, moderate to high densities of SRIF binding sites are detected in all zones of cerebral cortex, most parts of the limbic system (septum, hippocampus, bed nuclei of the stria terminalis and of the stria medullaris, habenula, interpeduncular nucleus, and amygdala), caudate putamen, several thalamic and hypothalamic areas, and selective brain stem nuclei. In addition, the higher resolution provided by the

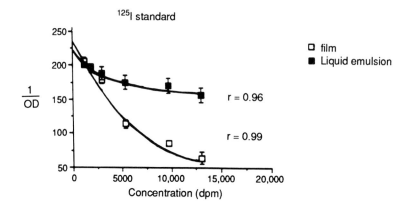

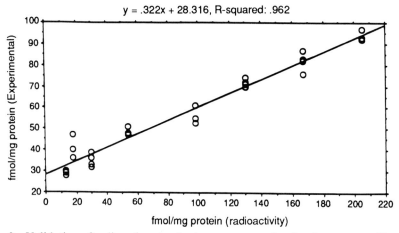

FIG. 2   Validation of radioactive standard measurement by densitometry on film and liquid emulsion autoradiography. The same standard sections, prepared as described in the text, were first apposed under tritium-sensitive films and then dipped in liquid emulsion. Radioactivity was evaluated by computer-assisted densitometry. Although the sensitivity of the standard curve is lower in liquid emulsion than in mere film autoradiography, the accuracy of the curve fitting is only moderately affected (top), and the reproducibility of the measurements, as evaluated in triplicate, is conserved (bottom).

present approach makes it possible to detect specific binding in several areas where it had not been hitherto reported, including the anterior olfactory nucleus, taenia tecta, septofimbrial nucleus, anterodorsal thalamic nucleus, Barrington's nucleus, and spinal trigeminal nucleus. It also allows for a finer assessment of intraregional patterns of specific SRIF labeling, namely, in the olfactory bulbs, the amygdaloid complex, the diencephalon, the colliculi, and the caudal brain stem.

In most areas, specifically labeled SRIF binding sites are rather uniformly distributed between neuronal perikarya and surrounding neuropil, making it difficult to distinguish somatodendritic from axonal labeling. The overall sparing of myelinated fiber bundles tends to suggest that only a small proportion of the labeled SRIF binding sites are on the membrane of, or in transit within, myelinated axons. There are a few examples, however, of apparent fiber labeling such as in the lateral geniculate body, the intramedullary trajectory of the glossopharingeal nerve, and the spinal trigeminal tract (31). Moreover, in several instances, the distribution of SRIF binding sites strikingly parallels that of the cell bodies of origin and terminal arborizations of specific projection systems. Thus, the pattern of SRIF binding in the interpeduncular nucleus is reminiscent of the terminal labeling observed after injection of an anterograde tracer in the medial habenula (32), while, in turn, the distribution of SRIF binding sites in the medial habenula is comparable to that of neurons retrogradely labeled after injection of horseradish peroxidase (33) in the interpeduncular nucleus.

In addition, SRIF binding densities are occassionally associated with a number of interconnected relay nuclei, forming apparent polysynaptic "somatostatinoceptive" pathways. The ascending olfactory system is thus heavily labeled with ^{125}I-labeled [Tyr^0,DTrp^8]SRIF$_{14}$ at the level of the external plexiform layer, anterior olfactory nucleus, pyriform and entorhinal cortices, nucleus of the lateral olfactory tract, and basolateral amygdaloid nucleus, with each of these nuclei projecting onto one another (34). Specific labeling densities are likewise observed within the optic tectum, vestibular and trigeminal sensory nuclei, inferior colliculus, and medial geniculate body, which all serve in the integration of opticoacoustical reflexes (35). Finally, key relays of the limbic system (i.e., septum, bed nucleus of stria terminalis, dentate gyrus, Ammon's horn of hippocampus, mammillary area) and interconnected centers responsible for central autonomic regulation (i.e., nucleus of the solitary tract and A1 catecholaminergic cell groups, subfornical organ, and hypothalamic supraoptic and paraventricular areas) all contain dense accumulations of SRIF binding sites.

In a few areas, including the periventricular and arcuate nucleus of the hypothalamus, the septofimbrial nucleus, and the caudal brain stem, SRIF binding is selectively concentrated over neuronal perikarya. Such labeled

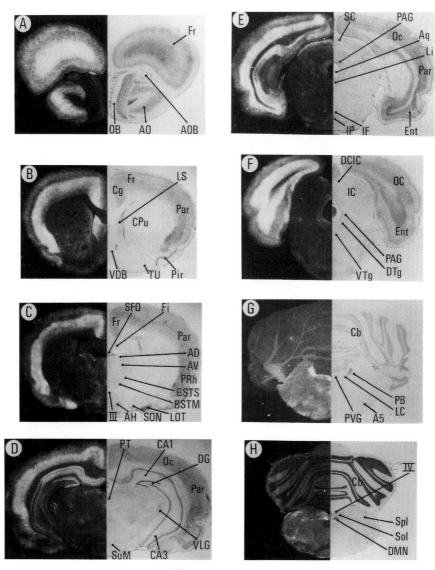

FIG. 3   Regional distribution of ^{125}I-labeled [Tyr^0,DTrp^8]SRIF$_{14}$ on coronal sections along the rostrocaudal axis of rat brain. Twenty-micron-thick sections are labeled from rostral (A) to caudal (H). Approximate corresponding levels in the atlas of Paxinos and Watson ("The Rat Brain" second edition, Academic Press, Sydney, 1986) are (A) A + 5.2, (B) A + 1.2, (C) A − 1.4, (D) A − 4.8, (E) A − 6.7, (F) A − 8.3, (G) A − 10.3, (H) A − 12.7. Dark-field (left) and bright-field (right) sections

cells presumably correspond to somatostatinoceptive neurons, that is, nerve cells that express SRIF receptors. The fact that these cells stand out much more clearly against the surrounding neuropil in these than in other regions could reflect a higher density of membrane-bound receptors and/or of intracytoplasmic receptor synthesis, or it may merely be due to weaker labeling of the neighboring parenchyma. Within the arcuate nucleus, the distribution of ^{125}I-labeled SRIF cells is strikingly reminiscent of that of growth hormone releasing hormone (GHRH)-immunoreactive perikarya previously reported in this area (36). This observation prompted us to compare quantitatively the distribution of ^{125}I-labeled SRIF and GHRH-immunoreactive cells in the rat arcuate nucleus and to conclude that there is a likely association of SRIF binding sites with GHRH neurons in this species (37). In fact up to 55–60% of GHRH-containing neurons visualized either by immunohistochemistry (38) or *in situ* hybridization (39) do carry SRIF receptors. In the brain stem noradrenergic cell groups, numerous ^{125}I-labeled SRIF cells correspond to tyrosine hydroxylase-immunoreactive neurons identified in adjacent sections (31). This overlap is most striking in the locus coeruleus, in which the entire catecholaminergic cell grouping is in perfect register with ^{125}I-tagged SRIF labeling, but this is also observed in the A5 and A1 cell groups of the ventral brain stem, where individual ^{125}I-labeled SRIF cells are also identified as tyrosine hydroxylase positive on adjacent sections (31).

---

are compared. Nonspecific binding was homogeneous throughout and accounted for 15–20% of total binding (not shown). Magnifications: ×1.8 (A) and ×3 (B–H). Abbreviation are as follows: **A5**, groups of noradrenaline cells; **AD**, anterodorsal nucleus; **AH**, anterior hypothalamic area; **AO**, anterior olfactory nucleus; **AOB**, accessory olfactory bulb; **Aq**, aqueduct of Sylvius; **AV**, anteroventral thalamic nucleus; **BSTM**, bed nucleus of stria terminalis, medial division; **BSTS**, bed nucleus of stria terminals, supracapsular division; **CA1 and CA3**, fields of Ammon's horn of the hippocampus; **Cb**, cerebellum; **Cg**, cingulate cortex; **CPu**, caudate putamen; **DCIC**, dorsal complex of inferior colliculus; **DG**, dentate gyrus; **DMN**, dorsal motor nucleus of vagus; **DTg**, dorsal tegmental nucleus; **Ent**, entorhinal cortex; **Fi**, fimbria; **Fr**, frontal cortex; **IC**, inferior colliculus; **IF**, interfascicular nucleus; **IP**, interpeduncular nucleus; **LC**, locus coeruleus; **Li**, linear Raphe nuclei; **LOT**, nucleus of the lateral olfactory tract; **LS**, lateral septum; **OB**, olfactory bulb; **Oc and OC**, occipital cortex; **PAG**, periaqueductal gray; **Par**, parietal cortex; **PB**, parabrachial nucleus; **Pir**, pyriform cortex; **PRh**, perihinal cortex; **PVG**, periventricular gray; **PT**, paratenial thalamic nucleus; **SC**, superior colliculus; **SFO**, subfornical organ; **Sol**, nucleus of the solitary tract; **SON**, supraoptic hypothalamic nucleus; **Spl**, spinal trigeminal nucleus, interpolar part; **SuM**, supramammillary complex of the hypothalamus; **TU**, tuberculum olfactorium; **VDB**, diagonal band of Broca, vertical limb; **VLG**, ventral lateral geniculate body of the thalamus; **VTg**, ventral tegmental nucleus. **III and IV**, third and fourth cerebral ventricles.

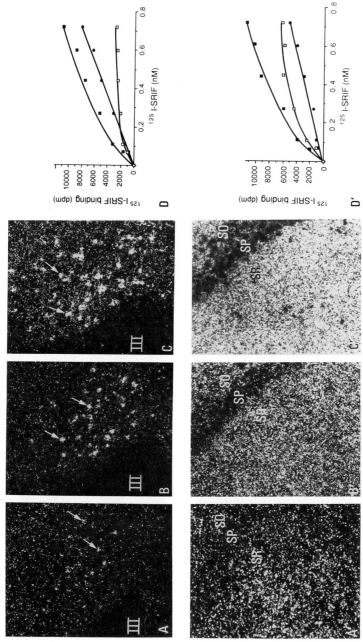

FIG. 4 Kinetics at equilibrium of ^{125}I-labeled [Tyr^0,DTrp^8]SRIF$_{14}$ binding on cell bodies in the arcuate nucleus of the hypothalamus and on neurophil in the CA1 of the hippocampus. Sections were incubated for 45 min at 25°C with increasing amounts of ^{125}I-labeled [Tyr^0,DTrp^8]SRIF$_{14}$. Visualization of the sections under dark-field illumination (magnifications: ×12.5) at the level of the arcuate is shown in A, B, and C for 0.07, 0.27, and 0.60 nM of tracer, respectively. Arrows point to representative highly labeled arcuate perikarya; III, third ventricle. Quantification of the data are shown in D, namely, total binding (■), nonspecific binding (●), and specific binding (□). Data are the means of six determinations. The same sections are visualized at the level of CA1 in A', B', and C', with quantification of the presented data in D'. SR, Stratum radiatum; SP, stratum pyramidal; SO, stratum oriens. (Reprinted with permission from Ref. 39.)

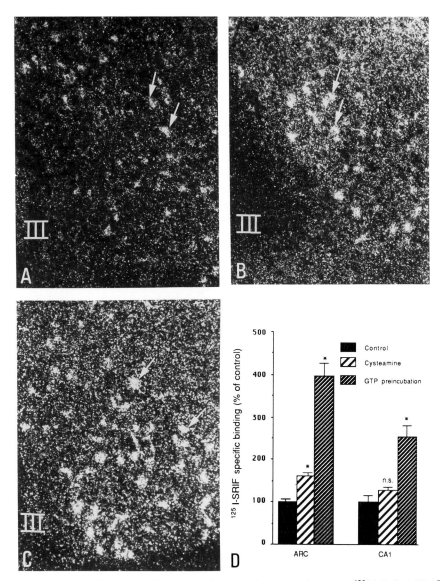

FIG. 5 Effect of *in vivo* and *in vitro* desaturation procedures on ^{125}I-labeled [Tyr^0 DTrp^8]SRIF$_{14}$ binding in the arcuate nucleus of the hypothalamus and the CA1 of the hippocampus. For *in vitro* endogenous SRIF desaturation, sections were preincubated at room temperature for 30 min in complete binding buffer supplemented with GTP (0.1 mM). They were then rinsed twice in the same buffer in absence of GTP. For *in vivo* desaturation, a single dose of cysteamine (300 mg/kg body weight) was injected subcutaneously 4 hr prior to decapitation. This results in a 56% decrease in brain SRIF concentrations. GTP-preincubated and cysteamine-treated sections were

Quantification of ^{125}I-Labeled [Tyr^0, DTrp^8]SRIF$_{14}$ Binding Sites by Light Microscopic Autoradiography

The resolution attained with light microscopic autoradiography allows one to define the kinetics of ^{125}I-labeled [Tyr^0,DTrp^8]SRIF$_{14}$ binding on both neuropil and perikarya in stratum radiatum of the CA1 of the hippocampus and arcuate nucleus, respectively (39). In both regions, ^{125}I-labeled [Tyr^0,DTrp^8]SRIF$_{14}$ binding is time and temperature dependent, saturable and specific, and reversible (Fig. 4). Saturation experiments indicate comparable affinities in the CA1 and the arcuate (0.2 and 0.6 nM, respectively). In both regions, displacement curves obtained with SRIF$_{14}$ and SRIF$_{28}$ are monophasic but shallow, whereas a somatostatin agonist, octreotide (Sandoz, Basel, Switzerland), induces biphasic displacement curves, thus suggesting that high (SSA) and low (SSB) affinity binding sites are present on both nerve terminals and perikarya. In both regions binding is GTP dependent, thereby confirming that the coupling of SRIF receptors to guanine nucleotide transducing proteins is still functional on brain sections.

Desaturation procedures (*in vivo* by cysteamine pretreatment and *in vitro* by GTP preincubation) result in an apparent increase in the number of measurable binding sites (Fig. 5). The increase is more marked in the arcuate nucleus than in CA1, indicating that the rate of occupancy of the binding sites by endogenous SRIF peptides is higher in arcuate nucleus than in the CA1. This suggests that SRIF could tonically inhibit GHRH metabolism, and we have recently found that SRIF depletion induced by cysteamine pretreatment results in a 2-fold increase in GHRH mRNA in the arcuate nucleus (40).

Conclusion

Optimizing the resolution of autoradiographic labeling of ^{125}I-labeled [Tyr^0,DTrp^8]SRIF$_{14}$ binding sites permits cytoarchitectonic accuracy in their

---

then incubated as indicated in the text. Visualization of representative sections at the level of the arcuate is provided in A–C under dark-field illumination (magnifications: ×12.5). (A) Control conditions; (B) cysteamine pretreatment; (C) GTP preincubation. Arrows point to representative perikarya; III, third, ventricle. (D) Histograms indicate the percentage increase in ^{125}I-labeled [Tyr^0,DTrp^8]SRIF$_{14}$ binding obtained with the two desaturation techniques in both regions. $p < 0.05$ vs respective controls. (Reprinted with permission from Ref. 39.)

localization and greatly improves the anatomical knowledge of the somato-statinoceptive brain elements. In addition, quantification of the data allows for more functional studies, including the assessment of the regional differences in endogenous SRIF occupancy of the receptors under various physiological conditions.

Acknowledgments

We would like to acknowledge the cooperation of Drs. Emmanuel Moyse and Alain Beaudet, which was of major importance in the localization studies, as well as the constant and skillful technical assistance of Ms. Catherine Videau and the photographic work of Gérard Delrue and Joël Grelier. These studies were supported by INSERM.

References

1. P. Brazeau, W. Vale, R. Burgus, N. Ling, M. Butcher, J. Rivier, and R. Guillemin, *Science* **179,** 77 (1973).
2. O. Johansson, T. Hökfelt, and R. E. Elde, *Neuroscience (Oxford)* **13,** 265 (1984).
3. L. Pradayrol, H. Jorvall, V. Mutt, and A. Ribet, *FEBS Lett.* **109,** 55 (1980).
4. R. H. Goodman, D. C. Aron, and B. A. Roos, *J. Biol. Chem.* **258,** 5570 (1983).
5. L. S. Shen and W. S. Rutter, *Science* **274,** 168 (1984).
6. P. E. Sawchenko, C. Arias, and J. Bettencourt, *J. Comp. Neurol.* **291,** 269 (1990).
7. J. Epelbaum, *Basic Clin. Aspects Neurosci.* **17,** 28 (1992).
8. J. Epelbaum, L. T. Arancibia, C. Kordon, and A. Enjalbert, *J. Neurochem.* **38,** 1515 (1982).
9. C. B. Srikant and Y. C. Patel, *Proc. Natl. Acad. Sci. U.S.A.* **78,** 3930 (1981).
10. A. Enjalbert, R. Rasolonjanahary, E. Moyse, C. Kordon, and J. Epelbaum, *Endocrinology (Baltimore)* **113,** 822 (1983).
11. V. T. Tran, M. F. Beal, and J. B. Martin, *Science* **228,** 492 (1985).
12. J. C. Reubi, *Neurosci. Lett.* **49,** 259 (1984).
13. K. Raynor and T. Reisine, *J. Pharmacol. Exp. Ther.* **251,** 510 (1989).
14. H. T. Wang, C. Bogen, T. Reisine, and M. Dichter, *Proc. Natl. Acad. Sci. U.S.A.* **86,** 9616 (1989).
15. W. N. Raby, C. W. Bourque, R. Benoit, and L. P. Renaud, *Ann. Endocrinol.* **50,** 46N (1989).
16. Y. Yamada, S. R. Post, H. W. Tager, G. I. Bell, and S. Seino, *Proc. Natl. Acad. Sci. U.S.A.* **89,** 251 (1992).
17. J. Epelbaum, A Enjalbert, M. Dussaillant, C. Kordon, and W. Rostene, *Peptides* **11,** 21 (1985).
18. K. Gulya, J. K. Wamsley, D. Gehlert, J. T. Pelton, S. P. Duckles, V. J. Hruby, and H. I. Yamamura, *J. Pharmacol. Exp. Ther.* **235,** 254 (1985).

19. S. Krantic, J. C. Martel, D. Weissman, and R. Quirion, *Brain Res.* **498,** 267 (1989).
20. Ph. Leroux, R. Quirion, and G. Pelletier, *Brain Res.* **347,** 74 (1985).
21. R. MacCarthy and L. M. Plunkett, *Brain Res. Bull.* **18,** 29 (1987).
22. J. C. Reubi and R. Maurer, *Neuroscience (Oxford)* **15,** 1183 (1985).
23. G. R. Uhl, V. Tran, S. H. Snyder, and J. B. Martin, *J. Comp. Neurol.* **240,** 288 (1985).
24. C. A. Whitford, J. M. Candy, C. R. Snell, B. H. Hirsty, A. E. Oakley, M. Johnson, and J. E. Thompson, *Eur. J. Pharmacol.* **138,** 327 (1987).
25. A. M. Rosier, Ph. Leroux, H. Vaudry, G. A. Orban, and F. Vandesande, *J. Comp. Neurol.* **310,** 189 (1991).
26. Ph. Leroux, B. Gonzales, A. Laquerriere, M. C. Bodenant, and H. Vaudry, *Neuroendocrinology* **567,** 533 (1988).
27. B. Gonzales, Ph. Leroux, A. Laquerriere, D. H. Coy, M. C. Bodenant, and H. Vaudry, *Dev. Brain Res.* **40,** 154 (1988).
28. E. Hamel and A. Beaudet, *Nature (London)* **312,** 155 (1984).
29. M. Herkenham and S. McLean, *in* "Quantitative Receptor Autoradiography" (C. A. Boast *et al.,* eds.), p. 137. Alan R. Liss, New York, 1986.
30. E. Moyse, A. Slama, C. Videau, P. D'Angela, C. Kordon, and J. Epelbaum, *Regul. Pept.* **26,** 225 (1989).
31. E. Moyse, A. Beaudet, J. Bertherat, and J. Epelbaum, *J. Chem. Neuro-anat.* **4,** in press.
32. M. Herkenham and W. J. Nauta, *J. Comp. Neurol.* **187,** 19 (1979).
33. E. R. Marchand, J. N. Riley, and R. V. Moore, *Brain Res.* **193,** 339 (1978).
34. J. De Olmos, G. F. Alheid, and C. A. Beltramino, *in* "The Rat Nervous System" (G. Paxinos, ed.), Vol. 1. p. 223. Academic Press, Sydney, 1985.
35. R. W. Rhoades, S. E. Fish, N. L. Chiaia, C. Bennett-Clarke, and R. D. Mooney, *J. Comp. Neurol.* **289,** 641 (1989).
36. B. Bloch, P. Brazeau, N. Ling, P. Bohlen, F. Esch, W. B. Wehrenberg, R. Benoit, F. Bloom, and R. Guillemin, *Nature (London)* 607 (1983).
37. J. Epelbaum, E. Moyse, G. S. Tannenbaum, C. Kordon, and A. Beaudet, *J. Neuroendocrinol.* **1,** 109 (1989).
38. J. Bertherat, P. Dournaud, A. Berod, E. Normand, B. Bloch, W. Rostene, C. Kordon, and J. Epelbaum, *Neuroendocrinology* in press (1992).
39. J. Bertherat, A. Slama, C. Kordon, C. Videau, and J. Epelbaum, *Neuroscience (Oxford)* **41,** 571 (1991).
40. J. Bertherat, A. Berod, E. Normand, B. Bloch, W. Rostene, C. Kordon, and J. Epelbaum, *J. Neuroendocrinol.* **3,** 115 (1991).

[18] Neurotensin Receptors in the Central Nervous System: Anatomical Localization by *in Vitro* Autoradiography and *in Situ* Hybridization Histochemistry

F. Javier Garcia-Ladona, Guadalupe Mengod, and José M. Palacios

Introduction

Neurotensin (NT) is a 13-amino acid neuropeptide first isolated from bovine hypothalamus (Fig. 1). It has been localized in several brain regions using radioimmunoassay and immunohistochemical techniques (1–4). NT in brain acts as a classic neurotransmitter and is involved in many brain functions (3, 5). In addition a wealth of experimental data suggest that NT, as other neuropeptides, may also act as a neurotrophic factor in the central nervous system (CNS) (6, 7). NT binds to specific membrane receptors (NTR) in brain that have been extensively characterized in membrane particulate fractions (8–11). Two classes of NT binding sites with different affinities have been reported; the high-affinity NT binding site is displaced by NT, and the low-affinity component is especially sensitive to levocastine (an antihistaminic drug) (11, 12). NT acting on its NTR triggers cell signaling pathways (13–16). A rat gene coding for NTR has been cloned and sequenced. The predicted receptor protein has a molecular weight of 47,000 and belongs to the G-protein-coupled receptor family (17).

In this context, the ability to determine the precise anatomical localization of NTR in the brain will undoubtedly help in understanding the role of NT in normal brain function (18), brain diseases (19, 20), as well as in plasticity phenomena (i. e., development or chronic treatments (7, 21–23). In this chapter, we present and discuss protocols for ligand binding autoradiography and *in situ* hybridization histochemistry which have been found useful for localizing and characterizing NTR in brain.

Methods in Neurosciences, Volume 12

| pGlu | Leu | Tyr | Glu | Asn | Lys | Pro | Arg | Arg | Pro | Tyr | Ile | Leu |

FIG. 1   Amino acid sequence of neurotensin.

Light and Electron Microscopic Localization of Neurotensin Binding Sites by *in Vitro* Autoradiography

The use of *in vitro* autoradiography to visualize NT was first reported by Young and Kuhar in 1981 (24). These results showed a rather restricted localization of NTR in rat brain. A high density of NT binding sites is present in the substantia nigra pars compacta and ventral tegmental area (Fig. 2A) (7, 24, 25). High levels of NT binding are also seen in the cingulate cortex, olfactory bulb, olfactory tubercle, islands of Calleja, vertical limb of diagonal band, amygdala, pre- and prosubiculum, ventral part of dentate gyrus, anterior dorsal thalamic nucleus, suprachiasmatic nucleus, and zona incerta (24–27). In some areas, NT binding sites correspond to NTR localized in the presynaptic terminals; this has been documented for these sites in caudate–putamen, olfactory tubercle, and nucleus accumbens (25, 28, 29).

So far only peptide derivatives, partial sequences of NT, and xenopsin (a neuropeptide isolated from *Xenopus*) (10, 17) have been described as agonists acting at NTR. These molecules have been used as pharmacological tools to characterize NTR by *in vitro* autoradiography. The pharmacological char-

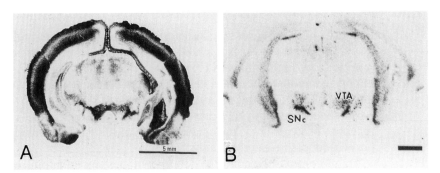

FIG. 2   Photographs from autoradiograms illustrating the localization of NTR in rat brain (**A**) [³H]Neurotensin binding sites in rat substantia nigra and ventral tegmental area of a 15-day-old animal. (**B**) *In situ* hybridization histochemistry of NTR mRNA in rat mesencephalon of the same age. The sections were incubated as indicated in the respective protocols described in the text. VTA, Ventral tegmental area; SNc, substantia nigra pars compacta. Bar in B = 2 mm.

TABLE I Autoradiographic Localization of Neurotensin Binding Sites in the CNS: Summary of Different Assay Conditions[a]

Radioligand	Buffer	Inhibitor	Time (min)	Temperature (°C)	Ref.[b]
0.1 nM ^{125}I-NT, 1 μM NT	50 mM Tris-HCl, pH 7.4	20 μM bacitracin, 500 μM ORPH, 100 μM PMSF	60	4	a
4 nM [^{3}H]NT, 1 μM NT	50 mM Tris-HCl, pH 7.4	20 μM bacitracin	60	4	b
4nM [^{3}H]NT, 5 μM NT	170 mM Tris-HCl, pH 7.6	100 μM bacitracin	60	4	c
0.1 nM ^{125}I-NT, 0.5 μM NT	50 mM Tris-HCl, pH 7.4	50 μM bacitracin	60	4	d
0.1 nM ^{125}I-NT, 1 μM NT	50 mM Tris-HCl, pH 7.4	50 μM bacitracin	60	4	e
5 nM [^{3}H]NT, 1 μM NT	170 mM Tris-HCl, pH 7.6	20 μM bacitracin	30	Room temperature	d
4 nM ^{125}I-NT, 5 μM NT	100 mM Tris-HCl, pH 7.6	100 μM bacitracin	60	4	f
0.1 nM ^{125}I-NT, 0.5 μM NT	100 mM Tris-HCl, pH 7.6	20 μM bacitracin	60	4	f

[a] Abbreviations: NT, neurotensin; ORPH, o-phenanthroline; PMSF, phenylmethylsulfonyl fluoride.

[b] References: (a) P. Mailleux, D. Pelaprat, and J. J. Vanderhaeghen, Brain Res. 508, 345 (1990); (b) R. Quirion, P. Gaudreau, S. St. Pierre, F. Rioux, C. B. Pert, Peptides 3, 757 (1982); (c) W. S. Young and M. J. Kuhar, Brain Res. 206, 273 (1981); (d) E. Moyse, W. Rostene, M. Vial, K. Leonard, J. Mazella, P. Kitabgi, J. P. Vincent, and A. Beaudet, Neuroscience (Oxford) 22, 525 (1987); (e) E. Szigethy, R. Quirion, and A. Beaudet, J. Comp. Neurol. 297, 487 (1990); (f) J. M. Palacios, A. Pazos, M. M. Dietl, M. Schlumpf, and W. Lichtensteiger, Neuroscience (Oxford) 25, 307 (1988).

acteristics of NT binding sites detected using this technique agree, in general, with previous data obtained using membrane preparations. NT binds in tissue sections to a single population of high-affinity binding sites (K_d 9 nM), and, among different peptides containing a segment of the NT sequence, NT(8–13) seems to be the most potent in displacing NT binding (25).

The autoradiograph localization of NT binding sites in brain has been achieved by using either [^{3}H]neurotensin or ^{125}I-labeled [Tyr^3] neurotensin (see Table 1). The iodination of NT has the advantage of producing a ligand of high specific activity (~2000 Ci/mmol) with low quenching effects as compared with the tritiated ligand (30). This allows for the use of low concentrations of radioligand in the incubations, which helps avoid autoradiographic artifacts. Labeled peptide is used at a concentration between 0.1 and 5 nM. Nonspecific binding is determined by incubating some sections in the presence of a high concentration (1000-fold that of the radiolabeled ligand) of unlabeled NT (see Table I). The incubation buffers should contain protease inhibitors in order to prevent the degradation of labeled and unlabeled neuro-peptides; the most commonly used is the antibiotic bacitracin at concentrations ranging between 20 and 100 μM (see Table I).

Light Microscopic Localization of Neurotensin Receptors

The following protocols have been used in our laboratory to localize high-affinity NT binding sites in brain tissue (Fig. 3) (7). The methods have also been used to detect NTR changes in human normal and pathological brain tissues (19, 20) as well as in primates (31), guinea pigs (32), and mice (33). Similar protocols have been applied to cat brain (34).

Preparation of Tissue

Rats are sacrificed by decapitation, and brains are rapidly dissected, frozen in powdered dry ice, and stored at −80°C until use. Cryostat sections (10 μm) are thaw-mounted in gelatin-precoated slides. Mounted sections are allowed to equilibrate at 4°C before being subjected to binding incubations. Some authors have also used tissues fixed by perfusion with mild fixatives prior to freezing. The latter method does not induce significant modifications of binding constants and may improve the general histological quality, especially if light microscopic resolution is required (see below).

Labeling of Neurotensin Receptors in Tissue Sections

Tissue sections are incubated for 60 min at 4°C with [³H]NT (63.4 Ci/mmol, available from New England Nuclear, Boston, MA, or Amersham, Arlington Heights, IL), in 100 mM Tris-HCl, pH 7.6, containing 0.05% bovine serum albumin (BSA) and 100 μM bacitracin. Consecutive sections are incubated under the same conditions but in the presence of 5 μM NT in order to quantify nonspecific binding. After incubation, the slides are washed 4 times for 2 min in 50 mM Tris-HCl, pH 7.6, at 4°C.

Autoradiogram Generation

After washing, the slides are dried with cold air. Labeled tissues can be kept (in a refrigerator and under low humidity conditions) until exposed to emulsion.

FIG. 3   Localization of [³H]neurotensin binding in the developing rat brain cortex. On the left side layers of the cortex are visualized by cresyl violet staining. The right side shows consecutive sections of the neocortex as labeled by [³H]NT. M, Molecular cell layer; CP, cortical plate; CC, corpus callosum; GD, gestational day; PN, postnatal. (From Ref. 7.)

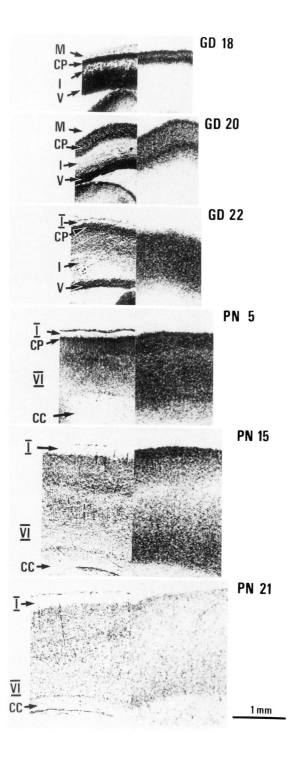

GD 18
M
CP
I
V

GD 20
M
CP
I
V

GD 22
I
CP
I
V

PN 5
I
CP
VI
CC

PN 15
I
VI
CC

PN 21
I
VI
CC

1 mm

Dry Emulsion: Light Microscopic Autoradiography

Dry tissue sections are exposed to ^{3}H-sensitive films together with appropriate radioactive standards, either commercial polymers or homemade radioactive brain mash. After 1 month of exposure, films can be developed (usually 1 week for ^{125}I-labeled NT).

Wet Emulsion Autoradiography

In some studies a higher anatomical resolution is desirable. In such situations, the film autoradiographic images cannot provide enough anatomical information, and the use of wet nuclear emulsions is necessary.

The slides containing the labeled brain sections are coated with photographic emulsions [i.e., Ilford (Cheshire, UK) K5 or Kodak (Rochester, NY) NTB3]. This process may induce the dissociation of radioactive NT; therefore, an additional cross-linking step is necessary in order to irreversibly link radioactive NT bound to its receptor, as was also demonstrated for other neuropeptides (35). The slides are immersed in aldehyde solutions made up in phosphate buffer; incubation in 4% glutaraldehyde in 50 mM phosphate buffer for 30 min at 4°C has been successfully used and retains around 70% radioactivity (25).

Electron Microscopic Localization of Neurotensin Receptors

Beaudet and collaborators have developed a technique which allows the visualization at the electron microscopic level of several neuropeptide receptors. NTR has been investigated in detail. Here, we summarize the main steps involved in electron microscopic visualization of NTR according to Dana *et al.* (36).

Tissue Preparation

Rat brains are perfused with and an ice-cold mixture of 1% tannic acid, 0.75% paraformaldehyde, and 0.1% glutaraldehyde in 0.1 M phosphate buffer. They are rapidly removed, placed on ice, and sliced on a vibrating microtome (Vibratome).

Ligand Binding

Tissue slices (70–75 μm) are immediately incubated with 0.1 nM monoiodo ^{125}I-labeled [Tyr^3]NT (2000 Ci/mmol) in 50 mM ice-cold Tris-HCl, pH 7.4, 5 mM MgCl$_2$, 0.2% BSA, 20 μM bacitracin, 0.25 mM sucrose for 1 hr at 4°C. For nonspecific binding, some sections are incubated with the above buffer but in the presence of nonradioactive NT. The sections are then washed 4 times for 2 min each in ice-cold buffer, fixed in 4% glutaraldehyde in 50 mM phosphate buffer for 30 min at 40°C, and postfixed for 1 hr at room temperature in 2% OsO$_4$, dehydrated in ethanols, and flat-embedded in Epon between two plastic coverslips.

Autoradiography

Epon-embedded slices are trimmed, reembedded in Beem capsules and cut on a ultramicrotome. Serial sections (80 nm) are deposited on parlodion-coated slides, stained with uranyl acetate and lead citrate, coated with carbon, and dipped into Ilford L4 emulsion. After the appropriate exposure time, sections are developed and examined with an electron microscope.

Combination of Receptor Autoradiography with Histochemical or Immunohistochemical Procedures

Once the anatomical localization of NTR is known, the next logical procedure is the determination with specific histochemical markers of the transmitters used by labeled cells. The combination of immunohistochemical techniques with autoradiography has allowed the chemical characterization of neurons containing NTR in brain sections (37–39). This double labeling can be done in the same section or in consecutive sections. The inconvenience of the first method is that most of the time the postfixation procedures necessary for the successful visualization of either the histological marker or receptor binding are found to be incompatible.

Beaudet and co-workers (37) applied the double-labeling methodology to chemically identify neurotensinoceptive neurons in the midbrain ventral tegmentum, where a large proportion of cells are known to be dopaminergic. They combined ^{125}I-labeled NT receptor "wet" autoradiography with immunohistochemistry employing tyrosine hydroxylase (TH), an enzyme specific for catecholamine synthesis and a reliable marker for dopamine-containing cells, and on serial, adjacent, 5–10 μm frozen sections of the substantia nigra and ventral tegmental area. They found that 95–100% of TH-positive neurons in the substantia nigra and 80–90% of TH-positive neurons in the ventral tegmental area contained NT receptors.

The same group compared the distribution of ^{125}I-labeled NT binding sites with that of acetylcholinesterase (AChE) reactivity in rat (38) and human (39) basal forebrain. A similar experimental approach was used. Consecutive 10 μm sections were processed for either autoradiography as described or histochemistry with AChE, a selective marker for cholinergic neurons. They concluded that the majority of ^{125}I-labeled NT cells in the rat basal forebrain were cholinergic. In humans, owing to technical limitations, it is impossible to draw such quantitative conclusions, yet there were regions such as the diagonal band and nucleus basalis of Meynert in which ^{125}I-labeled NT cells could be clearly identified as AChE-positive in adjacent sections.

Localization of Neurotensin Receptors in Cultured Nerve Cells

The localization and characterization of NT binding sites have also been performed in cultured nerve cells. The neuroblastoma cell line N1E-115 and primary cultures from neocortex and mesencephalon have been extensively used (14, 40–42).

The autoradiography of NT binding sites using photographic nuclear emulsions revealed the presence of a homogeneous population of NTR in primary neuronal cell cultures. Sites are present in the cell soma as well as in long processes and neurites (43). These NT binding sites have the pharmacological characteristics of high-affinity NTR observed in tissue, being insensitive to levocastine. ^{125}I-labeled NT was the radioligand of choice in the studies in which microscopic localization of NTR was performed. Culture medium is removed and replaced with the incubation buffer containing the radioligand and protease inhibitors. The incubation conditions are similar to those used in tissue and include a cross-linking step with 3.5% glutaraldehyde in phosphate buffer. Petri dishes or coverslips containing attached cells, labeled with radioactive NT, are dipped in Ilford K5 or Kodak NTB3 nuclear emulsions and exposed in the dark.

Localization of Neurotensin Receptors in Human Meningiomas

Autoradiographic techniques have been used to analyze NTR in tissues obtained postmortem from patients dying from neurodegenerative diseases and control cases (19, 20). NTR appear to survive postmortem changes well. These receptors have been also visualized and characterized in other pathological tissues such as human meningiomas (44).

Receptor binding ligand autoradiography was performed in 12 meningiomas obtained at the time of surgery and kept frozen until used. Twenty-micron sections were cut in a cryostat. Slide-mounted sections were labeled with [^{3}H]NT (80.7 Ci/mmol) for receptor binding autoradiography as described above. Ten of the meningiomas showed specific [^{3}H]NT binding, with a pharmacological specificity for [^{3}H]NT similar to neurotensin receptors. In all 12 meningiomas, the presence of [^{3}H]NT binding was closely related to tumor histology. [^{3}H]NT binding was preferentially located in syncytial areas, whereas no binding could be observed in fibrous areas. All meningothelial and transitional meningiomas showed specific binding, whereas only 2 of the 4 fibroblastic meningiomas presented measurable NT binding sites, suggesting a correlation between the presence of NTR and syncytial areas.

Localization of Neurotensin Receptor mRNA in the Central Nervous System

The NTR gene has been cloned and sequenced (17). The use of *in situ* hybridization histochemistry on brain sections allows one not only to support previous data on the autoradiographic localization of NTR, but also to follow possible changes in the levels of expression of receptor. We have used this technique to localize NTR (45) (Fig. 4) as well as other neurotransmitter and neuropeptide receptor mRNAs in brain (46, 47). Similar procedures (see below) have been used by Elde *et al.* (48) to visualize NTR mRNA.

Preparation of Tissue

We routinely use tissue prepared under the same conditions as for receptor autoradiography, in order to use both techniques for comparative studies. Dissected brains are rapidly frozen in powdered dry ice and cut (10 μm) in a microtome cryostat ($-20°C$). Precautions to avoid RNase contamination are taken throughout all manipulations (i.e., use of gloves and sterilized glass material). Mounted sections are brought to room temperature. Tissue sections are treated as previously described (47). Briefly, they are fixed for 10 min in 4% paraformaldehyde in phosphate-buffered saline (PBS: 2.6 mM KCl, 1.4 mM $KH_2 PO_4$, 136 mM NaCl, 8 mM Na_2HPO_4). After 3 washes in PBS, the sections are treated with predigested pronase (24 units/ml). The reaction is stopped by immersion in glycine–PBS buffer and PBS, then the slides are dehydrated in a graded series of ethanol and dried.

Synthesis and Labeling of Oligonucleotide Probes

We have used two different oligonucleotides complementary to bases 1100–1147 and to bases 1350–1397 of the rat NTR gene sequence (17). They are synthesized on an Applied Biosystems (Foster City, CA) Model 380B DNA synthesizer. Following a purification step on a 20% acrylamide/8 M urea preparative sequencing gel, the probes (2 pmol/μl) are labeled (1–4 $\times$ 10^4 Ci/mmol) at their 3' ends by incubation with terminal deoxynucleotidyl-transferase (Boehringer, Mannheim, Germany) and [α-^{32}P]dATP (3000 Ci/

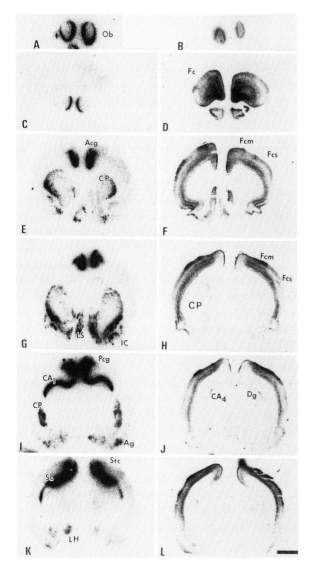

FIG. 4   Autoradiographic visualization of mRNA coding for neurotensin (left column) and neurotensin receptor (right column) in consecutive sections of a 10-day-old rat. Images are photographs from autoradiograms generated by hybridizing tissue sections with ^{32}P-labeled oligonucleotide probes complementary to selected regions of the neurotensin and neurotensin receptor genes. Dark areas are rich in hybridization signals. Acg, Anterior cingulate cortex; Ag, amygdaloid nuclei; CA$_1$, CA1 field of Ammon's horn; CA$_4$, CA4 field of Ammon's horn; CP, caudate–putamen; Dg, dentate gyrus; Fc, frontal cortex; Fcm, frontal cortex motor area; Fcs, frontal cortex sensory area; IC, inferior colliculus; LH, lateral hypothalamic area; LS, lateral septum; Ob, olfactory bulb; Pcg, posterior cingulate cortex; Sb, subiculum; Stc, striate cortex.

mmol) (Du Pont–NEN, Boston, MA). The radioactive probes are purified by chromatography through NACS PREPAC columns (BRL, Bethesda, MD).

In Situ Hybridization Histochemistry

Hybridization is carried out by incubating the sections with labeled probes diluted to a final concentration of $1–3 \times 10^7$ cpm/ml in the following buffer: 600 mM NaCl, 10 mM Tris-HCl, pH 7.5, 1 mM EDTA, 1× Denhardt's solution (0.02% Ficoll, 0.02% polyvinylpyrrolidone, 0.02% BSA), 500 μg/ml yeast tRNA, and 50% formamide. The sections are covered with Nescofilm and placed in a humid chamber at 42°C overnight. Sections are washed 4 times for 1 hr in 600 mM NaCl, 20 mM Tris-HCl, pH 7.5, 1 mM EDTA at 60°C, dehydrated in 95% ethanol/0.3 M ammonium acetate, and dried.

The slides are apposed to X-Omat film (Kodak) overnight at −80°C in order to check the intensity of the labeling. Then some slides are dipped in Kodak NTB3 photographic emulsion for 4 weeks and the rest apposed to βmax films (Amersham) for 15 days at 80°C.

With method we have, for example, localized the presence of NTR mRNA synthesizing cells in several brain regions during development (45) (Fig. 4). Figure 4 also illustrates the possibility of visualizing transmitter and receptor mRNAs in consecutive tissue sections.

Elde *et al.* (48) used a mixture of four 48-mer oligonucleotides complementary to different regions of the rat NTR mRNA as hybridization probes to visualize cells containing this messenger. The mixture of oligonucleotides is labeled at the 3′ end with [α^{35}S]dATP and terminal deoxynucleotidyltransferase and purified with NENsorb 20 columns. Thawed sections are directly hybridized for 15–20 hr with labeled probes at 42°C, rinsed in 1× SSC (1 × SSC: 15 mM sodium citrate, 150 mM NaCl) at 55°C for 1 hr, dehydrated, and apposed to Hyperfilm-βmax (Amersham) for 6–12 days. In general, the results obtained with both protocols are comparable.

Combination of in Situ Hybridization and Ligand Binding

The combination of *in situ* hybridization with receptor autoradiography is favored by the compatibility of tissue manipulation requirements for both techniques. As starting material, frozen brain sections can be used in both cases, thus allowing the visualization of mRNA and receptor proteins in consecutive sections of the same animal. Furthermore, both techniques can produce the same kind of visual support for data (films or emulsion-coated slides). *In situ* hybridization permits the visualization of the perikarya which synthesize receptor mRNAs, whereas receptor autoradiography reveals the final localization of the receptor proteins themselves. When combining both techniques, this fact can result in (1) an overlap of the signal patterns when receptors found in a given brain nucleus are synthesized by cells intrinsic to this nucleus or (2) a complementary rather than overlapping signal pattern if receptors visualized in one nucleus are synthesized by afferents to this nucleus whose perikarya are located in distant brain areas.

References

1. D. Kahn, G. M. Abrams, E. A. Zimmerman, R. Carraway, and S. E. Leeman, *Endocrinology* (*Baltimore*) **107,** 47 (1980).
2. L. Jennes, W. E. Stumpf, and P. W. Kalivas, *J. Comp. Neurol.* **210,** 211 (1982).
3. P. W. Kalivas, L. Jennes, C. B. Nemeroff, and A. J. Prange, Jr., *J. Comp. Neurol.* **210,** 225 (1982).
4. S. Inagaki, K. Shinoda, Y. Kubota, S. Shiosaka, T. Matsuzaki, and M. Tohyama, *Neuroscience* (*Oxford*) **8,** 487 (1983).
5. P. W. Kalivas, S. K. Burgess, C. B. Nemeroff, and A. J. Prange, Jr., *Neuroscience* (*Oxford*) **8,** 495, (1983).
6. M. R. Hanley, *Nature* (*London*) **315,** 14 (1985).
7. J. M. Palacios, A. Pazos, M. M. Dietl, M. Schlumpf, and W. Lichtensteiger, *Neuroscience* (*Oxford*) **25,** 307 (1988).
8. G. R. Uhl and S. H. Snyder, *Eur. J. Pharmacol.* **41,** 89 (1977).
9. M. Goedert, K. Pittaway, B. J. Williams, and P. C. Emson, *Brain Res.* **304,** 71 (1984).
10. J. Mazella, C. Poustis, C. Labbe, F. Checler, P. Kitabgi, C. Granier, J. Van Rietschoten, and J. P. Vincent, *J. Biol. Chem.* **258,** 3476 (1983).
11. J. Mazella, J. Chabry, P. Kitabgi, and J. P. Vincent, *J. Biol. Chem.* **263,** 144 (1988).
12. P. Kitabgi, W. Rostene, M. Dussaollant, A. Schotte, P. M. Laduron, and J. P. Vincent, *Eur. J. Pharmacol.* **140,** 285 (1987).
13. M. Goedert, R. D. Pinnock, C. P. Downes, P. W. Mantyh, and P. C. Emson, *Brain Res.* **323,** 193 (1984).
14. J. A. Gilbert and E. Richelson, *Eur. J. Pharmacol.* **99,** 245 (1984).
15. J. C. Bozou, S. Amar, J. P. Vincent, and P. Kitabgi, *Mol. Pharmacol.* **29,** 489 (1986).
16. M. Sato, S. Shiosaka, and M. Tohyama, *Dev. Brain Res.* **58,** 97 (1991).
17. K. Tanaka, M. Masu, and S. Nakanishi, *Neuron* **4,** 847 (1990).
18. G. R. Uhl, *in* "Handbook of Chemical Neuroanatomy" (A. Bjorklund, T. Hökfelt, and M. J. Kuhar, eds.), Vol. 9, p. 443, Elsevier, Amsterdam 1990.
19. G. Chinaglia, A. Probst, and J. M. Palacios, *Neuroscience* (*Oxford*) **39,** 351 (1990).
20. J. M. Palacios, G. Chinaglia, M. Rigo, J. Ulrich, and A. Probst, *Synapse* **7,** 114 (1991).
21. B. Levant, G. Bissette, E. Widerlov, and C. B. Nemeroff, *Regul. Pept.* **32,** 193 (1991).
22. C. B. Nemeroff, *Biol. Psychiatry* **15,** 283 (1980).

23. G. R. Uhl and M. Kuhar, *Nature (London)* **309,** 350 (1984).
24. W. S. Young and M. J. Kuhar, *Brain Res.* **206,** 273 (1981).
25. E. Moyse, W. Rostene, M. Vial, K. Leonard, J. Mazella, P. Kitabgi, J. P. Vincent, and A. Beaudet, *Neuroscience (Oxford)* **22,** 525 (1987).
26. C. Kohler, A. C. Radesater, H. Hall, and B. Winblad, *Neuroscience (Oxford)* **16,** 577 (1985).
27. R. Quirion, P. Gaudreau, S. St. Pierre, F. Rioux, and C. B. Pert, *Peptides* **3,** 757 (1982).
28. J. M. Palacios and M. J. Kuhar, *Nature (London)* **294,** 587 (1981).
29. R. Quirion, C. C. Chiueh, H. D. Everist, and A. Pert, *Brain Res.* **327,** 385 (1985).
30. M. J. Kuhar and J. R. Unnerstall, *Trends Neurosci.* **8,** 49 (1985).
31. R. Quirion, S. Welner, S. Gauthier, and P. Bedard, *Synapse* **1,** 559 (1987).
32. C. Waeber and J. M. Palacios, *Brain Res.* **528,** 207 (1990).
33. J. M. Palacios, F. J. Garcia-Ladona, L. Triarhou, and G. Mengod, to be published.
34. M. Goedert, S. P. Hunt, P. W. Mantyh, and P. C. Emson, *Dev. Brain Res.* **20,** 127 (1985).
35. A. Beaudet and E. Szigethy, *in* "Brain Imaging" (N. A. Sharif and M. Levis, eds.), p. 186. Horwood, Chichester, 1989.
36. C. Dana, M. Vial, K. Leonard, A. Beauregard, P. Kitabgi, J. P. Vincent, W. Rosténe, and A. Beaudet, *J. Neurosci.* **9,** 2247 (1989).
37. E. Szigethy and A. Beaudet, *J. Comp. Neurol.* **279,** 128 (1989).
38. E. Szigethy and A. Beaudet, *Neurosci. Lett.* **83,** 47 (1987).
39. E. Szigethy, R. Quirion, and A. Beaudet, *J. Comp. Neurol.* **297,** 487 (1990).
40. B. Cusack, T. Stanton, and E. Richelson, *Eur. J. Pharmacol.* **206,** 339 (1991).
41. F. Checler, J. Mazella, P. Kitabgi, and J. P. Vincent, *J. Neurochem.* **47,** 1742 (1986).
42. J. Chabry, F. Checler, J. P. Vincent, and J. Mazella, *J. Neurosci.* **10,** 3916 (1990).
43. C. Dana, D. Pelaprat, M. Vial, A. Brouard, A. M. Lhiaubet, and W. Rostène, *Dev. Brain Res.* **61,** 259 (1991).
44. S. Przedborski, M. Levivier, and J. L. Cadet, *Ann. Neurol.* **30,** 650 (1991).
45. F. J. Garcia-Ladona, G. Mengod, and J. M. Palacios, submitted for publication.
46. M. T. Vilaró, M. I. Martinez-Mir, M. Sarasa, M. Pompeiano, J. M. Palacios, and G. Mengod, *Curr. Aspects Neurosci.* **3,** 1 (1991).
47. J. M. Palacios, G. Mengod, M. Sarasa, M. T. Vilaró, M. Pompeiano, and M. I. Martinez-Mir *J. Rec. Res.* **11,** 459 (1991).
48. R. Elde, M. Schalling, S. Ceccatelli, S. Nakanishi, and T. Hökfelt. *Neurosci. Lett.* **120,** 134 (1990).

[19] Combined *in Situ* Hybridization and Immunocytochemical Studies of Neurotransmitter Receptor RNA in Cultured Cells

Sherry Bursztajn, Xing Su, and Stephen A. Berman

Introduction

Molecular cloning of the genes which code for neurotransmitter receptors or their subunits, together with the production of monoclonal antibodies specific for certain receptors or other cellular epitopes, has added a new dimension to our ability to probe synaptic biology. The nicotinic acetylcholine receptors (nAChRs) are one of the first neurotransmitter receptors to be isolated, characterized, cloned, and sequenced. In the central and peripheral nervous systems both nicotinic and muscarinic cholinergic receptors are expressed by neurons, in culture and *in vivo,* and both receptors respond to acetylcholine (ACh) released from nerve terminals (1, 6, 12). Our studies have focused primarily on the regulation of nAChRs which are expressed in high concentrations at the neuromuscular junctions and in nerve–muscle cocultures. The nicotinic receptors are expressed in skeletal muscle and cholinergic neurons, and both are ligand-gated ion channels, exhibiting multiple subtypes which differ in subunit composition and other biochemical and electrophysiological properties.

The accumulation of nicotinic acetylcholine receptors serves as a hallmark of the cholinergic neurotransmission at the neuromuscular junction. These receptors are integral membrane proteins that traverse the bimolecular leaflet and bind the neurotransmitter acetylcholine. Biochemical studies have shown that the nAChRs are composed of five subunits with a stoichiometry of $\alpha_2\beta\gamma\sigma$ forming an ion channel (20, 26, 33). The structural and functional properties of nAChRs have been well characterized during stages of muscle development and in the adult neuromuscular synapse. Experiments using recombinant DNA and patch clamp techniques suggest that bovine fetal muscle expresses transient nAChR subtypes composed of the $\alpha, \beta, \gamma,$ and σ subunits, and in adult neuromuscular junction stable forms composed of $\alpha, \beta, \sigma,$ and ε subunits appear (27). Hybridization analysis of rat muscle RNA with nAChR

Methods in Neurosciences, Volume 12

subunit-specific cDNA probes shows that a switch between γ and ε subunits takes place during postnatal development of the neuromuscular junction (17, 41). The amount of mRNA of the nAChR α subunit expressed appears to be under the control of cholinergic motor neurons. Sensory neurons do not have the capacity to regulate the nAChR message levels (9). In adult rat skeletal muscle the motor neuron regulates the expression of two functionally different nAChR subtypes by differential expression of subunit-specific mRNAs (41).

In contrast to the muscle nAChRs, the neuronal nAChRs are less well characterized. Neuronal nAChRs, like muscle nAChRs, are ligand-gated ion channels, containing multiple subtypes, whose functional properties are beginning to be better understood. Nine different genes are known to encode the neuronal nAChRs. Six of the genes, coding for $\alpha2$, $\alpha3$, $\alpha4$, $\alpha7$, $\beta2$, and $\beta4$, are known to generate functional receptors (13, 24). The function of the genes that code for $\alpha5$, $\alpha8$, and $\beta3$ has not yet been elucidated (5, 36). The proteins encoded by these genes have homologous extracellular, transmembrane, and cytoplasmic domains classified as α or β (or non-α) subunits. Expression studies in *Xenopus* oocytes have shown that each of the subunits ($\alpha2$, $\alpha3$, and $\alpha4$) forms functional receptors with either $\beta2$ or $\beta4$ (4, 13, 40). However the $\alpha7$ subunit forms an nAChR gated ion channel in the absence of the other subunits (13). Although there is extensive homology between the α nAChR subunits of muscle with the α neuronal nAChRs as well as between the β subunits of muscle with the β subunits of neuronal nAChRs, the arrangements of the subunits to produce a functional receptor differ.

Hydrophobicity profiles deduced from primary structural data suggest that the brain nicotinic receptors fold through the membrane in a manner identical to the *Torpedo* electric organ nAChR, that is, each subunit spans the membrane four times, placing both the N and C termini outside the cell (11, 25). Electrophysiological analysis of nAChRs ($\alpha2\beta2$, $\alpha3\beta2$, $\alpha4\beta2$) has demonstrated that each subtype has unique biophysical properties that can be activated by acetylcholine and nicotine and is not blocked by α-bungarotoxin (32). Though it is known that the $\alpha4\beta2$ subtypes are blocked by neuronal α-bungarotoxin, but $\alpha2\beta2$ and $\alpha3\beta4$ are not, the detailed pharmacology of these receptors has yet to be elucidated (14–16, 40).

The availability of nucleotide sequences which code for specific gene products and procedures for labeling of RNA and DNA clones should permit *in situ* hybridization studies which will lead to a better understanding of where in the brain nAChRs are distributed, what population of cells express the mRNA coding for each of the subunits, and how each of the subunits is regulated. Combining *in situ* hybridization with immunocytochemistry provides a powerful tool for simultaneous detection of both the message for a particular protein and the translation product of that message.

Basic Principle of *in Situ* Hybridization

In Situ *Hybridization*

The major goal of *in situ* hybridization is to visualize specific messenger RNA (mRNA) in precise group of cells on a single cell level. A specific mRNA codes for one protein through a unique sequence of nucleic acids. Cells contain many mRNA species, some coding for common structural proteins which are expressed in great abundance and others coding for specific proteins such as receptors or neurotransmitters specific for a cell type. To visualize a specific mRNA species, labeled complementary nucleic acid (either DNA or RNA) is used. Under specific conditions (described in the text; see also Fig. 1) two DNA strands (or one DNA and one RNA, or two RNA strands) can hybridize, that is, become double-stranded. Each strand forms double-stranded structures stabilized by hydrogen bonding (Fig. 1). Complementary base matching over regions of sufficient length can give rise to highly specific labeling analogous to antibody identification of specific protein or peptide epitopes. In our studies we have used intron probes to visualize pre-mRNA inside the nucleus and exon probes to visualize mature mRNA (Fig. 1). In both cases we have used the primer extension

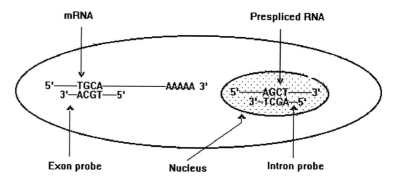

FIG. 1 Schematic diagram of *in situ* hybridization. Cultured cells are fixed (see text) before being hybridized to a probe. The probe can be single-stranded DNA or RNA which has been labeled by a low-energy isotope such as ^{35}S or ^{3}H. The probe is designed so that it hybridizes to the prespliced RNA in the nuclei (intron probe) or both mature mRNA in the cytoplasm and prespliced RNA in the nuclei. The probe forms stable base-paired hybrids with the target RNA in the cells, and the unhybridized probe is washed from the cells after hybridization. After exposing the cells to photographic emulsion (see text for details), the regions having a high density of silver grains (above background) will indicate that the cells underneath the grains expressed the gene of interest.

method to make the probes with ^{35}S- or ^{3}H-labeled deoxynucleotides (see below).

The labeling reaction is similar to a sequencing reaction: a primer is annealed to a DNA template, and a DNA polymerase extends the primer by incorporating nucleotides. The reaction is normally divided into two steps. First the primer is extended in the presence of one or two radioactive nucleotides, and radiolabeled short DNA strands are made. The DNA strands are then further extended by adding cold nucleotides at increased concentrations in order for the synthesizing DNA strands to pass through a desired restriction enzyme site. The DNA molecule is then linearized by the enzyme, and the labeled DNA probe is isolated after denaturing and separating the DNA strands by gel electrophoresis. Because all the commercially available DNA polymerases have either strand-displacement activity or $3' \rightarrow 5'$ exonuclease activity, the labeled DNA strands can be either displaced from their templates or digested by the polymerase if the polymerization is allowed to go too far. This is a common problem associated with failure of labeling. It is thus critical not to use excess nucleotides in a labeling reaction.

Although single-stranded DNA template is the best choice, double-stranded plasmid DNA can also be used. The plasmid DNA should have an insert of 200–500 nucleotides which is part of a gene of interest. If single-stranded DNA is required, the insert DNA should be the sense strand. The primer should anneal to the 3' end of the insert and allow the synthesis of the antisense strand.

Detection of Low-Abundance Sequences

The nAChR mRNAs in muscle cells are expressed in low abundance compared to mRNAs for many other proteins such as actin or myosin. The described *in situ* hybridization techniques allow us not only to detect nAChR sequences within cells, but also to determine both intracellular domains where these sequences are concentrated (8) and how they are compartmentalized within nuclei (2). It is generally accepted that membrane proteins and proteins for export are synthesized on membrane-bound ribosomes, whereas cytoskeletal and cytoplasmic proteins are synthesized on free polyribosomes (31). Studies on synaptogenesis in the cerebellar cortex have shown a close association between polyribosomes and postsynaptic membrane (29, 30), suggesting that components of the postsynaptic membrane may be synthesized close to the site of expression. To substantiate this hypothesis, the use of *in situ* hybridization in combination with immunocytochemistry can provide a powerful tool. The described probe labeling techniques and the combined *in situ* hybridization and immunocytochemistry procedures allow us to answer important questions regarding mRNA and protein processing on subcellular levels using a mild fixation approach (21, 37). Combining a high specific activity labeling method based on primer extension (8) with immunofluorescent/dark-field microscopy, we have achieved a new *in situ*/immunocytochemical method that is extremely effective for detection of messages expressed in low abundance and the detection of neurofilament protein in nerve/muscle coculture.

Immunocytochemistry and *in Situ* Hybridization

A combined immunocytochemical localization of genes coding for a particular message and its protein product is particularly useful in studies of synapse formation. Receptors for acetylcholine are present at high concentration at the neuromuscular junction (34). In the central nervous system (CNS), for instance, the glycine receptors are presumably concentrated at synaptic sites (38). Thus the synthesis of message for a particular molecule, its transport out of the nucleus, and its translation into a protein product destined for specific expression on the cell surface require mechanisms which determine not only how the molecules are synthesized but also how they are delivered to the different cellular domains. Furthermore, *in situ* hybridization in combination with immunocytochemistry can detect changes in both quantity and location of mRNA and protein at a single-cell level. This approach should be particularly valuable in the CNS where myriads of neuronal interconnections create networks by which important information is transmitted. To identify neuronal processes while, at the same time, localizing mRNA for the nAChR, we utilized a neurofilament monoclonal antibody and a DNA clone for the nAChR α subunit to examine both gene expression and protein content at the single-cell level.

Protocol

Preparation of Coverslips

Coverslips on which cells are grown must be treated to allow the cells to adhere firmly to the glass and at the same time prevent the probe from sticking excessively to the substrate. For cleaning glass coverslips, we place them in concentrated hydrochloric acid (HCl) for 30 min, followed by a 30-min wash in ethanol, and steam autoclave with 0.1 M EDTA. Glass coverslips are rinsed in sterile distilled water and allowed to dry. Plastic coverslips are treated the same way except the HCl is omitted. Cells are plated on collagen freshly prepared from a rat tail (3).

Fixation of Cells

At the desired time of cellular development, cells are fixed in 4% paraformaldehyde in phosphate-buffered saline (PBS) and 5 mM MgCl$_2$ for 15 min (21, 37) or in 2% paraformaldehyde in PBS containing 1.8% lysine and 0.21% (w/v) sodium periodate for 30 min (10). Both fixatives give good morphological preservation of cells and good retention of intracellular antigens. These fixatives do not extensively cross-link protein and provide excellent RNA retention, which is especially important when the RNAs to be hybridized are expressed in low abundance. Since these fixatives do not create a barrier to diffusion of the probes, we have found no need to use dilute acids, detergents, alcohols, or proteases in our prehybridization protocol.

Immunocytochemistry

The localization and distribution of receptor proteins in a neuronal cell population can be readily detected using monoclonal or polyclonal antibodies. Cells or tissue sections are fixed in 2 or 4% paraformaldehyde, washed with PBS, blocked with PBS containing 0.1% bovine serum albumin (BSA), and incubated with a primary antibody at appropriate dilution. For antigens which are expressed intracellularly, cells are permeabilized with 0.1% Triton X-100 for 5 min after fixation. For antigens which are expressed on cell surfaces, such as the acetylcholine receptors, permeabilization is not necessary. Cells are incubated with the antibody for 1 hr at 37°C or overnight at 4°C. The distribution of nAChR protein after antibody incubation can be readily visualized with electron-dense or fluorescent secondary antibodies.

An example of nAChR distribution on ciliary neurons is shown in Fig. 2. Neurons after fixation were stained with MAb35 (a monoclonal antibody generated against the nAChR α subunit), and the antigen was visualized with an anti-rat immunoglobulin G (IgG) conjugated to horseradish peroxidase (HRP). The dense reaction product can be readily seen in the soma of neurons as well as in certain neuronal precursors (Fig. 2a). For most of our combined immunocytochemistry–*in situ* hybridization studies we used fluorescent chromophores. This allowed us to readily localize the nAChR protein, the neurofilament protein, as well as the distribution of mRNAs on a single-cell level.

Choosing Radiolabeled Deoxynucleotide

We radiolabel the probes with two [^{35}S]thiodeoxynucleoside triphosphates. In some cases we label the probes with [^{32}P]deoxynucleoside triphosphates (dNTPs). The ^{32}P-labeled probes have a higher specific activity and yield

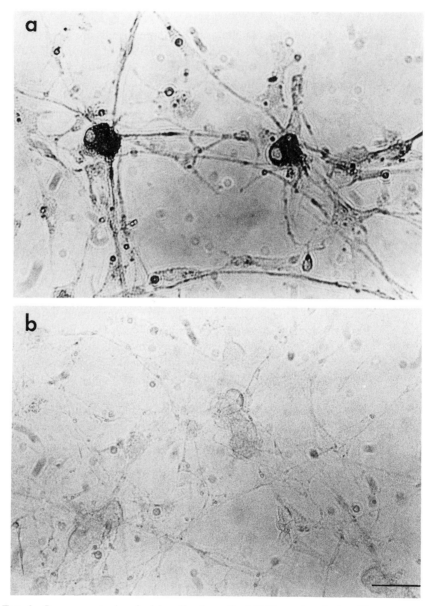

FIG. 2   Immunocytochemical localization of nAChR α subunits in ciliary neurons. (a) Neurons were incubated with a monoclonal antibody to nAChR α subunit, and the antigen was visualized with anti-rat IgG conjugated to horseradish peroxidase. The dense reaction product is present in the cell soma and neuronal processes. Bar, 50 μm. (b) Neurons were processed as described above except that the primary antibody was omitted. Reaction product was not observed. Bar, 50 μm.

shorter autoradiographic exposure times (1–3 days). However, the long path length of the energetic β emission results in an anatomical resolution inadequate for single-cell and intracellular localization. ^{32}P-Labeled probes are useful for regional localization using X-ray film. The highest resolution, lowest background, and best anatomical detail is obtained using ^{3}H-labeled probes. However, the drawback with ^{3}H-labeled probes is their low specific activity that requires up to 3 months of exposure time. This is even more disadvantageous when detecting sequences, such as the nAChRs, that are present in low abundance. For these reasons, we routinely use ^{35}S-labeled probes which yielded higher specific activities, greater signals, and shorter exposure times (7–12 days) than ^{3}H-labeled probes, while still having excellent resolution on a single-cell level.

Probe Labeling

A variety of protocols have been utilized in *in situ* hybridization procedures, all of which depend on rendering the cells permeable to the probe without the loss of RNA and with the retention of cellular morphology. The proper probes and the conditions for labeling are critical for the specificity of the *in situ* hybridization technique. We have found that hybridizing with single-stranded DNA probes (2, 8) gives higher specificity and less background than RNA single-stranded probes. We have used a primer extension method for synthesis and labeling of the single-stranded DNA (2, 8). Other methods that utilize nick translation or random priming give, in our preparation, high backgrounds with unacceptable levels of nonspecific binding. The procedure described below permits the investigator to make, in a 10-μl reaction volume, enough probe to hybridize at least 10 coverslips. Several labelings can proceed simultaneously.

Reagents

Single-stranded DNA at 200 ng/μl (500 ng/μl for M13 or other double-stranded plasmid DNA)
Primer (about 17-mer) at 20 ng/μl
10× DNA polymerase buffer (250 mM Tris-HCl, pH 7.5, 500 mM NaCl, 70 mM MgCl$_2$)
0.1 M dithiothreitol (DTT)
dGTP and dTTP mixture (1 mM each)
dATP and dCTP mixture (100 μM each)
^{35}S-Labeled dATP and dCTP at about 1400 Ci/mmol (NEN, Boston, MA)

DNA polymerase large fragment (Klenow from BRL, Gaithersburg, MD) at 6 units/μl

A restriction enzyme (RE) and its 10× buffer

[^{32}P]dATP or [^{32}P]dCTP (used for tracing).

Glycogen at 10 μg/μl

Annealing

1. Prepare 2.5× polymerase buffer by mixing 7.5 μl water and 2.5 μl of 10× polymerase buffer

2. Mix 1 μl of 2.5× polymerase buffer, 1 μl DNA, and 0.5 μl primer

3. Incubate at 65°C for 5 min then 37°C for 20 min. Spin briefly to collect the liquid in the bottom.

Labeling Reaction

During annealing, set up the following reaction mixture (for one 10-μl reaction). An accurate pipetman such as P2 pipette is recommended.

1 μl of 1 mM dGTP plus dTTP

0.75 μl of 10× polymerase buffer

0.5 μl of 0.1 M DTT

2 μl ^{35}S-labeled dATP

2 μl ^{35}S-labeled dCTP

0.25 μl [^{32}P]dATP (or [^{32}P]dCTP)

Add the annealed DNA to this mixture followed by 1 unit Klenow polymerase in about 1 μl (The enzyme can be diluted in 10 mM Tris-HCl, pH 7.5, 5 mM MgCl$_2$, and 0.5 μg/μl BSA). Mix well, spin briefly, and incubate at 37°C for 1 hr.

Fill-in Reaction

Add 1 μl of 100 μM cold dATP plus dCTP to the reaction. Continue the incubation for another 30 min.

Restriction Enzyme Digestion

1. Prepare 40 μl of 1× restriction enzyme (RE) buffer containing 5 units of the appropriate RE.

2. Add the above solution directly to the labeling reaction, mix, and incubate for 1 hr at 37°C (or recommended temperature).

3. Stop the reaction and precipitate DNA by adding 1 μl (10 μg) of glycogen, 5 μl of 3 M sodium acetate, and 125 μl of ethanol. (Salt can be omitted if there is 0.1 M NaCl in the 1× RE buffer.) Keep the tube on dry ice for 20 min or −20°C for several hours.

Procedure for Isolating Labeled Probe

1. Prepare a small 1.5% agarose gel (i.e., 20 ml in a 5 × 7 × 0.5 cm) in [Tris base (1 M, w/v), Boric acid (1 M, w/v), EDTA (20 mM, w/v)] 1× TBE plus ethidium bromide (1 μg/ml).

2. Spin down the DNA in a microcentrifuge for 15 min, then remove the ethanol completely. Dry the DNA briefly (just enough to remove the ethanol).

3. Use 3 μl water to solubilize the DNA.

4. Add 7 μl formamide to the DNA, and make sure that all the DNA is dissolved by moving the liquid with a pipette tip over the area where DNA may stay. Then spin the tube for a few seconds to collect the liquid in the bottom of the tube. The DNA standard can be 0.3 μg of 1-kb ladder (BRL) in 3 μl, mixed with 7 μl formamide.

5. Boil the DNA for 2 min.

6. Load the 10-μl sample into a slot on the gel, and run the gel at 10 V/cm. About 10 ng of a single-stranded DNA band (>200 nucleotides) can be visualized with the help of a UV light box (the gel can be photographed as usual).

7. Cut the desired DNA band out of the gel with a razor blade. Check the radioactivity with a counter to be sure that the band cut has the highest counts versus other bands that may show up.

8. Purify the DNA using the glass powder method (i.e., GeneClean II kit from Bio101, La Jolla, CA).

9. Elute the DNA in total 50 μl of water. Use 1 or 2 μl of the solution to determine the total counts per minute. Calculate the specific activity after determining the counts and estimating the amount of DNA which has been purified (compare the band intensity with those of a standard 1-kb ladder in the photograph of the gel).

10. Save the probe at −20°C and use it within 1 or 2 days.

Immunocytochemistry

1. After fixation, wash the cells 3 times with PBS.

2. Permeabilize the cells with 0.1% Triton X-100 for 5 min.

3. Wash the cells 2 times for 5 min each with PBS containing 0.1% BSA plus 2% goat serum (PBSS buffer).

4. Stain the cells with monoclonal antibody to neurofilament protein (or an antibody of your choice; concentration must be titrated).

5. Wash the cells twice with PBSS buffer.

6. Incubate the cells with biotinylated antibody (1:50).

7. Incubate the cells with streptavidin–Texas red (1:200) or streptavidin–fluorescein isothiocyanate (FITC) (1:75).

8. Wash the cells with PBS or leave in 70% ethanol until ready for hybridization.

9. If cultures are kept in ethanol, they must be rehydrated in PBS containing 5 mM MgCl$_2$ for 10 min before hybridization.

Hybridization

1. Rehydrate in PBS if necessary.

2. To reduce background, acetylate coverslips with 0.1 M ethanolamine containing 0.25% acetic anhydride (final), 0.2 M Tris-HCl and 1 M glycine, pH 8.0.

3. Prehybridize coverslips in 50% deionized formamide in 2× standard saline citrate (1× SSC is 0.3 M NaCl, 30 mM sodium citrate, pH 7.2). Incubate at 65°C for 10 min.

4. Prepare the hybridization mixture consisting of 4× 10^5 cpm of DNA probe, 40 μg sheared salmon sperm DNA, 40 μg *Escherichia coli* tRNA, and 0.4% BSA, 20 mM vanadyl ribonucleoside inhibitor, 20% dextran sulfate, 1 μg/μl heparin, and 20 mM DTT, heat the mixture at 90°C for 10 min, and incubate on the coverslip for 4 hr at 37°C.

5. Wash the coverslips in 10-ml Coplin jars with 2× SSC containing 50% formamide of increasing stringencies at 37°C. All washes must contain 20 mM DTT in order to reduce background.

6. Wash the coverslips in 1× SSC until no radioactivity is detected as monitored with a minimonitor.

7. Attach coverslips to slides and coat with Kodak (Rochester, NY) NTB-3 emulsion.

8. Store the slides in dark boxes at 4°C for 7–12 days and develop with a Kodak developer and fixer. The short exposure time of ^{35}S is particularly advantageous when a combined approach of *in situ* hybridization and immunocytochemistry is utilized. The fluorescent chromophores fade readily after a long exposure time, which is usually needed when using tritium, and therefore it may not have been possible to detect the protein localization in our double-label techniques.

Examples of Detection of Nicotinic Acetylcholine Receptor RNA by *in Situ* Hybridization

In our previous studies using immunocytochemistry, we have examined the relationship of nAChR clusters to nuclei in multinucleated myotubes (7). To determine the location of the nuclei, we stained the cultures with bisbenzamide, which binds specifically to DNA (Fig. 3). By alternating among fluorescent filters which excite and emit at specific wavelengths, we readily obtained double or triple exposures of the same microscopic field. Figure 3a shows the spatial orientation of nAChR clusters as compared to nuclei in a myotube. The aggregation of the nuclei beneath the nAChR clusters is apparent. Alternating the fluorescent filters allows us to determine which nuclei are associated with nAChR clusters. These observations led us to test the hypothesis whether every nucleus in the multinucleated myotube actively expresses nAChR mRNA. Figure 3b shows cell hybridized with the intron–exon probe to the nAChR subunits. Our results show that only certain nuclei actively express the message and others are either inactive or express the RNA at a very low level. Control cells hybridized with M13 bacteriophage without the insert show few scattered grains (Fig. 3c), indicating that our hybridization is specific.

Is the protein or peptide synthesized in the specific cells or intracellular compartment in which the immunoreactivity appears? The presence of immunoreactivity in intracellular compartments might be the result of protein transport to the compartment rather than demonstrating the actual site of synthesis. Furthermore, the absence of immunoreactivity in a given compartment may not indicate lack of synthesis, but rather a rapid degradation or exocytosis. *In situ* hybridization in combination with immunocytochemistry can resolve these questions.

In our own case, we combined immunocytochemistry with *in situ* hybridization for still other reasons. We wished to determine whether innervation influenced the heterogenous expression of nAChR subunit RNAs in the nuclei of a multinucleated myotube (8), how the distribution of neurons on muscle cells affected the expression of nAChR RNAs, and whether only cholinergic innervation would induce myotube nuclei to actively express nAChR RNAs (9). An example of our combined immunocytochemistry and *in situ* hybridization approach is shown in Fig. 4.

Figure 4a shows ciliary neurons, cocultured with muscle cells, stained with a monoclonal neurofilament antibody (labeled with Texas red) and hybridized with a ^{35}S-labeled intron/exon probe for nAChR. The fluorescence of neurons is readily localized, and, using a dark-field condenser, both autoradiographic silver grains and neurons are easily seen in the same field (Fig. 4a). By interchanging the fluorescent filters, double-labeled exposures are obtained. These show the distribution of the neurons and the nAChR α subunit RNA as judged by the distribution of silver grains and the nuclei, which are stained with bisbenzamide. This approach allowed us to show that when ciliary neurons are uniformly distributed throughout the myotubes, the majority of nuclei actively express nAChR subunit mRNA (9). However, when myotubes are plated a distance away from the source of neurons, only

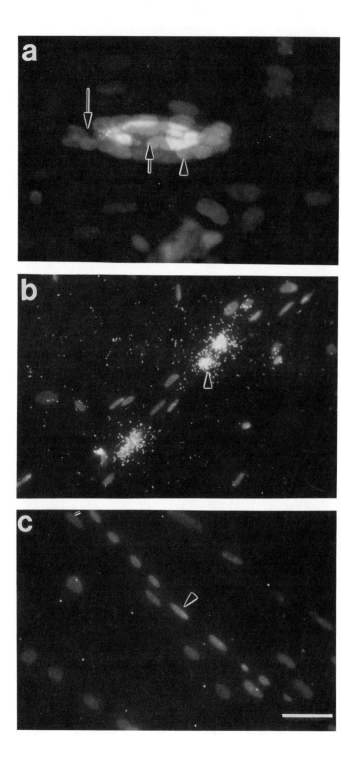

certain myotube nuclei actively express nAChR subunit RNA (9), whereas others have only a few grains associated with nuclei (Fig. 4b). Dorsal root ganglion cells (DRG) cocultured with myotubes and subjected to *in situ* hybridization and immunocytochemistry show that only very few nuclei actively express nAChR subunit RNA, which is the basal level of expression (Fig. 4c). The DRG neurons were stained with a neurofilament monoclonal antibody, and protein was visualized using streptavidin–Texas red. Typical controls consisted of myotubes cocultured with ciliary neurons that were stained with a neurofilament antibody and hybridized with M13 bacteriophage lacking the nAChR subunit insert (Fig. 4d). There is only a low background of occasional silver grains (Fig. 4d).

Application of *in Situ* Hybridization to Studies of Neuronal Receptors

Proteins primarily involved in cellular maintenance are usually present in every cell type, whereas others, such as neurotransmitters and receptors, are specific for certain cell types and are usually present in low abundance. Each of the proteins is translated from a specific mRNA. Recent developments in cloning techniques have made available a large number of probes which can be used in the detection of such messages by *in situ* hybridization. This approach can detect low-abundance messages if the probe is of sufficiently high specific activity.

Our approach that combines *in situ* hybridization with immunocytochemistry allows us to localize both the message and the protein for which a particular gene is coding within a single cell. This is of particular importance when working with a heterogeneous cell population, since it allows us to demonstrate directly the cells responsible for the synthesis of the mRNA

FIG. 3 Double-label immunofluoresence of muscle cells and autoradiographs of cellular RNA by *in situ* hybridization. (a) Double-label exposures of myotubes incubated with α bungarotoxin-tetramethyl rhodamine BTX-TMR, which binds to nAChRs (arrows), and nuclei (arrowhead) stained with bisbenzamide. Note the accumulation of nuclei around the nAChR cluster. (b) Muscle cells hybridized with a ^{35}S-labeled nAChR α subunit intron/exon probe show that the silver grains, denoting RNA localization, are present over certain nuclei and that the other nuclei present in the same cell are either inactive or express RNA at low levels. (c) Control hybridizations to bacteriophage M13mp19 lacking any insert show few scattered grains. Bar, 50 μm.

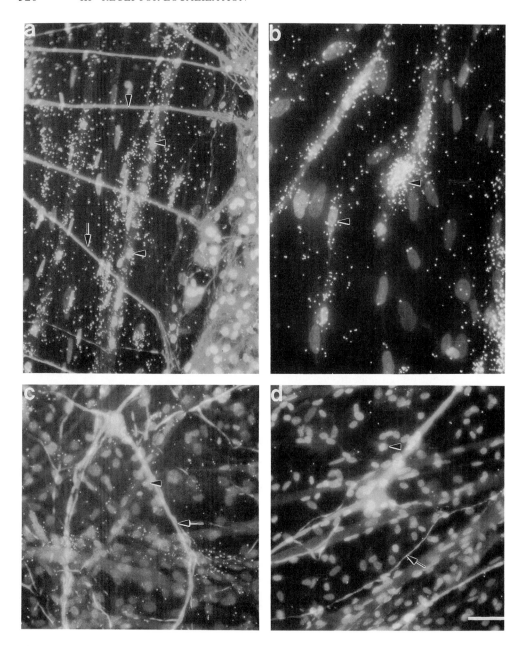

and whether there are specific compartments within the cells where the message is produced. Such experiments provide more information than a Northern analysis. In the combined detection, the protein is readily visualized using a biotinylated monoclonal antibody followed by incubation with streptavidin conjugated to Texas red or FITC. This staining withstands the subsequent treatments of *in situ* hybridization. The distribution of the protein or epitopes is delineated by a red or green fluorescence and the specific message by the autoradiographic silver grains. The location of the nuclei of all cells in a culture is readily detected by staining the nuclear DNA with bisbenzamide after hybridization. By alternating the fluorescence filters and using a dark-field condenser, we have obtained a triple label of cells in the same microscopic field.

Utilizing *in situ* hybridization in combination with immunocytochemistry will provide insight into the mechanisms involved in synthesis and posttranslational processing of key protein constituents at synaptic contacts. This approach should allow one to answer the following questions: Are all the messages and the proteins for which they code synthesized in the same cellular compartments? What alterations in gene and protein expression occur during synapse development, maintenance, and senescence? Is there compartmentalization at the synapse of specific mRNAs whose protein products determine whether a particular synapse is to be maintained or eliminated? More generally, this method permits one to localize messenger and protein within any cellular system.

FIG. 4 Combined *in situ* hybridization and immunocytochemistry in nerve–muscle cell cocultures with nAChR intron/exon probe. Cultures were observed with fluorescent and dark-field microscopy. Neurons in all cultures were stained with the neurofilament antibody and visualized with streptavidin–Texas red, showing the cell bodies and neuronal processes (arrows). Nuclei were visualized with bisbenzamide (arrowhead) and the grains with dark-field illumination (white dots). (a) *In situ* hybridization of ciliary neuron muscle cells. (b) Coculture of spinal cord explant with muscle cells, five explant diameters away from spinal cord. Certain nuclei actively express nAChR α subunit message. (c) Dorsal root ganglia cocultured with muscle cells and subjected to *in situ* hybridization and immunocytochemistry. Muscle cells express much less nAChR α subunit message than those innervated with cholinergic neurons. (d) Ciliary neurons cocultured with muscle cells and hybridized with M13mp8 bacteriophage without insert. Few background grains are seen scattered throughout the culture. Bar, 20 μm.

References

1. J. H. Ashe and C. A. Yarosh, *Neuropharmacology* **23**, 1321 (1984).
2. S. A. Berman, S. Bursztajn, B. Bowen, and W. Gilbert, *Science* **247**, 212 (1990).
3. M. B. Bornstein, *Lab. Invest.* **7**, 134 (1958).
4. J. Boulter, J. Connolly, E. Deneris, D. Goldman, S. Heinemann, and J. Patrick, *Proc. Natl. Acad. Sci. U.S.A.* **84**, (1987).
5. J. Boulter, M. Hollmann, A. O'Shea-Greenfield, M. Hartley, E. Deneris, C. Maron, and S. Heinemann, *Science* **249**, 1033 (1990).
6. D. A. Brown and A. A. Selyanko, **365**, 365 (1985).
7. J. M. Bruner and S. Bursztajn, *Dev. Biol.* **115**, 35 (1986).
8. S. Bursztajn, S. A. Berman, and W. Gilbert, *Proc. Natl. Acad. Sci. U.S.A.* **86**, 2928 (1989).
9. S. Bursztajn, S. A. Berman, and W. Gilbert, *J. Neurobiol.* **21**, 387 (1990).
10. S. Bursztajn and G. D. Fischback, *J. Cell Biol.* **98**, 498 (1984).
11. T. Claudio, M. Ballivet, J. Patrick, and S. Heinemann, *Proc. Natl. Acad. Sci. U.S.A.* **80**, 1111 (1983).
12. A. E. Cole and P. Shinnick-Gallagher, Nature (*London*) **307**, 270 (1984).
13. S. Couturier, L. Erkman, S. Valera, D. Rungger, S. Bertrand, J. Boulter, M. Ballivet, and D. Bertrand, *J. Biol. Chem.* **265**, 17560 (1990).
14. E. S. Deneris, J. Boulter, J. Connolly, E. Wada, K. Wada, D. Goldman, L. W. Swanson, J. Patrick, and S. Heinemann, *Clin. Chem.* **35**, 731 (1989).
15. E. Deneris, J. Boulter, L. Swanson, J. Patrick, and S. Heinemann, *J. Biol. Chem.* **264**, 6268 (1989).
16. D. Goldman, E. Deneris, A. Kochhar, J. Patrick, and S. Heinemann, *Cell* (*Cambridge, Mass.*) **48**, 965 (1987).
17. G. D. Fischback and S. M. Schuetze, *J. Physiol.* (*London*) **303**, 125 (1980).
18. A. T. Hasse, ed., *Curr. Top. Microbiol. Immunol.* **143**, 9 (1989).
19. H. Hofler, *Pathol. Res. Pract.* **182**, 421 (1987).
20. A. Karlin, R. Cox, R. R. Kaldany, P. Lobel, and E. Holtzman, *Cold Spring Harbor Symp. Quant. Biol.* **48**, 1 (1983).
21. J. B. Lawrence and R. H. Singer, *Nucleic Acids Res.* **13**, 1777 (1985).
22. J. Lindstrom, R. Schoepfer, and P. Whiting, *Mol. Neurobiol.* **1**, 281 (1987).
23. R. H. Loring and R. E. Zigmond, *J. Neurosci.* **72**, 153 (1987).
24. C. W. Luetje, J. Patrick, and P. Seguela, *FASEB J.* **4**, 2753 (1990).
25. P. D. McCrea, J. Popot, and D. M. Engleman, *EMBO J.* **6**, 3619 (1987).
26. A. K. Mitra, M. P. McCarthy, and R. M. Stroud, *Cell Biol.* **109**, 755 (1989).
27. M. Mishina, T. Takai, K. Imoto, M. Noda, T. Takahashi, S. Numa, C. Methfessel, and B. Sakmann, *Nature* (*London*) **321**, 406 (1986).
28. H. Mullink, J. M. M. Walboomers, T. M. Tadema, D. J. Janse, and C. J. L. M. Meijer, *J. Histochem. Cytochem.* **37**, 603 (1989).
29. E. L. Palacios-Peru, L. Miranda-Contreras, R. V. Mendoza, and E. Zambrano, *Neuroscience* (*Oxford*) **24**, 111 (1988).
30. E. L. Palacios-Peru, L. Palacios, and R. V. Mendoza, *J. Submicrosc. Cytol.* **13**, 145 (1981).

31. G. Palade, *Science* **189,** 347 (1965).
32. R. L. Papke, J. Boulter, J. Patrick, and S. Heinemann, *Neuron* **3,** 589 (1989).
33. M. A. Raftery, M. W. Munkapiller, C. D. Strader, and L. E. Hood, *Science* **208,** 1454 (1980).
34. M. M. Salpeter, and R. H. Loring, *Prog. Neurobiol.* **25,** 297 (1985).
35. J. Schmidt, *Int. Rev. Neurobiol.* **30,** 1 (1988).
36. R. Schoepfer, P. Whiting, F. Esch, R. Blacher, S. Shimasaki, and J. Lindstrom, *Neuron* **1,** 241 (1988).
37. R. H. Singer, J. B. Lawrence, and C. Villnave, *BioTechniques* **4,** 230 (1986).
38. A. Triller, F. Cluzeavd, F. Pfeiffer, H. Betz, and H. Korn, *J. Cell Biol.* **101,** 683 (1985).
39. K. L. Valentino, J. H. Eberwine, and J. D. Barchas, eds., "*In Situ* Hybridization: Applications to Neurobiology." Oxford Univ. Press, New York, 1987.
40. W. Wada, M. Ballivet, and J. Boulter, *Science* **240,** 330 (1988).
41. V. Witzemann, E. Stein, B. Barg, T. Konno, M. Koenen, W. Kues, M. Criado, M. Hofmann, and B. Sakmann, *J. Biochem. (Tokyo)* **194,** 437 (1990).

[20] Autoradiographic Localization of Putative Neuronal Nicotinic Receptors Using Snake Venom Neurotoxins

Ralph H. Loring and David W. Schulz

Introduction

Postsynaptically acting snake venom neurotoxins, especially α-bungarotoxin (αBTX) from the venom of *Bungarus multicinctus*, have proved to be invaluable in localizing and quantitating muscle nicotinic receptors (1). αBTX is a significant fraction of the total protein in *B. multicinctus* venom, is readily radiolabeled with ^{125}I, and binds and blocks muscle nicotinic receptors of many species virtually irreversibly ($t_{1/2}$ for dissociation from rat muscle exceeds 10 days; Ref. 2). However, although αBTX binds to many types of neurons, in most cases it has little effect on neuronal nicotinic receptor function (3). The number of known subtypes of neuronal nicotinic receptors is large and is growing. At least seven neuronal α subunits and four neuronal β subunits have been cloned, and from these subunits at least seven combinations have been shown to produce functional receptors when expressed in *Xenopus* oocytes (4). Neuronal bungarotoxin (NBTX), a minor neurotoxin from the venom of *B. multicinctus,* blocks some of these subtypes (4, 5). In the rat and chick nervous systems, NBTX (variously called bungarotoxin 3.1, toxin F, or κ-bungarotoxin) is believed to bind and block the $\alpha_3\beta_2$ subtype of receptor with high affinity (6), and it possibly blocks several other subtypes at higher concentrations (4, 7). The use of ^{125}I-labeled NBTX is complicated by the fact that it readily binds to some sites recognized by αBTX in addition to receptor subtypes that do not recognize αBTX (5).

Interest in neuronal αBTX binding sites has recently increased due to several developments. A purified αBTX binding site from the chick optic lobe has been shown to form a functional receptor when reconstituted into lipid membranes (8), and two genes for αBTX-binding proteins (α_7 and α_8) have now been cloned (9), one of which (α_7) forms a functional receptor that is blocked by αBTX when expressed by itself in oocytes (10). Therefore, this chapter describes the use of ^{125}I-labeled αBTX and ^{125}I-labeled NBTX for autoradiographic localization and quantitation of putative neuronal nicotinic receptors in brain slices, at the light and electron microscopic levels.

Methods in Neurosciences, Volume 12

Sources of Snake Venom Neurotoxins

Pure snake venom toxins are necessary for both iodination and for blocking the binding of radiolabeled toxins. Purification of αBTX (11–13) or NBTX (14–18) requires biochemical expertise. Commercial sources of purified αBTX include Biotoxins, Inc. (St. Cloud, FL), Sigma Chemical Co. (St. Louis, MO), Molecular Probes (Eugene, OR), and Miami Serpentarium (Punta Gorda, FL). Users should be warned that, in the past, commercial lots of αBTX have been contaminated with other neurotoxins, including NBTX (19), so that additional purification may be necessary. To our knowledge, the only current commercial source of NBTX is Biotoxins, Inc.

Iodination of Snake Venom Neurotoxins

^{125}I-Labeled αBTX is commercially available from a variety of sources, including Amersham (Arlington Heights, IL), ICN (Costa Mesa, CA), and NEN/Du Pont (Boston, MA). It is also quite easy to prepare. In contrast, ^{125}I-labeled NBTX is not commercially available, its production is generally more difficult, and the labeled product tends to radiodecompose more quickly. Both chloramine-T (e.g., Sigma) and Iodogen (Pierce Chemical, Rockford, IL) can be used for iodinating NBTX, but, in our hands, chloramine-T seems to work well only when used to iodinate larger amounts (20 versus 10 μg). Therefore, two iodination protocols for NBTX are presented, both of which work equally well for αBTX. Both protocols use high concentration, low pH Na^{125}I (e.g., NEN/Du Pont NEZ 033L, or Amersham IMS.300). Radioiodinations should be performed behind lead shielding in a properly monitored fume hood equipped with charcoal filters with the approval of the institutional radiation safety officer. Toxin concentrations are most easily estimated using the absorbance at 280 nm of stock solutions [extinction coefficient of αBTX (0.1%) is 1.24, Ref. 20; NBTX, 0.82, Ref. 21].

Radiodination Protocol I for Iodinating 20 μg Toxin (22)

Stock Solutions
    0.15 *M* Tris-HCl buffer, pH 7.8
    3 mg/ml sodium metabisulfite in distilled water (made shortly before use)
    1 mg/ml chloramine-T in Tris-HCl (made just before use)

Buffer A: 5 mM sodium phosphate buffer, pH 7.5, with 1 mg/ml bovine serum albumin (BSA; e.g., Sigma fraction V)

Buffer B: 50 mM sodium phosphate buffer, pH 7.5, 3 mM NaCl, and 10 mg/ml BSA

Procedure

1. Prepare a disposable polypropylene column (e.g., 731-1550, Bio-Rad, Richmond, CA) packed with 1 ml CM-cellulose (Whatman, Clifton, NJ, CM52) made up in distilled water. Equilibrate with four 1-ml washes of buffer A.

2. Dissolve 20 μg toxin in 200 μl Tris-HCl. Add to 5 mCi Na[125]I in a suitable iodination chamber equipped with a charcoal trap. Mix well.

3. Dissolve chloramine-T to 1 mg/ml in Tris-HCl. Add 30 μl to the reaction and mix well.

4. Mix well every 2 min. After 10 min add 20 μl more chloramine-T. After a total of 15 min, stop the reaction by adding 100 μl sodium metabisulfite.

5. Add the reaction mixture (~0.45 ml) to 1 ml buffer A. Dilute 10 μl to 1 ml in buffer A. Count 10 μl in a γ counter to estimate the total radioactivity to be added to the column.

6. Add 1 ml more buffer A and pour the combined material over the CM-cellulose column.

7. Collect effluent in polypropylene tubes. Wash the column 7 times with 1 ml of buffer A, keeping each effluent separate. Then wash 5 times with 1 ml buffer B.

8. For each effluent, dilute 10 μl into 1 ml buffer A, and count 10 μl of that. More than 99% of the unreacted ^{125}I should wash through the column with buffer A, and the iodinated toxin should be eluted with buffer B (tubes 8–10).

Radiodination Protocol II for Iodinating 10 μg Toxin (Modified from Ref. 23)

Stock Solutions

0.5, 0.25, and 0.05 M sodium phosphate, pH 7.4

10 mg/ml BSA in 0.5 M sodium phosphate, pH 7.4

Saturated (~0.2 g/ml) tyrosine in 0.25 M sodium phosphate, pH 7.4

Procedure

1. Coat clean 12 × 75 mm glass tubes with Iodogen (Pierce) by mixing 8.6 mg with 10 ml methylene chloride and then, with a glass transfer tip, coating the bottom of each tube with 5–10 μl of the mixture. The methylene

chloride/Iodogen should dry clear with no hint of a precipitate. Several tubes can be coated and stored desiccated in the dark at 4°C. After storage, a spare tube can be tested by adding 0.1 ml of 0.1 M NaI in 0.25 M sodium phosphate; The solution should turn yellow within a few seconds.

2. Prepare a disposable polypropylene column (e.g., 731-1550, Bio-Rad) packed with 7 ml swelled and defined Sephadex G-10 (e.g., Sigma) in distilled water. Pretreat the column by passing through 3 ml of 10 mg/ml BSA in 0.5 M sodium phosphate and washing with at least 20 ml of 0.25 M sodium phosphate.

3. Dilute Na^{125}I to approximately 5 mCi/0.1 ml in 0.25 M sodium phosphate.

4. Add 10 μl of 1 mg/ml toxin in 0.25 M sodium phosphate, 30–50 μl 0.25 M sodium phosphate, and 40–60 μl (~2–3 mCi) Na^{125}I to a final volume of 100 μl in an Iodogen-coated tube. Stopper and mix well. Mix every 2–3 min.

5. After 15 min add 10 μl tyrosine, mix well, and let sit 2 min before loading material on the top of the Sephadex column. Elute with 0.05 M sodium phosphate (pH 7.4) and collect 900-μl fractions into polypropylene tubes containing 100 μl of 10 mg/ml BSA in 0.5 M sodium phosphate. Dilute 10-μl samples of fractions as in Protocol I for counting. Labeled toxin should elute in tubes 3–5, while free ^{125}I and ^{125}I-labeled tyrosine come out after tube 10.

"Good" iodinations using either protocol for NBTX will incorporate about 50% of the ^{125}I, or 80–90% for αBTX. After iodination, the labeled toxins can be stored at 4°C. If the material will not be used on living tissue, 0.02% sodium azide will discourage bacterial contamination. ^{125}I-Labeled αBTX has a storage life of 4–8 weeks, while that of ^{125}I-labeled NBTX is only 2–3 weeks.

Binding Assays and Determination of Specific Activity of Radiolabeled Neurotoxins

Before performing autoradiographic experiments, it is necessary to characterize radioligand binding in homogenates of the tissue in question. Binding assays allow estimations of saturating concentrations, affinity, rates of association and dissociation, and maximum site densities, all of which are useful in designing autoradiographic experiments. In addition, isotopic dilution experiments allow estimation of the specific activity and concentration of the ^{125}I-labeled toxin. In such an experiment, addition of precisely known (checked with the extinction coefficient) and progressively increasing amounts of unlabeled toxin dilutes the labeled material, decreasing the amount of radioactivity bound to a fixed number of tissue sites. Assuming that the labeled and unlabeled toxin bind equally well, the concentration dependence of the decrease in bound radioactivity is a measure of the concentration of the labeled toxin. From the known concentration, the specific activity is obtained by dividing the amount of radioactivity in a fixed volume by the molar mass of toxin.

We routinely perform an isotopic dilution assay within a few days of iodination, using either *Torpedo* membranes for ^{125}I-labeled αBTX (20) or chick retina homogenates for ^{125}I-labeled NBTX. For example, 12 frozen retinas (stored at $-80°C$) from 1- to 4-day-old chicks are homogenized in 2 ml of 20 mM sodium phosphate, pH 7.5, 170 mM NaCl, 1 mM EDTA, 1 mM EGTA, and 0.1 mM phenylmethylsulfonyl fluoride (in ethanol; final ethanol concentration 0.4%, v/v) using 20 strokes of a glass–Teflon homogenizer. An estimated 2 pmol of ^{125}I-labeled NBTX is mixed in quadruplicate samples in 1.5-ml plastic centrifuge tubes with 0, 0.5, 1, 2, 4, or 100 pmol nonradioactive NBTX in final volumes of 50 μl homogenization buffer containing 2 mg/ml BSA. To prevent NBTX binding to αBTX binding sites, 100 pmol αBTX is also included in all tubes. Fifty microliters of homogenate is added to each tube, mixed well, and incubated at room temperature for 2 hr. Nonbound ^{125}I-labeled NBTX is removed in the supernatant after three 1-ml washes and centrifugations (14,000 g, 3 min at 4°C) using 10 mM sodium phosphate buffer, pH 7.5, 1 mM MgCl$_2$, 100 mM NaCl, 320 mM sucrose, and 10 mg/ml BSA. The pellets are counted for radioactivity, and binding in the presence of 100 pmol NBTX is subtracted from the other samples to determine the specific binding. To confirm that the binding sites are saturated under these conditions, other tubes containing an estimated 3 pmol ^{125}I-labeled NBTX are mixed with either 0 or 100 pmol NBTX; specific binding in these samples should not be more than 20% higher. A convenient way to analyze the data is to plot the reciprocal of the specific counts versus the competing unlabeled toxin concentration (e.g., Ref. 20). The data should give a straight line with the x-intercept being the negative of the ^{125}I-labeled NBTX concentration.

Filtration assays may be used to study the binding of ^{125}I-labeled αBTX (24–26) or ^{125}I-labeled NBTX to brain tissue homogenates. In this example, rat caudate–putamen (striatum) was used since NBTX is known to block nicotine-evoked dopamine release in striatal slices and synaptosomes (27).

Filtration Assay

Stock Solutions

Buffer A: 120 mM NaCl, 50 mM Tris (pH 7.4), and 2 mg/ml BSA
Buffer B: 120 mM NaCl, 50 mM Tris (pH 7.4), and 0.1 mg/ml BSA

Procedure

1. Immerse a sufficient number of Whatman GF/C filters in incubation buffer; maintain at 4°C until filtration.

2. Decapitate adult male Sprague-Dawley rats and rapidly dissect striata on ice. Weigh the tissue and place it in incubation buffer for immediate use, or freeze the tissue for later experiments. We found no differences in binding characteristics between fresh striata and tissue that had been stored at −20°C for 2 months.

3. Using a Brinkmann Polytron (Westbury, NY) at setting 6, homogenize the tissue in 20 volumes of buffer A for 20 sec at 4°C. Centrifuge at 25,000 g for 15 min. Resuspend the pellet in the same volume of buffer A and repeat the centrifugation step. Suspend the final pellet at a tissue concentration of 60 mg original wet weight/ml in cold buffer A and maintain on ice.

4. Aliquot 50-μl volumes of buffer or competing compounds into 12 × 75 mm polypropylene culture tubes at 4°C.

5. Add 50 μl of ^{125}I-labeled NBTX at three times the desired final concentration. Initiate the incubation by adding 50 μl of tissue suspension, vortexing, and placing samples in a water bath at 37°C.

6. Incubate for 1 hr; shorter time periods may be used if nonequilibrium conditions are desired. Terminate the reaction by rapidly diluting with 4 ml cold buffer A, followed by immediate filtration onto Whatman GF/C filters. Rinse the filters twice with 4-ml aliquots of cold buffer B. (*Note:* These experiments used a 12-unit Millipore filtration apparatus, which allows for convenient buffer switching. If an automated apparatus using a single reservoir of buffer is employed, buffer B is recommended throughout the rinse procedure.) Count filters in a γ counter.

The relevant findings for the design of autoradiographic experiments were that ^{125}I-labeled NBTX binding to rat striatal membranes is saturable with a K_d of 4.6 nM. The binding reaches equilibrium by 60 min at 37°C, and the toxin dissociates with a $t_{1/2}$ of approximately 20 min at 37°C (28).

Localization of Receptors in Tissue Slices

It is beyond the scope of this chapter to give a complete overview of all autoradiographic techniques (tissue preparation, fixation, choice of emulsion, etc.) at the tissue slice, light, or electron microscopic levels. Interested readers are referred to books by Rogers (29), Baker (30), and Leslie and Altar (31). Localization of [^{3}H]nicotine or [3H]acetylcholine nicotinic binding sites in the rat brain has been described in detail (32–34). The distribution of these high-affinity agonist binding sites closely parallels that observed when immunohistochemical (35) or *in situ* hybridization (36, 37) approaches are used. In contrast, the distribution of ^{125}I-labeled αBTX binding in rat brain correlates very poorly with that of ^{3}H-labeled agonists, with very few regions displaying high densities of both types of binding sites (38, 39). NBTX clearly has some structural homology with αBTX, yet it differs with respect to its ability to block nicotine-mediated responses in neuronal tissue. In this example, the anatomical distribution of ^{125}I-labeled NBTX in rat brain was determined (28).

Autoradiography Using Rat Brain Tissue Sections

Stock Solutions

Isotonic sucrose: 0.32 M sucrose, buffered with 50 mM sodium phosphate, pH 7.4
Buffer A: 120 mM NaCl, 50 mM Tris (pH 7.4), and 2 mg/ml BSA
Buffer B: 120 mM NaCl, 50 mM Tris (pH 7.4), and *no* BSA

Preparation of Tissue Sections

1. Dip glass microscope slides in a 50°C solution of 5 g/liter gelatin and 0.5 g/liter CrKSO$_4$ · 12H$_2$O. Allow 1–2 days to dry.
2. Anesthetize an adult male Sprague-Dawley rat with chloral hydrate (35%, 0.14 ml/100 g body weight). Perfuse intracardially with 100 ml 4°C isotonic sucrose. Remove the brain and freeze onto a cryostat chuck using O.C.T. compound. Sections may be cut immediately, or the frozen brain and chuck may be stored at −20°C.
3. Cut 20-μm sections and thaw-mount onto glass slides previously coated with gelatin. It is desirable to obtain sets of 10–20 consecutive sections at various levels of the brain. Store slides desiccated at −20°C; use within 2 months.

Incubation of Tissue Sections

1. Preincubate sections at room temperature for 30 min with buffer A by covering the section with 200 μl of buffer and maintaining in a humidified chamber. Include nonradioactive competitive agents, if desired.
2. Incubate the slides at 37°C for 1 hr in buffer A containing ^{125}I-labeled NBTX plus appropriate competing agent(s). In experiments where only a single concentration of ^{125}I-labeled NBTX is required, 2–3 nM is optimal.

This level is below the K_d, yet yields a workable range of optical density readings among various brain regions following 1 week of exposure to film (assuming a specific activity of $\sim$1000 Ci/mmol).

3. Wash sections in buffer A at 4°C for 30 min, then transfer slides to buffer B for an additional 30 min at 4°C. These conditions were optimal for minimizing nonspecific binding. Rinse slides in distilled water at 4°C for about 5 sec, shake off excess, and blow dry at ambient temperature.

4. Place slides in a standard film cassette, appose to LKB Ultrafilm (Piscataway, NJ), and expose for 5–10 days. The duration of exposure will depend on the concentration and specific activity of ^{125}I-labeled NBTX used. An additional factor that influences exposure time is the selection of brain regions to be analyzed, since optical density values in the range of 0.1–0.8 are optimal. If quantitative receptor density is desired, a set of ^{125}I standards should be exposed with the sections. Either brain paste standards containing ^{125}I (e.g., Ref. 40) or commercially available standards such as Amersham ^{125}I microscales can be used. If the latter option is selected, a comparison should be made between brain paste and the polymer standards, since there may be differences in the relationship between optical density and radioactivity for some film types over certain durations of exposure (40).

5. Develop film using products such as Kodak D-19 and Kodak Rapid Fix (Rochester, NY).

Quantitative Analysis

Autoradiograms may be analyzed using one of the many available computerized image analysis systems, such as the one we use from Imaging Research Inc. (St. Catherines, ON). Most of these systems provide the option of either sampling multiple square-shaped areas within a given brain region or actually tracing the region of interest. To avoid the difficulty of expressing receptor density in moles per mg protein, we expressed results in units of ligand bound per unit area (28, 40, 41).

^{125}I-Labeled NBTX binds heterogeneously to various regions of rat brain (Fig. 1A). When 0.5 μM αBTX is included in the incubation medium (Fig. 1B), specific labeling disappears entirely in some regions but is only slightly decreased in others. The difference in binding in Fig. 1A versus Fig. 1B represents the component of ^{125}I-labeled NBTX binding that is also recognized by αBTX. Since αBTX fails to block nicotinic activity in most mammalian neuronal preparations studied thus far, the areas labeled by ^{125}I-NBTX in the presence of nonradioactive αBTX are likely to contain agonist recognition sites. In Fig. 1C, binding in the presence of both αBTX and 100 μM nicotine (or other nicotinic agents such as dihydro-β-erythroidine or d-tubocurare)

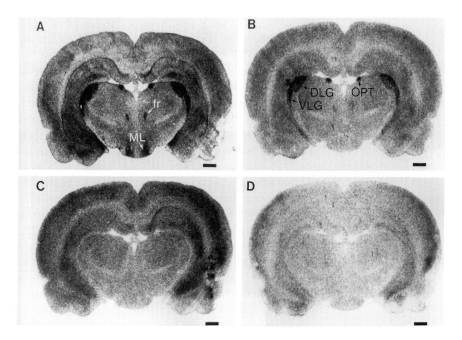

FIG. 1   Localization of ^{125}I-labeled NBTX (2 nM) binding to sections from frozen rat brain (20 μm thick). ^{125}I-labeled NBTX binding was assayed in the presence of the following connecting ligands: (A) none, (B) 0.5 μM unlabeled αBTX, (C) 0.5 μM unlabeled αBTX plus 100 μM nicotine, and (D) 0.5 μM unlabeled αBTX, 100 μM nicotine, and 0.5 μM unlabeled neuronal bungarotoxin. fr, Fasciculus retroflexus; ML, the lateral part of the medial mammillary nucleus of the hypothalamus; OPT, olivary pretectal nucleus; DLG, dorsal lateral geniculate nucleus; VLG, ventral lateral geniculate nucleus. The apparent increase in optical density over the cortex and hippocampus in (C) relative to (B) reflects experimental variability. Bars, 1 mm. [Modified with permission from Schulz et al. (28).]

leaves no distinct labeling of brain regions. Thus, a second component of ^{125}I-labeled NBTX binding is represented by the difference between Fig. 1B and Fig. 1C and comprises the binding not recognized by αBTX that is displaceable by nicotine. A third component of ^{125}I-labeled NBTX binding is represented by the difference between Fig. 1C and Fig. 1D, consisting of sites that are competed for only by nonradioactive NBTX. These sites are unlikely to have functional significance, since they are homogeneously distributed in the brain and are not displaced by a series of 15 different compounds having high affinity for various neurotransmitter receptors (28).

Localization of Receptors in Light Microscopic Autoradiographs

Light microscopic autoradiographs are useful as a guide for determining exposures of electron microscopic autoradiograms and for localizing receptors in relatively large pieces (e.g., 1–2 mm^3) of tissue. For example, intact chick ciliary ganglia are incubated in ^{125}I-labeled toxin in as little as 50 μl of Eagle's minimal essential medium (GIBCO, Grand Island, NY) at 37°C in a humidified atmosphere of 95% O_2/5% CO_2 for up to 4 hr without serious morphological damage (22, 42). After incubation in the presence or absence of competing ligands, excess labeled toxin is washed away with 100 μl fresh medium (10 min), and then the ganglia are immersed overnight at room temperature in a fixative consisting of 2% paraformaldehyde (freshly prepared) and 2% glutaraldehyde (electron microscopy grade), buffered with 0.12 M sodium cacodylate, pH 7.4. (Paraformaldehyde is prepared in a fume hood the day of use by heating ingredients for an aqueous 10% solution to about 70°C and adding 10 N NaOH dropwise with stirring until the paraformaldehyde is dissolved. After cooling, the white floccular material is filtered; the filtrate is neutralized and then diluted to the final concentration with the other ingredients of the fixative.) Ganglia are postfixed in 1% OsO_4 in 1.2 M cacodylate buffer, pH 7.4, at room temperature for 70 min, followed by rinsing in 0.9% NaCl. The ganglia are rinsed with buffer and then dehydrated at 4°C by passing through increasing ethanol solutions (50, 75, 90, and 95% for 5 min each, 2 times for 10 min each in 100% ethanol, and then 2 times for 30 min each in propylene oxide at room temperature). The tissue is then embedded in Epon plastic by placing in 1 : 1 Epon/propylene oxide for 30 min, leaving overnight in Epon, and transferring the next day to fresh Epon before hardening in an oven.

We describe here the flat substrate method for emulsion coating (43) using Ilford L4, which is also used for electron microscopic autoradiography. Experimenters only interested in light microscopic autoradiograms should consider using emulsions with higher sensitivity, such as Kodak NTB-2 or Ilford K5. Tissue sections (0.5 μm) are mounted on clean microscope slides (75 × 25 mm) with a frosted end on one side. A drop of distilled water is placed on the same side as the frosting, but approximately 2 cm from the unfrosted end, near the middle of the slide. Three to four sections are transferred to the drop, and brief heating on a hotplate (80–90°C) evaporates the water and fixes the sections to the slide. The back of the slide can be etched with a diamond scribe to locate the sections, and the sections identified by marking the frosting with a pencil. Three to six slides are made for each tissue sample. It is important that the sections be placed in the same location on each slide.

Coating Slides with Ilford L4

Materials

Two 250-ml beakers, one 100-ml beaker

Ilford L4 emulsion (Polysciences, Warrington, PA)

Light-tight slide boxes and black vinyl electrical tape

Silica gel capsules or 5–8 g Drierite wrapped in Kimwipes

Drying racks for slides, conveniently made from clothespins separated by sections of soda straws and hung on coat hanger wire attached to a frame

Procedure

1. In the darkroom under a safelight, weigh 4–5 g Ilford L4 emulsion on a beam balance into one of the 250-ml beakers, and place both 250 ml beakers in a water bath (45°C) for 15 min. Carefully add 20 ml distilled water and slowly swirl without creating bubbles. Gently pour the emulsion from one heated beaker to the other, until the emulsion is completely melted (i.e., nothing solid remains at the bottom after transfer). Then transfer to the 100-ml beaker and let cool to room temperature.

2. Using a medicine dropper, cover a clean microscope slide with emulsion from the unfrosted end to about 5–10 mm below the frosting. Gently drain off excess emulsion into the beaker, avoiding bubbles at all times. Blot the bottom edge (farthest from the frosting) and hang to dry. This test slide is examined outside the darkroom. When viewed from the back to see interference colors, the emulsion should have a uniform blue cover over the area where the sections are mounted on the specimen slides. Dilute the emulsion accordingly and retest. (A small amount of undiluted emulsion can be saved separately in case of overdilution.)

3. After achieving a blue color (corresponding to a bilayer of silver halide crystals in Ilford L4), carefully coat the specimen slides. Retest the emulsion every 30 min and adjust if necessary.

4. Dry the slides on the drying racks for at least 1 hr and place the slides in the slotted light-tight boxes containing desiccant. Tape the boxes closed with the black tape, and expose at 4°C.

After suitable exposure (1–3 days), one slide from each tissue sample should be developed. Two slides can be developed at once, back to back, in a series of 30-ml beakers. All photochemicals should be made in distilled water. A typical development sequence is 4 min in Kodak D-19 at 20°C, a dip in distilled water, 30 sec in 2% acetic acid, a dip in fresh distilled water, 2 min in 25% sodium thiosulfate, two dips in distilled water, 2 min in distilled water, and then air drying. Chemicals should be changed after two sets of slides, water after every set. Autoradiograms can be easily examined under dark-field microscopy (e.g., Fig. 2). Depending on the grain density, the exposure for the remaining slides can then be estimated.

Localization and Quantitation of Receptors in Electron Microscopic Autoradiographs

Electron microscopic autoradiograms are conveniently prepared using the Salpeter and Bachmann procedure (43) by modifying a few steps used above to make light microscope autoradiograms. The tissue blocks should be stained with uranyl acetate (2%, 2 hr) to increase the tissue contrast before embedding in plastic. The sections are then mounted on a thin layer of collodion, instead of directly on glass slides. Collodion-coated slides are prepared by carefully washing the microscope slides in detergent, rinsing several times in water and then distilled water, and carefully drying so as not to leave any lint on the slides. Collodion (~0.5% in amyl acetate) is filtered through Whatman No. 1 filter paper into a Coplin staining jar, and the slides are dipped to 2–3 mm below the frosting and then dried in a lint-free area. The collodion solution should be diluted or enriched, if necessary, to give a silver interference color when the collodion layer is stripped onto a water surface.

Sections for the electron microscope (pale gold interference color, ~100 nm) are mounted in the same area of the collodion-coated slides as was used above for the light microscope autoradiograms. However, usually two or three drops of water are used, and 3–6 sections, depending on size, are mounted in each drop. To do this, one drop of water is applied at a time, and the sections are transferred to the drop using a platinum wire loop; the sections are anchored by maneuvering them with an eyelash to touch the collodion, and the drop of water is then drawn away with a pointed stick or filter paper. Care must be taken to not tear or touch the collodion during this step. After drying, the sections can be visualized by lightly breathing on the slide, which is then enscribed on the back. Again, 4–7 slides from each tissue block are prepared. Care should be taken to determine that sufficient areas of interest (crisp synaptic profiles, neuronal processes, etc.) are present on each section to make analysis possible (see below).

The sections are stained again with 2% uranyl acetate (3–6 hr) to improve image contrast and carefully washed with distilled water. The dried slides are then coated with evaporated carbon about 5 nm thick in a vacuum chamber. The slides are then coated with Ilford L4, but diluted to give a purple interference color (corresponding to a monolayer of silver halide

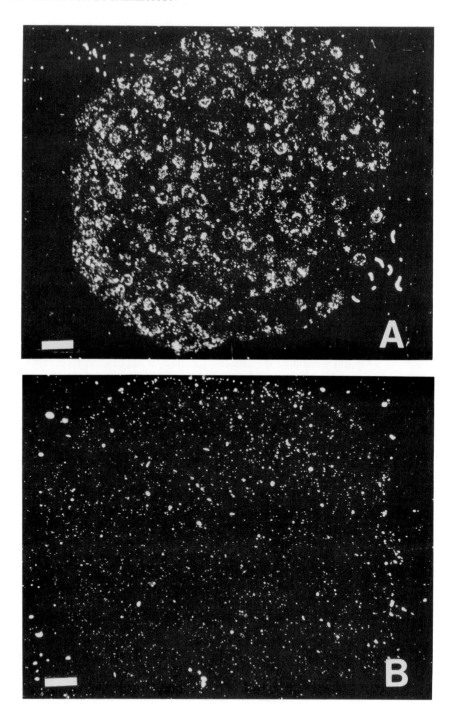

crystals) to achieve the best possible resolution. Several blank collodion-coated slides are also carbon coated and used as test slides for the final dilutions of the emulsion.

After exposure, the slides are developed in Kodak D-19 as above, but for 2 min at 20°C. After development, the slides are soaked in distilled water until the collodion layer begins to detach; the edges of the slide are scraped to cut the collodion, and the collodion is carefully stripped onto a water surface (a staining dish approximately 7 cm high, 8 cm wide, and 15 cm long is convenient). If the collodion sticks to the slide, a 1% HF solution applied near an edge can often help, but the HF should not be allowed near the area with the sections, or dirt particles will result. The sections should be visible as a bronze interference color on the floating silver collodion, and cleaned copper grids can be carefully placed over the sections with tweezers. A convenient way to recover the grids is to wrap the slide with an unstretched layer of Parafilm and, touching a part of the collodion away from the grids, push the slide under the surface of the water. The collodion will stick tightly to the Parafilm, and, after drying, the grids can be detached from the rest of the collodion layer and examined in the electron microscope.

To allow quantitation, the sensitivity of the emulsion to ^{125}I decay must be determined. Companion emulsion-coated slides (without sections) are pressed for a fixed time period against a source with a known amount of ^{125}I per unit area. Other slides not exposed to radioactivity serve to determine the intrinsic background of the emulsion. After subtracting background, the sensitivity of the emulsion is determined as the number of autoradiographic grains per unit area versus radioactive decays per unit area.

It is beyond the scope of this chapter to discuss the theory of various techniques for analyzing electron microscopic autoradiograms (see Refs. 29, 30, 44, and 45). However, it is important to realize that at the electron microscopic level, autoradiographic grains can be some distance from the site of ^{125}I decay (46). Correction for this radiation spread is the major challenge of quantitative autoradiographic analysis. What follows is a first-approximation analysis for determining the density of toxin binding sites on membranes that approximate a straight line.

FIG. 2 Light microscopic localization of ^{125}I-labeled αBTX binding to chick ciliary ganglia examined under dark-field microscopy. (A) Overexposed autoradiogram of a ganglion incubated for 4 hr in 20 nM ^{125}I-labeled αBTX. (B) Similar ganglion preincubated for 1 hr in 1 μM labeled αBTX before incubation with 20 nM ^{125}I-labeled αBTX and 1 μM unlabeled αBTX for 4 hr. The 0.5-μm sections were coated with Ilford L-4 and exposed for 21 days. The silver grains appear as white specks and in (A) can be seen to ring individual cells. Bars, 100 μm. [Modified with permission from Loring et al. (42).]

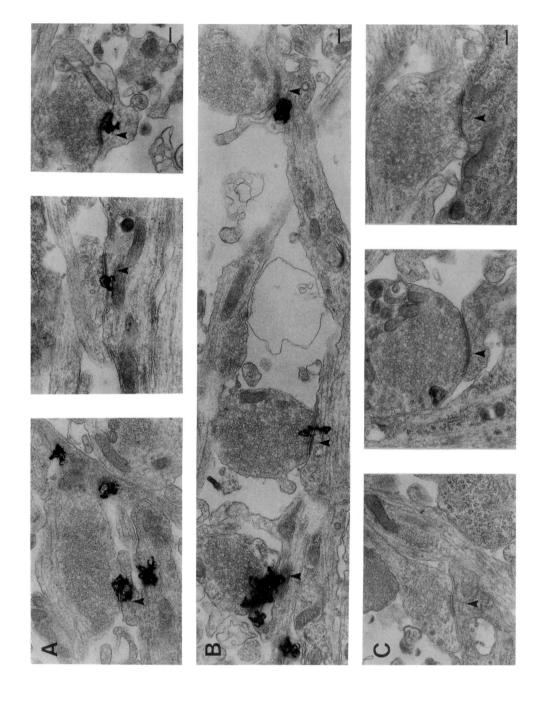

Randomly selected areas of tissue are photographed to a final print magnification of ×15,000, and crisp cytoplasmic membranes are marked using fine-tipped colored pens to encode different areas of interest (e.g., dense synaptic membranes, somal cytoplasmic membrane). The lengths of the lines are measured, and all grains whose centers lie within 0.25 μm of a membrane are counted. The center of a grain is defined as the center of the smallest circle that completely surrounds it. The 0.25 μm distance (or 3.75 mm at ×15,000, magnification) includes approximately 85% of all grains arising from a linear source (46). Figure 3B illustrates some of the difficulties encountered during this phase. The right-most synaptic profile (arrowhead) has a region that is not entirely crisp, and that portion is not used, since the width of the membrane cannot be accurately determined (and is greater than the thickness of the section). The left-most synaptic profile in Fig. 3B is overexposed, as more than one grain is superimposed, making grain counting difficult (we estimate three grains). This illustrates the need to develop the autoradiograms at the proper exposure. (Highly overexposed autoradiograms are still quite useful for qualitative "instant" localizations of binding sites; See Refs. 1 and 22). Assuming proper counting of the grains and measurements of the membranes, the average grains per square micron of membrane for each category is calculated based on the number of grains, the length of the membranes, and the thickness of the section. From this, the number of sites per square micron can be calculated using equations found in Ref. 47, which requires the specific activity of the toxin at the time the sections were coated with emulsion, the length of exposure, and the emulsion sensitivity. A more precise analysis can also be performed that makes fewer assumptions about the shape and radiation spread of various sources in the tissue (45). In the example illustrated in Fig. 3, the uncorrected site density for NBTX binding was about 4800 sites/μm^2 at synapses in the absence of competing ligands or in the presence of αBTX, but the density was near 0 in the presence of unlabeled NBTX (48). The calculations arbitrarily assume that all sites are postsynaptic, since electron microscopy does not have sufficient resolution to separate pre- from postsynaptic membranes.

FIG. 3 Synaptic localization of ^{125}I-labeled NBTX binding to cultured neurons from the rat superior cervical ganglion under electron microscopy. (A, top row) Three examples of synaptic profiles from a culture incubated in ^{125}I-labeled NBTX (80 nM) in the absence of competing ligands. (B, middle row) Synaptic profiles from cultures incubated in the presence of 80 nM ^{125}I-labeled NBTX and 2 μM unlabeled αBTX. (C, bottom row) Synaptic profiles of cultures incubated in 80 nM ^{125}I-labeled NBTX and 2 μM unlabeled NBTX. Arrowheads point to synaptic profiles. Autoradiographic grains appear as black squiggles. Bars, 0.25 μm. [Reprinted with permission from Loring *et al.* (48).]

Acknowledgments

We thank Dr. Richard E. Zigmond, in whose laboratory most of this work was performed, for continued advice and support. R.H.L. is supported in part by grants from the National Institutes of Health (NS22472) and the Smokeless Tobacco Research Council. D.W.S. was a Pharmaceutical Manufacturers Association Foundation Fellow in Pharmacology–Morphology.

References

1. H. C. Fertuck and M. M. Salpeter, *Proc. Natl. Acad. Sci. U.S.A.* **71,** 1376 (1974).
2. S. Heinemann, J. Merlie, and J. Lindstrom, *Nature (London)* **274,** 65 (1978).
3. D. A. Brown and L. Fumagalli, *Brain Res.* **129,** 165 (1977).
4. E. S. Deneris, J. Connolly, S. W. Rogers, and R. Duvoisin, *Trends Pharmacol. Sci.* **12,** 34 (1991).
5. R. H. Loring and R. E. Zigmond, *Trends Neurosci.* **11,** 73 (1988).
6. C. W. Luetje, K. Wada, S. Rogers, S. N. Abramson, K. Tsuji, S. Heinemann, and J. Patrick, *J. Neurochem.* **55,** 632 (1990).
7. J. Boulter, J. Connolly, E. Deneris, D. Goldman, S. Heinemann, and J. Patrick, *Proc. Natl. Acad. Sci. U.S.A.* **84,** 7763 (1987).
8. C. Gotti, A. E. Ogando, W. Hanke, R. Schlue, M. Moretti, and F. Clementi, *Proc. Natl. Acad. Sci. U.S.A.* **88,** 3258 (1991).
9. R. Schoepfer, W. G. Conroy, P. Whiting, M. Gore, and J. Lindstrom, *Neuron* **5,** 35 (1990).
10. S. Couturier, D. Bertrand, J.-M. Matter, M.-C. Hernandez, S. Bertrand, N. Millar, S. Valera, T. Barkas, and M. Ballivet, *Neuron* **5,** 847 (1990).
11. C. Y. Lee, S. L. Chang, S. T. Kaw, and S. H. Luh, *J. Chromatogr.* **72,** 71 (1972).
12. R. A. Love and R. M. Stroud, *Protein Eng.* **1,** 37 (1986).
13. P. A. Kosen, J. Finer-Moore, M. P. McCarthy, and V. J. Basus, *Biochemistry* **27,** 2775 (1988).
14. P. R. Ravdin and D. K. Berg, *Proc. Natl. Acad. Sci. U.S.A.* **76,** 2072 (1979).
15. V. A. Chiappinelli, *Brain Res.* **277,** 9 (1983).
16. R. H. Loring, V. A. Chiappinelli, R. E. Zigmond, and J. B. Cohen, *Neuroscience (Oxford)* **11,** 989 (1984).
17. R. H. Loring, D. Andrews, W. Lane, and R. E. Zigmond, *Brain Res.* **385,** 30 (1986).
18. J. J. Fiordalisi, C. H. Fetter, A. TenHarmsel, R. Gigowski, V. A. Chiappinelli, and G. A. Grant, *Biochemistry* **30,** 10337 (1991).
19. V. A. Chiappinelli, J. B. Cohen, and R. E. Zigmond, *Brain Res.* **211,** 107 (1981).
20. R. H. Loring, S. W. Jones, J. Matthews-Bellinger, and M. M. Salpeter, *J. Biol. Chem.* **257,** 1418 (1982).
21. V. A. Chiappinelli and J. C. Lee, *J. Biol. Chem.* **260,** 6182 (1985).

22. R. H. Loring and R. E. Zigmond, *J. Neurosci.* **7**, 2153 (1987).

23. V. A. Chiappinelli, K. M. Wolf, J. A. DeBin, and I. L. Holt, *Brain Res.* **402**, 21 (1987).

24. B. J. Morley, J. F. Lordon, G. B. Brown, G. E. Kemp, and R. J. Bradley, *Brain Res.* **134**, 161 (1977).

25. M. Segal, Y. Dudai, and A. Amsterdam, *Brain Res.* **148**, 105 (1978).

26. S. Hunt and J. Schmidt, *Brain Res.* **157**, 213 (1978).

27. D. W. Schulz and R. E. Zigmond, *Neurosci. Lett.* **98**, 310 (1989).

28. D. W. Schulz, R. H. Loring, E. Aizenman, and R. E. Zigmond, *J. Neurosci.* **11**, 287 (1991).

29. A. W. Rogers, "Techniques of Autoradiography," 3rd Ed., Elsevier, Amsterdam, 1979.

30. J. R. J. Baker, "Autoradiography: A Comprehensive Overview." Oxford Univ. Press, Oxford, 1989.

31. F. M. Leslie and C. A. Altar, eds., *in* "Receptor Biochemistry and Methodology" (J. C. Venter and L. C. Harrison, series eds.), Vol. 13. Alan R. Liss, New York, 1988.

32. R. D. Schwartz, R. McGee, and K. J. Kellar, *Mol. Pharmacol.* **22**, 56 (1982).

33. P. B. S. Clarke, C. B. Pert, and A. Pert, *Brain Res.* **323**, 390 (1984).

34. E. D. London, S. B. Waller, and J. K. Wamsley, *Neurosci. Lett.* **53**, 179 (1985).

35. L. W. Swanson, D. M. Simmons, P. J. Whiting, and J. Lindstrom, *J. Neurosci.* **7**, 3334 (1987).

36. D. Goldman, D. Simmons, L. W. Swanson, J. Patrick, and S. Heinemann, *Proc. Natl. Acad. Sci. U.S.A.* **83**, 4076 (1986).

37. E. Wada, K. Wada, J. Boulter, E. Deneris, S. Heinemann, J. Patrick, and L. W. Swanson, *J. Comp. Neurol.* **284**, 314 (1989).

38. M. J. Marks and A. C. Collins, *Mol. Pharmacol.* **22**, 554 (1982).

39. P. B. S. Clarke, R. D. Schwartz, S. M. Paul, C. B. Pert, and A. Pert, *J. Neurosci.* **5**, 1307 (1985).

40. A. P. Davenport and M. D. Hall, *J. Neurosci. Methods* **25**, 75 (1988).

41. C. R. Clark and M. D. Hall, *Trends Biochem. Sci.* **11**, 195 (1986).

42. R. H. Loring, L. M. Dahm, and R. E. Zigmond, *Neuroscience (Oxford)* **11**, 989 (1985).

43. M. M. Salpeter and L. Bachmann, *J. Cell Biol.* **22**, 469 (1964).

44. M. M. Salpeter and F. A. McHenry, *in* "Advanced Techniques in Biological Electron Microscopy" (J. K. Koehler, ed.), p. 113. Springer-Verlag, New York, 1973.

45. M. M. Salpeter, F. A. McHenry, and E. E. Salpeter, *J. Cell Biol.* **76**, 127 (1978).

46. M. M. Salpeter, H. C. Fertuck, and E. E. Salpeter, *J. Cell Biol.* **72**, 161 (1977).

47. J. A. Matthews-Bellinger, and M. M. Salpeter, *J. Physiol. (London)* **279**, 197 (1978).

48. R. H. Loring, D. W. Y. Sah, S. C. Landis, and R. E. Zigmond, *Neuroscience (Oxford)* **24**, 1071 (1988).

[21] Determination of Binding Sites for Neuropeptide Y Using Combined Autoradiography and Immunocytochemistry

Eveline P. C. T. de Rijk and Eric W. Roubos

Introduction

Neuropeptide Y (NPY), a peptide consisting of 36 amino acids, was discovered by Tatemoto *et al.* (1) in porcine brain as a member of the pancreatic polypeptide family. The discovery was followed by many publications describing the localization of NPY in the brain and pituitary of various mammals, amphibians, and fish (e.g., Refs. 2–4).

NPY acts not only within various areas of the central nervous system (5–7) but also in peripheral systems such as the circulatory system and in reproductive organs (8–10). The functions of NPY seem to be widespread, as it is involved, for instance, in the control of circadian rhythmicity (11), growth (e.g., Refs. 12–14), and background adaptation (15–17). On the other hand, in many systems the locations of NPY receptors and the functional significance of NPY are unknown. To solve this problems, the method of autoradiography is commonly used. In some cases, *in vivo* labeling with radiolabeled receptor ligands has been applied. This method has rendered information about the location of the receptors, both in the central nervous system and in peripheral organs. Nevertheless, the *in vivo* method has some serious drawbacks: NPY does not readily penetrate the blood–brain barrier, high concentrations of radiolabeled NPY are needed to attain a sufficiently high concentration of the ligand near the binding site, and this concentration is hard to control. These drawbacks may be overcome by using *in vitro* autoradiography of brain sections. Particularly in mammals such as the rat, cat, and monkey, this approach has yielded detailed information about the locations of NPY receptors in various brain areas, including cortex, septum, hippocampus, thalamus, and mammiliary nuclei (18, 19), and in peripheral structures, such as the vas deferens (20) and cardiac blood vessels (21). However, this *in vitro* approach has the disadvantage that the tissue preparation (fixation and/or freezing and thawing) may severely affect receptor binding characteristics. This problem does not occur when the ligand is

Methods in Neurosciences, Volume 12

applied to living tissue. In this chapter we describe such an approach, studying the presence of NPY binding sites in the intermediate lobe of the pituitary of the clawed toad *Xenopus laevis*.

Little is known about the NPY binding sites in the pituitary. Because NPY is able to influence pituitary hormone release in rats (22, 23), fish (24), and amphibians (15, 17), the pituitary seems to be an important target for NPY. The *in vivo* autoradiography technique described here permits the localization of NPY binding sites of individual cells in the pituitary. This "dispersed cell labeling method" involves incubation of dispersed, living cells with the specific NPY ligand ^{125}I-Bolton–Hunter-labeled NPY (^{125}I-BH-NPY) followed by autoradiography. To recognize endocrine melanotrope cells of the intermediate lobe, autoradiography is combined with immunocytochemistry using an anti-αMSH serum.

Preliminary Choices

Pars Intermedia of Xenopus Laevis

The intermediate lobe of the pituitary of the amphibian *Xenopus laevis* has been studied in detail. It contains endocrine melanotrope cells which release α-melanophore-stimulating hormone (αMSH), a peptide that induces pigment dispersion in the melanophores of the skin when the animal is placed on a dark background. In addition to melanotrope cells another cell type is present, the folliculostellate cell. This cell type is a glial-like cell and contacts the melanotrope cells very tightly with slender processes (16).

Superfusion experiments with neurointermediate lobes have shown that αMSH release is inhibited by NPY (15). By light and electron microscopic immunocytochemistry, NPY has been demonstrated in nerve endings in the intermediate lobe (16, 25). These results indicate that NPY plays an important role in the physiological process of background adaptation of *Xenopus laevis*. Therefore, *Xenopus laevis* is regarded as a suitable object to study NPY binding sites in the intermediate lobe of the pituitary.

Ligands Specific for Neuropeptide Y Receptors

Various ligands for NPY receptors have been used to localize NPY binding sites, namely [³H]NPY (19, 26, 27), iodinated NPY (28, 29), and the iodinated NPY-specific ligand ^{125}I-BH-NPY (18, 30, 31). The use of ^{125}I-BH-NPY has several advantages, such as (1) the higher degree of labeling specificity than [³H]NPY and ^{125}I-labeled NPY (18), (2) higher stability to autoradiation breakdown (18), (3) higher autoradiographical resolution (18), (4) absence of quenching, which is often seen after labeling with tritiated compounds (32), and (5) much shorter exposure time compared to autoradiography with a tritiated ligand (6 days of iodinated NPY versus 6 weeks for tritiated NPY) (18). Moreover, because the distributions of radioactivity in rat brain sections after incubation with [³H]NPY and ¹²⁵I-BH-NPY are similar (18), as are their respective biological activities (30), we choose to use ¹²⁵I-BH-NPY for NPY receptor localization in the pituitary of *Xenopus laevis*.

Labeling Method

Light microscopic autoradiography is one of the most extensively described methods for neurotransmitter and neuropeptide receptor mapping [e.g., see review by Kuhar (33)]. In general, autoradiography can be readily applied, and, in the case of ligands labeled with isotopes that strongly radiate, results can be obtained rapidly. In the case of localizing peptide binding sites on individual cells in the pituitary, the traditional methods of *in vivo* and *in vitro* labeling of tissue sections proved to be unsatisfactory (see above). Therefore, the dispersed cell labeling method was chosen. A problem that shows up when investigations are carried out to obtain precise receptor localization is the identification of the receptor-containing cells. A generally used method for cell identification is immunocytochemistry. We show that a combination of autoradiography and immunocytochemistry can be used to investigate accurately the locations of NPY binding sites of dissociated cells. In addition, by measuring the degree of radioactive labeling of individual cells, an estimation of the amount of the receptor present may be obtained.

Results are presented of an experiment in which living, dispersed cells of the intermediate pituitary lobe (melanotropes and folliculostellate cells) were labeled with ¹²⁵I-BH-NPY. After the cells had attached to glass slides, they were treated immunocytochemically with a specific anti-αMSH serum.

Materials and Methods

Cell Dissociation

Specimens of *Xenopus laevis* are perfused intracardially with a Ringer's solution containing 112 mM NaCl, 2 mM KCl, 2 mM CaCl₂, and 15 mM HEPES, pH 7.4 (Calbiochem, La Jolla, CA; Ultrol grade) for 10 min. Then, neurointermediate pituitary lobes are dissected out and washed in complete medium (CM) consisting of 6.7 ml L-15 medium (GIBCO, Grand Island,

NY), 3 ml Milli-Q water, 100 μl kanamycin (GIBCO 5160), 100 μl antimycotic solution (GIBCO 5240), 0.8 mg $CaCl_2 \cdot H_2O$, and 2 mg glucose (pH 7.4), sterilized by ultrafiltration (0.22 μm, Millex-GV; Millipore, Bedford, MA). Subsequently, the lobes are transferred to 2 ml fresh CM and dissociated for 45 min in 0.5 mg/ml collagenase type V (Sigma, St. Louis, MO) and 10 mg/ml protease type IX (Sigma) in CM at 22°C. Next, they are carefully suspended by 10 passes through a siliconized Pasteur pipette, transferred to a syringe, and filtered through a nylon filter (pore size 150 μm) by air pressure. Neural lobe tissue remains on the filter. Dispersed cells are washed in CM containing 10% fetal calf serum (GIBCO). Finally, cells are collected by centrifugation (5 min, 50 g, 22°C), and the cell yield is determined using a hemocytometer and the trypan blue exclusion test (for details, see Ref. 34).

Labeling

Cells are cultured for 2 days in a centrifuge tube containing 0.75 ml CM with fetal calf serum in a shaking water bath at 22°C (medium is changed after 24 hr). Then, 50 μl of the cell suspension is placed on a poly(L-lysine)-coated glass slide, and cells are allowed to attach for 1 hr at 20 °C. For the determination of autoradiographic background labeling, two slides are immediately processed for immunocytochemistry (see below). Six slides are incubated with 100 μl of a solution of ^{125}I-BH-NPY (specific activity 2000 Ci/mmol; Amersham International, Amersham, England) in Ringer's solution, three with a concentration of 25 pM and three with 100 pM, for 45 min. As a control, for each concentration two slides are incubated in the same way, but an excess of unlabeled porcine NPY (1 μM) is added. After the incubations, slides are washed thoroughly in Ringer's solution for 3 min.

Immunocytochemistry and Autoradiography

Slides are stained by immunocytochemistry using an anti-αMSH serum raised in rabbits (35, 36). After immersion in 4% paraformaldehyde, slides are treated, sequentially, with 20% normal goat serum in phosphate-buffered saline (PBS), pH 7.6, for 10 min, anti-αMSH (diluted 1 : 500 in PBS) for 90 min, and goat anti-rabbit immunoglobulin G and rabbit-peroxidase-anti-peroxidase (rabbit-PAP) (Nordic Immunology, Tilburg, The Netherlands, 1 : 500 in PBS) for 1 hr. Finally, slides are treated with 0.02% 3,3'-diaminobenzidine, 0.6% nickel ammonium sulfate, and 0.005% H_2O_2 in PBS for 10 min. After drying, the slides are dipped in liquid autoradiographic L4 emulsion

(Ilford, Cheshire, UK). After exposure for 7 days at 4°C, the emulsion is developed in Kodak (Rochester, NY) D-19b developer (5 min at 20°C) and fixed in sodium thiosulfate (5 min).

Measurements and Statistics

Silver grains are counted using bright-field microscopy (magnification ×1000). On each slide at least 50 αMSH-positive cells and at least 15 stellate cells are examined. For each cell type, the number of labeled cells is expressed as a percentage of the total number of the counted cells, and the numbers of grains are expressed per cell. Measurements of control slides routinely show that the degree of background labeling is never more than 5 grains per cell. Therefore, in the final calculations only cells revealing more than 5 grains are considered.

A random sampling procedure is maintained throughout the experiment. The data are analyzed with a one-way analysis of variance ($a = 5\%$; Ref. 37), followed by the multiple range test of Duncan (38). The analysis is preceded by tests for the homogeneity of variance (Bartlett's test; Ref. 37) and for the joint assessment of normality (39).

Results and Discussion

Cell Yield

To avoid contamination of the dispersed cell preparations with red blood cells, animals were perfused intracardially with Ringer's solution. After dissociation of neurointermediate lobes it was necessary to determine the cell yield, so that an appropriate amount of cells can be brought onto a glass slide. As was shown by the trypan blue exclusion test, the cell yield after lobe dissociation was approximately 10,000 cells, which is about 15% of the total number of cells present in the lobe (36). Using a concentration of approximately 150 cells per 50 μl per slide, more than 90% of the cells appeared to lie separately, whereas the rest had formed small aggregates. The ratio of the two main cell types present in the pars intermedia, the melanotropes and the folliculostellate cells, was estimated at 10 to 1. Cells of other types, such as endothelial and muscle cells, were extremely scarce. Except for some small aggregates consisting of a few cells, dissociation was nearly complete. Differences occurred between dissociated cells and aggregated cells; both dissociated melanotropes and folliculostellate cells were rounded up (Figs. 1B and 2), whereas aggregated cells had retained their normal shape (melanotropes, polygonal; folliculostellate cells, with long slender processes; Figs. 1A and 3).

Specific Labeling of Cells

Before submitting the cells to autoradiography, cells were cultured for several days in order to enable them to synthesize new receptors. After 2 days of culture, folliculostellate cells could be readily distinguished from melanotropes by their larger size (20–25 μm), smaller, irregularly shaped nucleus that was excentrically located, and the αMSH-immunonegative cytoplasm (Figs. 1 and 2). After labeling, grains were found on the whole surface of the folliculostellate cell. Grains were preferentially located in the cell periphery and tended to be grouped together (Figs. 2 and 3). After incubation with 25 pM ^{125}I-BH-NPY, more than 75% of the cells showed silver grains, with a density of approximately 10 grains per cell. A similar labeling density was found after labeling with 100 mM ^{125}I-BH-NPY (Figs. 5 and 6). Apparently, the binding sites were already saturated at a concentration of about 25 pM. However, to gain more information about the binding capacities of the NPY binding sites, further investigations are needed in which a dose–response relationship is determined between label concentration and grain density.

In autoradiograms, melanotropes could be clearly recognized by their size (mean diameter of about 15 μm), their large, round nucleus, and strongly αMSH-immunopositive cytoplasm (Fig. 1). After labeling with 25 pM ^{125}I-BH-NPY, about 15% of the cells showed more than 5 silver grains (cells with 5 grains or more were considered to be positive). On the average, approximately 2 grains were present per cell (Fig. 1B). When a high concentration of radioactive ligand was used (100 pM ^{125}I-BH-NPY), the percentage of cells labeled as well as the number of grains present per cell were not significantly different (20% of the cells were labeled, and the number of grains per cell was approximately 4; see Figs. 5 and 6).

Control incubations were carried out by adding excess unlabeled NPY. At a concentration of 1 mM cold NPY, total suppression of binding was observed. Treatment with excess NPY did not significantly influence the percentage of labeled melanotropes (Fig. 5) or the number of grains per melanotrope cell (Fig. 6). Therefore, it seems that melanotropes do not have binding sites with high affinity for NPY, suggesting that few if any NPY receptors are present. Adding 1 mM NPY very strongly decreased the percentage of labeled folliculostellate cells as well as the number of grains on these cells; the values were similar to those for the melanotropes (Figs. 4–6; from more than 75% to less than 25%, and from more than 10 to less than 3, respectively).

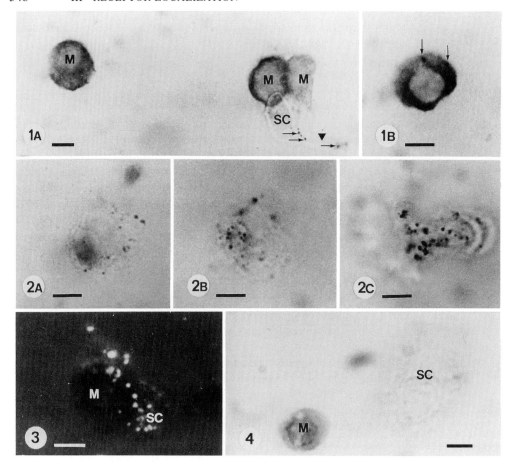

Fig. 1   Autoradiographed cell suspension of the pars intermedia of *Xenopus laevis*.
(A) Unlabeled isolated melanotrope cell (M), immunoreactive with αMSH, and an
aggregate of two melanotropes and labeled (arrows) folliculostellate cell (SC) with
slender process (arrowhead). (B) Melanotrope only slightly labeled (arrows) with
large, round nucleus and strongly immunopositive cytoplasm. Cells were incubated
in 25 pM ^{125}I-BH-NPY. Bar, 5 μm.

Fig. 2   Folliculostellate cells after labeling with 25 pM ^{125}I-BH-NPY for 45 min.
Note that silver grains are preferentially located at the periphery (A) and tend to be
grouped together (B, C). Bar, 5 μm.

Fig. 3   Aggregate of a melanotrope (M) and a folliculostellate cell (SC) labeled with
25 pM ^{125}I-BH-NPY for 45 min. Only the folliculostellate cell shows silver grains.
Bar, 5 μm.

Fig. 4   Unlabeled melanotrope (M) and folliculostellate cell (SC) after incubation
in 25 pM ^{125}I-BH-NPY and 1 μM NPY for 45 min. Note the absence of silver grains.
Bar, 5 μm.

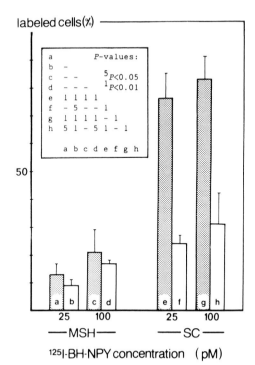

FIG. 5   Percentages (means ± SEM) of labeled αMSH-immunopositive melanotropes (MSH) and αMSH-immunonegative folliculostellate cells (SC) after incubation in 25 or 100 pM ^{125}I-BH-NPY without (▨) and with 1 μM NPY (□). Each mean value was statistically compared with all other mean values, using a one-way analysis of variance and Duncan's multiple range comparison test; significances are indicated at the 5 and 1% levels (see inset).

The following results are essential to conclude that the presence of ^{125}I-BH-NPY on folliculostellate cells is due to specific binding: (1) the almost complete abolishment of binding by incubation with excess NPY and (2) the absence of an effect of a strong increase in ligand concentration (from 25 to 100 pM) to increase the percentage of labeled folliculostellate cells or the labeling intensity per folliculostellate cell.

Concluding Remarks

Various autoradiographic techniques are available for localization of receptors for neuronal and endocrine messengers. The dispersed cell labeling approach presented in this chapter appears to extend the potentials of these

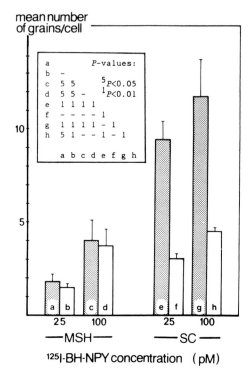

FIG. 6   Number of grains (means ± SEM) on αMSH-immunopositive melanotropes (MSH) and αMSH-immunonegative folliculostellate cells (SC) after incubation in 25 or 100 pM ^{125} I-BH-NPY, without (▨) and with 1 μM NPY (□). Means were statistically compared as described in Fig. 5.

methods, especially with respect to the accuracy of receptor detection at the cellular level. By studying single, living cells, binding sites can readily be localized and their density quantitatively determined. Moreover, the combined use of autoradiography and immunocytochemistry permits the identification of individual cell types under examination. It is expected that this method, used here for demonstrating NPY receptors of folliculostellate cells in the pituitary of *Xenopus laevis,* will be useful for receptor studies in various other tissues (cf. Ref. 40).

Acknowledgments

The authors are greatly indebted to Mr. P. M. J. M. Cruijsen for technical assistance and to Dr. B. G. Jenks for critically reading the manuscript. We also wish to thank Mr. R. J. C. Engels for the *Xenopus* husbandry.

References

1. K. Tatemoto, M. Carlquist, and V. Mutt, *Nature (London)* **296,** 659 (1982).
2. J. M. Danger, M. C. Tonon, B. G. Jenks, S. Saint-Pierre, J. C. Martel, A. Fasolo, B. Breton, R. Quirion, G. Pelletier, and H. Vaudry, *Fundam. Clin. Pharmacol.* **4,** 307 (1990).
3. J. M. Danger, J. Guy, M. Benyamina, S. Jegou, F. Leboulenger, J. Coté, M. C. Tonon, G. Pelletier, and H. Vaudry, *Peptides* **6,** 1225 (1985).
4. A. Pontet, J. M. Danger, P. Dubourg, G. Pelletier, H. Vaudry, A. Calas, and O. Kah, *Cell Tissue Res.* **255,** 529 (1989).
5. B. M. Chronwall, D. A. DiMaggio, V. J. Massari, V. M. Pickel, D. A. Ruggiero, and T. L. O'Donohue, *Neuroscience (Oxford)* **4,** 1159 (1985).
6. T. L. O'Donohue, B. M. Chronwall, R. M. Pruss, E. Mezey, J. Z. Kiss, L. E. Eiden, V. J. Massari, R. E. Tessel, V. M. Pickel, D. A. DiMaggio, A. J. Hotchkiss, W. R. Crowley, and Z. Zukowska-Grojec, *Peptides* **6,** 755 (1985).
7. C. Aoki and V. M. Pickel, *in* "Central and Peripheral Significance of Neuropeptide Y and Its Related Peptides" (J. M. Allen and J. I. Koenig, eds.), p. 186. Annals of the New York Academy of Sciences, New York, 1990.
8. A. Rudehill, A. Sollevi, A. Franco-Cereceda, and J. M. Lundberg, *Peptides* **7,** 821 (1986).
9. L. Edvinsson, R. Hakansson, C. Wahlestedt, and R. Uddman, *Trends Pharmacol. Sci.* **8,** 231 (1987).
10. B. Waeber, J. F. Aubert, R. Corder, D. Evequoz, J. Nussberger, R. Gaillard, and H. R. Brunner, *Am. J. Hypertens.* **1,** 193 (1988).
11. R. Y. Moore and J. P. Card, *in* "Central and Peripheral Significance of Neuropeptide Y and Its Related Peptides" (J. M. Allen and J. I. Koenig, eds.), p. 247. Annals of the New York Academy of Sciences, New York, 1990.
12. S. P. Kalra and W. R. Crowley, *Life Sci.* **35,** 1173 (1984).
13. L. Kerkérian, J. Guy, G. Lefevre, and G. Pelletier, *Peptides* **6,** 1201 (1985).
14. J. Guy, S. Li, and G. Pelletier, *Regul. Pept.* **23,** 209 (1988).
15. B. M. L. Verburg-van Kemenade, B. G. Jenks, J. M. Danger, and H. Vaudry, *Peptides* **8,** 61 (1987).
16. E. P. C. T. de Rijk, B. G. Jenks, H. Vaudry, and E. W. Roubos, *Neuroscience (Oxford)* **38,** 495 (1990).
17. H. P. de Koning, B. G. Jenks, W. J. J. M. Scheenen, E. P. C. T. de Rijk, R. T. J. M. Caris, and E. W. Roubos, *Neuroendocrinology* **54,** 68 (1991).
18. J. C. Martel, A. Fournier, S. Saint-Pierre, and R. Quirion, *Neuroscience (Oxford)* **36,** 225 (1990).
19. A. M. Rosier, G. A. Orban, and F. Vandesande, *J. Comp. Neurol.* **293,** 486 (1990).
20. J. C. Martel, A. Fournier, S. Saint-Pierre, Y. Dumont, M. Forest, and R. Quirion, *Mol. Pharmacol.* **38,** 494 (1990).
21. J. Wharton and J. M. Polak, *in* "Central and Peripheral Significance of Neuropeptide Y and Its Related Peptides" (J. M. Allen and J. I. Koenig, eds.), p. 133. Annals of the New York Academy of Sciences, New York, 1990.

22. J. I. Koenig, *in* "Central and Peripheral Significance of Neuropeptide Y and Its Related Peptides" (J. M. Allen and J. I. Koenig, eds.), p. 317. Annals of the New York Academy of Sciences, New York, 1990.

23. D. J. O'Halloran, P. M. Jones, J. H. Steel, M. A. Ghatei, J. M. Polak, and S. R. Bloom, *in* "Central and Peripheral Significance of Neuropeptide Y and Its Related Peptides" (J. M. Allen and J. I. Koenig, eds.), p. 329. Annals of the New York Academy of Sciences, New York, 1990.

24. C. Peng, Y. P. Huang, and R. E. Peter, *Neuroendocrinology* **52,** 28 (1990).

25. E. P. C. T. de Rijk, F. J. C. van Strien, and E. W. Roubos, *J. Neurosci.* **12,** 864 (1992).

26. J. C. Martel, S. Saint-Pierre, and R. Quirion, *Peptides* **7,** 55 (1986).

27. A. Westlind-Danielsson, A. Unden, J. Abens, S. Andell, and T. Bartfai, *Neurosci. Lett.* **74,** 237 (1987).

28. A. Härfstrand, K. Fuxe, L. F. Agnati, F. Benfenati, and M. Goldstein, *Acta Physiol. Scand.* **128,** 195 (1986).

29. M. W. Walker and R. J. Miller, *Mol. Pharmacol.* **34,** 779 (1988).

30. R. S. L. Chang, V. J. Lotti, D. J. Cerino, and P. J. Kling, *Life Sci.* **37,** 2111 (1985).

31. T. Nakajima, Y. Yashima, and K. Nakamura, *Brain Res.* **380,** 144 (1986).

32. M. Herkenham and L. Sokoloff, *Brain Res.* **321,** 363 (1984).

33. M. J. Kuhar, *in* "Neurotransmitter Receptor Binding" (H. I. Yamamura, S. J. Enna, and M. J. Kuhar, eds.), p. 153. Raven, New York, 1985.

34. E. Louiset, L. Cazin, M. Lamacz, M. C. Tonon, and H. Vaudry, *Neuroendocrinology* **48,** 507 (1988).

35. I. D. van Zoest, P. S. Heijmen, P. M. J. M. Cruijsen, and B. G. Jenks, *Gen. Comp. Endocrinol.* **76,** 19 (1989).

36. E. P. C. T. de Rijk, B. G. Jenks, and S. E. Wendelaar Bonga, *Gen. Comp. Endocrinol.* **79,** 74 (1990).

37. C. J. Bliss, *in* "Statistics in Biology," Vol. 1. McGraw-Hill, New York, 1967.

38. R. G. D. Steel and J. H. Torrie, *in* "Principles and Procedures of Statistics." McGraw-Hill, New York, 1960.

39. H. H. Shapiro and M. B. Wilk, *Biometry* **52,** 591 (1965).

40. S. James, C. J. S. Hassall, J. M. Polak, and G. Burnstock, *Cell Tissue Res.* **259,** 129 (1990).

[22] Progesterone and Estrogen Receptor Immunocytochemistry in the Central Nervous System

Maryvonne Warembourg and Pierre Poulain

I. Introduction

The ovarian steroid hormones estradiol and progesterone interact to influence regulation of female sexual behavior and reproductive physiology. The facilitation and inhibitory effects of these steroids are mediated by a direct action in the central nervous system (CNS) and involve interaction of estradiol and progesterone with neural intracellular receptors. Our knowledge concerning the anatomical distribution of steroid receptors in target organs was initially provided by autoradiographic studies using a radioactive steroid as a marker for the receptor. These procedures result in localization of radiolabeled ligand binding in steroid hormone target cells. Recently, development of specific monoclonal antibodies to estrogen receptor (ER) and to progesterone receptor (PR) has made it possible to detect the receptor molecules themselves utilizing immunocytochemical techniques. The localization of ER and PR is studied in the brain using light microscopic immunohistochemical methods. By means of a double immunocytochemical technique, it is possible to demonstrate simultaneously or sequentially in the same tissue section the coexistence of ER and PR. It is also possible to show the presence of a steroid receptor with a neuropeptide or a neurotransmitter. Immunohistochemical localizations of PR and ER are combined with other neuroanatomical procedures in order to obtain more information from estrogen receptor-immunoreactive (ER-IR) or progesterone receptor-immunoreactive (PR-IR) neurons. A single or a double immunocytochemical staining can be used in conjunction with the retrograde tracing technique to study the axonal projections of ER-IR or PR-IR neurons. In combined procedures, the general rule is that perfusion, fixation, and other tissue processing protocols should permit optimal visualization of immunohistochemically demonstrable compounds but also allow demonstration of neuronally transported fluorescent labels. This chapter gives a survey of available procedures and their applications in the CNS.

II. Gonadal Steroid Receptor Immunocytochemistry: Optimal Procedure for Use in Studies of the Central Nervous System

A. Monoclonal Antibodies

The purification of steroid hormone receptors has allowed the preparation of highly specific antireceptor monoclonal antibodies. Monoclonal mouse immunoglobulin G (IgG) antibodies (Mi 60, LET 126, and LET 64) raised against the rabbit PR are used for PR detection (1). Details of PR purification, immunization of mice, cell fusion, hybridoma cloning, and screening procedures have been described elsewhere (2). The characteristics and specificity of these antibodies have been studied by immunochemical and immunocytochemical methods (1–6). Moreover, among a selection of several well-characterized monoclonal antibodies the ability of these three antibodies to give the best immunocytochemical staining of PR has been tested in the uterus of various species (7). Comparative analysis has revealed that the three antibodies were the most sensitive monoclonals in guinea pigs, rabbits, and monkeys. Thus, our experiments are carried out in the guinea pig and occasionally in the monkey.

A monoclonal rat antibody to human ER is used for ER detection. This antibody (Abbott ER-ICA monoclonal kit) is purchased from the Abbott Diagnostic Products Laboratories (Rungis, France). The characterization and specificity of the antibody have been described previously (8, 9). The antibody reacts well with guinea pig and monkey tissues. The use of PR and ER antibodies as immunocytochemical reagents is described here.

B. Fixation and Perfusion

Fixation is immensely important in preparing the tissue for steroid receptor immunohistochemistry, as the morphological characteristics and the immunoreactivity of the desired antigen must be preserved while allowing the penetration of antibodies. Fixatives such as Carnoy's, Helly's, and ethanol preclude steroid receptor staining, whereas aldehyde fixatives (paraformaldehyde, glutaraldehyde) preserve the immunoreactivity of the receptor. Presently, the best procedure appears to be the use of 2% picric acid and 4% paraformaldehyde (PAF) in a 0.1 M sodium phosphate buffer, pH 7.4. In our experiments, localization of ER and PR in unfixed sections from frozen tissue has always failed. A comparison of results obtained after fixation by intracardiac perfusion and by immersion before or after sectioning of frozen brain tissue has revealed a marked superiority of the method involving fixation by perfusion. Perfusion of fixative through the vascular system is the method of choice for most areas of the brain. Nevertheless, in experiments done in our laboratory, it has been found that fixation by immersion in a PAF solution can be performed on 3- to 4-mm-thick slabs of neural tissue with little or no immunocytochemically detectable loss of steroid receptor activity (10). The perfusion of the animal starts with a washing solution consisting of an isotonic sodium chloride solution including heparin. Washing is important because blood cells have endogenous peroxidase activity that can give artifacts. In our experiments, 400-g guinea pigs are perfused for 20–25 min with a volume of 500 ml. The amount of fixative depends on the size of the animal. Following perfusion, the brain is immediately removed from the skull, cut into pieces, and immersed overnight in the same fixative at 4°C.

C. Preparation of Tissue Sections

Following postfixation, the brain pieces are soaked in a 10% sucrose solution in 0.1 M sodium phosphate buffer at 4°C until they sink to the bottom of the jar, usually after 24 hr. This treatment allows for removal of excess fixative and avoids formation of ice crystals during the freezing procedure. The tissue is then encased in Tissue tek O.C.T. compound (Miles Scientific, Naperville, IL) and frozen by immersion in liquid nitrogen-cooled isopentane. Fifteen-micrometer frontal sections are cut in a cryostat at −20°C and mounted on gelatin-coated slides. Often, frozen samples are kept at −80°C before sectioning.

Initial attempts to immunocytochemically identify progesterone receptors in brains fixed and processed routinely for paraffin embedding have met with little success. The use of cryostat or vibratome sections is necessary. This condition probably circumvents the problem of loss of receptors present inside target neurons at very low concentrations and not preserved in paraffin-embedded tissue sections.

D. Immunocytochemical Detection Systems

The sections are washed in 10 mM sodium phosphate-buffered saline (PBS), pH 7.4, for 30 min, three times for 10 min each, before processing for immunocytochemistry. Different immunocytochemical staining techniques can be used to visualize the primary antibody. In our laboratory, immunofluorescence and immunoperoxidase procedures for single PR and ER detection are used as follows.

1. Immunofluorescence

1. Sections are incubated with 5% nonimmune serum of the animal species in which the secondary antibody is raised (sheep in this procedure) in PBS for 20 min. This reduces the nonspecific binding of primary antibody.

2. Sections are treated with the mouse monoclonal anti-PR antiserum (2.5 µg/ml) in PBS or with the rat monoclonal anti-ER antiserum used at the dilution supplied for 18–64 hr at 4°C in a humid atmosphere.

3. The slides are washed with PBS three times for 5 min each, at room temperature. These washes are carried out after each subsequent incubation unless otherwise stated.

4. Sections are incubated with a 1:200 dilution in PBS of biotinylated sheep anti-mouse or anti-rat Ig (species-specific antibodies; Amersham International, Amersham, England) for 1.5 hr at 4°C.

5. Fluorescein isothiocyanate (FITC)–streptavidin diluted 1:80 or Texas red–streptavidin (Amersham) diluted 1:120 in PBS is applied to the section for 1.5 hr at 4°C.

6. Sections are coverslipped with glycerine–PBS (3:1, v/v).

7. Sections are examined under a Leitz Orthoplan fluorescent microscope equipped with the Ploemopak filter blocks I_2 (450–490 nm wavelengths) for identification of FITC and N_2 (530–560 nm) for identification of Texas red.

2. Immunoperoxidase Staining Procedures

a. Peroxidase–Antiperoxidase Method

1. Sections are incubated with 5% normal serum goat in PBS for 20 min. Steps 2 and 3 are the same as for the immunofluorescence method.

4. Sections are incubated in goat anti-mouse Ig (Nordic Immunological Laboratories, Tilburg, The Netherlands) diluted 1:60 or goat anti-rat Ig from the Abbott kit for 1 hr at room temperature.

5. Mouse peroxidase–antiperoxidase (PAP) (Sternberger Mayer, Jarrettsville, MD) at 1:100 dilution in PBS or rat PAP from the Abbott kit is applied to the section for 1 hr at room temperature.

6. The slides are flooded with 0.05% 3,3′-diaminobenzidine tetrahydrochloride (DAB) in Tris buffer, pH 7.6, and 0.05% H_2O_2 to reveal peroxidase staining. The appropriate time of incubation has to be judged by visual observation of the sections as they are turning brown; this usually takes 8–10 min. Recently, a simple method of nickel sulfate intensification of the DAB reaction product has been introduced to enhance the detection of steroid receptor immunoreactivity (11). According to this method, after rinsing in PBS, sections are washed twice in acetate–imidazole buffer (175 mM sodium acetate, 10 mM imidazole, pH 7.2) and are then incubated with 0.05% DAB

dissolved in a solution containing 100 mM NiSO$_4$ 125 nM sodium acetate, 10 mM imidazole at pH 6.5 to which H$_2$O$_2$ is added to a final concentration of 0.01%. The sections are incubated for 20 min, then transferred to fresh acetate–imidazole buffer. This step stops the reaction. Using this method, PR or ER immunoreactivity displays a dark blue color.

7. Sections are washed, dehydrated in ethanol followed by toluene, and covered with a synthetic mounting medium (Eukitt) under a coverslip. The material is routinely examined without counterstaining with a light microscope.

b. Biotinylated Antibody–Peroxidase–Streptavidin Method

Steps 1–4 are as described for the immunofluorescence method.

5. Streptavidin–horseradish peroxidase conjugate (Amersham) diluted 1:100 in PBS is applied to the section for 1 hr at room temperature.

6. Sections are processed in DAB as the chromogen, with or without intensification, and mounted as described (see Peroxidase–Antiperoxidase Method, steps 6 and 7).

c. Controls for Antibody Specificity

The specificity of the immunoreaction is checked by comparing sections stained with antireceptor antibodies and control antibody (mouse receptor-unrelated monoclonal antibody or normal rat immunoglobulin) used at the same concentration. Sections are also incubated with diluent only (PBS) or presaturated antiserum. In this study, only monoclonal anti-PR antibody is presaturated with highly purified PR (4) as previously described (5, 6); antibody to ER is not pretreated because purified ER is unavailable.

III. Applications and Advantages of Steroid Receptor Immunocytochemistry

A. Applications: Localization of Progesterone and Estrogen Receptors

Topographic maps on the distribution of gonadal steroid receptors have been obtained in the ovariectomized guinea pig brain by the techniques described in this chapter (6, 12). Adult female guinea pigs are ovariectomized 10–14 days before perfusion. Optimal immunocytochemical staining of PR is detected in ovariectomized guinea pigs that receive daily a subcutaneous injection of 10 μg estradiol benzoate, dissolved in 0.3 ml sesame oil, for 5 days. Estradiol priming induces PR. On the other hand, optimal staining of ER is obtained in ovariectomized guinea pig without estrogen treatment. In staining procedures, the PAP and biotin–streptavidin–peroxidase methods produce the most sensitive and reliable results under our experimental conditions. PR and ER immunoreactivities are present only in nuclei of target neurons; the nucleolus remains unstained. PR-IR neurons are mainly located in two specific regions: the preoptic area (periventricular preoptic and preoptic medial nuclei) and the mediobasal hypothalamus (periventricular arcuate, ventrolateral, and ventral premammillary nuclei).

Figure 1a is an example of the immunofluorescence method and shows the distribution of PR in the preoptic area. The localization of ER-IR cells in the guinea pig brain corresponds to that observed for PR in the preoptic area and the mediobasal hypothalamus. A striking difference in the immunocytochemical distribution is the occurrence of ER in the supraoptic nucleus and the amygdala. It should be noted, however, that the staining intensity in these regions is lower than that obtained in the hypothalamic areas. Figure 1b is an example of the PAP method without intensification of the DAB by nickel sulfate and shows estrogen receptors in the arcuate nucleus. Recently, we have studied the localization of PR-IR and ER-IR neurons in the monkey brain and pituitary. Results parallel those mentioned above for PR-IR and ER-IR neurons in guinea pig brain. Figure 1c is an example of the biotin–streptavidin–peroxidase method with intensification of DAB by nickel sulfate and shows the distribution of PR in the monkey pituitary.

B. Advantages of Immunocytochemical Technique

Until recently, autoradiography with radiolabeled steroid hormones has been used to localize the receptor; however, ligand-binding procedures have limitations. First, the labeled steroid fails to detect receptor that is already occupied by endogenous hormone. Second, from animals that have received a saturating tritiated dose prior to sacrifice, the autoradiographic technique is carried out on frozen, unfixed tissues in order to avoid the inability of fixed receptor protein to specifically bind steroid. As the association of hormone with receptor is noncovalent, the diffusion of the hormone should be prevented during experimental manipulations of preparations and long periods of emulsion exposure. Thus, the autoradiographic technique involves

FIG. 1   (a) Progesterone receptor immunoreactivity in a frontal section of the guinea pig preoptic area after immunofluorescence staining. (b) ER immunoreactivity in a frontal section of the guinea pig arcuate nucleus after PAP staining. (c) PR immunoreactivity in the monkey anterior pituitary after biotin–streptavidin–peroxidase staining with intensification of DAB by nickel sulfate. The third ventricle is indicated by asterisks. Bar, 150 μm for (a) and (b), 90 μm for (c).

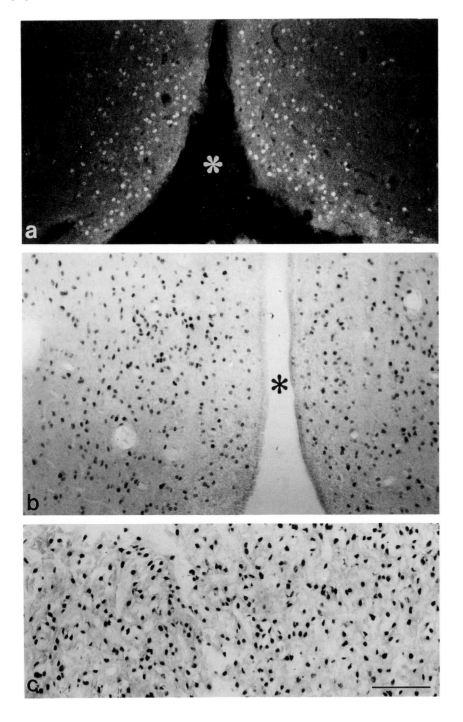

mounting frozen sections on dried photographic emulsions and excludes all fluids used in the classic histological techniques. Third, the presence within cells of nonreceptor steroid-binding proteins and local metabolism of steroids may also interfere. The immunocytochemical technique provides a new tool to initiate studies independently of steroid binding. With immunocytochemical demonstration, the PR or ER protein itself is detected. It becomes possible therefore to identify both unoccupied and occupied receptors under various physiological, pathological, and pharmacological conditions. Moreover, the immunocytochemical technique permits studies concerning induction and modulation of steroid hormone receptors. For the neuroanatomical localization of steroid receptors, the present immunocytochemical techniques are more sensitive than previously used autoradiographic studies (13–16). In guinea pigs, the concentration of PR-IR neurons is higher in the preoptic area and the mediobasal hypothalamus after immunocytochemistry than that obtained by autoradiography after injection of the ^{3}H-labeled progestin receptor ligand, R5020. The immunocytochemical technique detects receptors within individual neurons but does not give quantitative information about the concentration of receptors within those neurons. Nevertheless, differences in immunostaining intensity reflect variations in the content of receptors.

Past studies on the chemical characterization of neuroendocrine targets for steroid hormones have used a combination of autoradiography to detect steroid receptors and immunocytochemistry to mark specific cell types (neuropeptides, neurotransmitters) on the same tissue. Likewise, the methodological approach for the study of axonal projections of these targets was autoradiography associated with axonal retrograde tracing techniques. These procedures are technically difficult and often require compromises. Fixation is required for immunocytochemistry, since antigenicity is lost in the absence of fixative, and for axonal tracing. This fixation may, however, reduce the intensity of the steroid autoradiographic signal by removing some of the nuclear-bound steroid–receptor complexes. Moreover, the section thickness affects the efficiency of detection of the β particule arising from the ^{3}H-labeled steroid binding, and only a fraction of the population of positive cells is identified. These different factors may lead to an underestimation of the number of target cells and particularly hinder the appreciation of labeled neurons with lower levels of steroid receptor protein expression. In addition, long exposure times and photodevelopment of autoradiograms before immunocytochemical processing may decrease the intensity of immunostaining. It is also possible that neuropeptide-immunoreactive cells will not be detected by autoradiography if their nuclei are located too far from the emulsion, causing misinterpretations. At the present time, the preparation of specific antibodies to steroid receptors provides other approaches that permit one to avoid the use of autoradiography. Some of these combined procedures are considered.

IV. Applications of the Combined Techniques

The perfusion, fixation, and sectioning parameters, as well as the dilution of anti-ER and anti-PR antibodies and washes to be employed in these combined techniques, are those previously described, permitting optimal immunohistochemical visualization of steroid receptors.

A. Demonstration of Two Antigens

It is often desirable to demonstrate more than one antigen within the same neuron. Numerous combined methods are available for this. Below are described the protocols used in our laboratory to demonstrate the presence of PR and ER in the same neuron (12), taking advantage of the fact that anti-ER antibodies are raised in rats whereas anti-PR antibodies are raised in mice.

1. Double Immunofluorescence Method
The double staining procedure involves sequential incubations of tissue sections with mouse anti-PR antibody containing 5% nonimmune sheep serum for 24–64 hr at 4°C; sheep anti-mouse Ig conjugated to FITC (Amersham) diluted 1:50 for 1.5 hr at 4°C; rat anti-ER antibody for 24–48 hr at 4°C; biotinylated sheep anti-rat Ig diluted 1:200 for 1.5 hr at 4°C; and Texas red–streptavidin diluted 1:120 for 1.5 hr at 4°C. Sections are coverslipped with glycerine–PBS (3:1, v/v) and examined by fluorescence microscopy with either of the two standard filter systems for FITC and Texas red. The cells containing PR-IR show green fluorescence with FITC whereas the cells containing ER-IR in the same tissue exhibit red fluorescence with Texas red. By switching filters, coexistence can be directly established. There is no overlap in excitation wavelength between Texas red and FITC, and Texas red does not shine through when the FITC-induced immunofluorescence is analyzed. However, a convincing demonstration of colocalization of two receptors in the same neuron requires color photographs since both fluorophores occupy a ''nuclear'' position.

2. Immunofluorescence and Peroxidase–Antiperoxidase Method
For immunofluorescence, after incubation in the primary antiserum to the first antigen (either PR or ER) the sections are placed in biotinylated sheep

anti-mouse or anti-rat Ig and then in FITC–streptavidin or Texas red–streptavidin. Sections are wet-mounted as outlined above and observed by fluorescence microscopy. After photography, the coverslips are removed and the second antigen (either ER or PR) is visualized as a brown color (DAB) following all steps of the PAP method (Section II,D,2) and results compared to previous photographs (Fig. 2).

3. Double Immunoperoxidase Method

The first antigen (PR) is first demonstrated by the PAP method as above, but the peroxidase activity is in this case developed with a substrate giving another color than DAB (4-chloro-1-naphthol). The sections are treated with 0.06% 4-chloro-1-naphthol/0.02% H_2O_2 to yield a blush color and are wet-mounted. After bright-field photography, the coverslips are removed and the same sections are incubated overnight in sodium phosphate buffer containing 4% formaldehyde at 4°C to destroy the peroxidase activity of the first PAP (17). The sections are then washed 5 times in PBS, and the second antigen

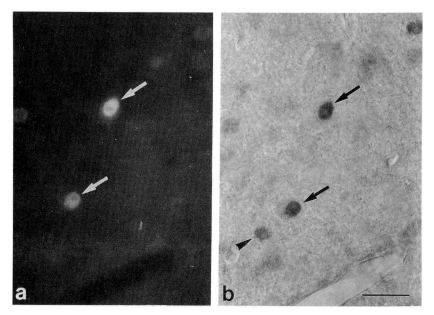

FIG. 2 Immunofluorescence (a) and immunoperoxidase staining (b). Photomicrographs were taken of the same frontal section through the guinea pig ventrolateral nucleus stained with antiserum to PR (a) and stained with antiserum to ER (b) after the first photography. Doubly labeled cell nuclei are indicated by arrows. The arrowhead indicates an ER-positive/PR-negative cell nucleus. Bar, 30 μm.

(ER) is visualized as a brown reaction product of the DAB by the PAP method. The dual immunoperoxidase method results in black-stained cell nuclei. This staining due to the overlap of the brown reaction precipitate on the blue product indicates that the two antigens coexist in the same neuron. Color mixing can be easily distinguished in the microscope but is, of course, difficult to see in black and white photomicrographs. This sequential immunoperoxidase staining is rather suitable for demonstrating two distinct and separate neuronal systems, when each tissue antigen can be labeled by a distinct color staining or be separated on the basis of cellular localization.

4. Controls

Controls for the sequential dual staining immunocytochemical procedures consist of replacing the primary antiserum of the first sequence with buffer, preabsorbed antiserum, or control antibody in the complete staining protocol; replacing the primary antiserum of the second sequence with buffer, preabsorbed antiserum, or control antibody; omitting each of the different constituents of the two linkages, one at a time; substituting the primary, secondary, and tertiary antiserum of the first linkage with the corresponding antiserum of the other linkage to test the absence of cross-reactivities between the individual immunoreagents.

B. Use of Pathway Tracing Procedures in Combination with Steroid Receptor Immunocytochemistry

Various combinations of steroid immunocytochemistry and tracing techniques can now be mastered without major pitfalls. In our studies, such combinations have been used to determine the projection of PR-IR and ER-IR neurons to different areas of the guinea pig brain. The axonal tracing is performed first, and, after a suitable survival time, the tissue is prepared and sections processed for immunofluorescence as mentioned above. The main obstacle to combining immunohistochemistry and retrograde tracers is the necessity for the dyes to be stable through the aqueous phase when incubation with antisera and rinsing procedures are performed. It is also necessary for the fluorescent dyes to be distinguishable from secondary immunostains.

Fluorescent axonal tracers used in our studies are Fluorogold (FG) (Fluorochrome, Inc., Englewood, CO), Lucifer Yellow CH (LY) (Fluka, Buchs, Switzerland), and Granular Blue (GB) (EMS Polyvoy, Gross Umstradt, Germany). All these tracers effectively survive the processing of the tissue for immunocytochemistry. The fluorescent marker for immunocytochemistry is Texas red. Demonstration of cell bodies labeled with the axonal tracer, with Texas red, or doubly labeled with the two fluorochromes is carried out by switching between filter systems. Filter block A (340–380 nm) was used to examine the fluorescence of FG- and GB-containing cell bodies and filter block H2 (390–490 nm) the fluorescence of LY-containing neurons. There was no interference between these tracers and Texas red under our conditions of observation.

1. Retrograde Axonal Tracing with Fluorogold

Fluorogold is regarded as the most intensely fluorescent of the tracers. From the site of administration, the dye is taken into the terminals by endocytosis and reaches the cell bodies by rapid retrograde axonal transport. However, the possibility of uptake and transport through axons of passage should be considered in the interpretation of experiments.

FG has been used in our experiments to demonstrate the projection of arcuate nucleus PR-IR neurons to the preoptic area (18). Iontophoretic injections in the brain are performed with micropipettes with 20–40 μm tip diameters, filled with a solution of FG. Animals are anesthetized and placed in a stereotaxic frame. Two solutions of FG are used. When using 1% FG in 0.1 M sodium acetate buffer, pH 3.3 (19), injections are successively achieved with positive currents of 1 μA up to 5 min (4 sec on, 4 sec off). When using 2% FG in 0.1 M sodium cacodylate trihydrate, pH 7.2 (20), stronger currents (5 μA up to 4 min) can be injected because clogging of the microelectrode tip is less frequent. Five days after the injection, animals are subjected to immunocytochemical analysis as described above. Spherical injection sites of 300–1000 μm in diameter are observed, according to the amount of current delivered. In every cases, a necrosis is apparent at the center of the injection site.

Cell bodies labeled with FG display fluorescence in the cytoplasm and the nucleolus. The nucleus is faintly fluorescent (Fig. 3a). The fluorescence of FG fades slightly during the immunocytochemical procedures. Nevertheless, silvered granules are always seen in the cytoplasm. Doubly labeled cells are easily identifiable (Fig. 3a,b).

FIG. 3   Fluorescent photomicrographs of frontal sections through the guinea pig arcuate nucleus demonstrating tracer-labeled cells (a, c, e) and PR-IR cells (b, d, f). Doubly labeled cells are indicated by arrows. In (a) the cell was retrogradely labeled after injection of FG in the preoptic area. In (c) cells were back-filled after application of a crystal of LY on the cut surface of the median eminence. In (e), cells were retrogradely labeled after intravenous injection of GB. Bar, 50 μm for (a) to (d), 23 μm for (e) and (f).

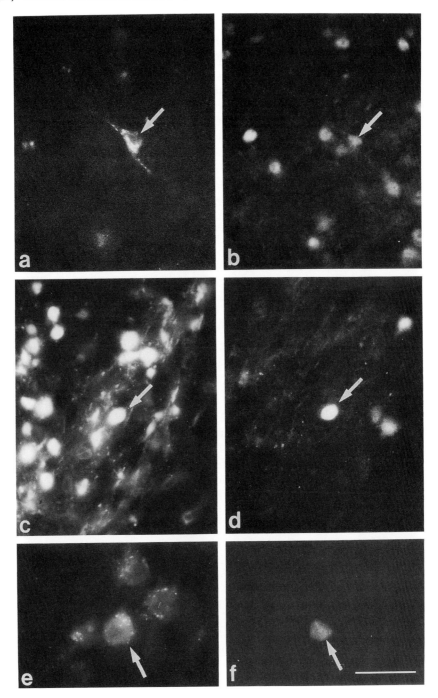

2. Neuron Back-filling with Lucifer Yellow CH

In our studies, LY has been used to fill neurons through their cut processes. Filling of the neuron is accomplished by diffusion of the tracer along the sectioned axon up to the cell body (back-filling) and does not involve active transport. A simple *in vitro* method has been developed that allows the labeling of hypothalamic neurons projecting directly to the median eminence. With this method, it has been demonstrated that only a small population of arcuate neurons, which contained PR receptor, project to the median eminence (10).

The brains are dissected and placed ventral side up. The median eminence is cut at the level of the pituitary stalk, and a small crystal of LY is placed on the cut surface for 1–2 min. After rinsing out superfluous LY with physiological saline, a piece containing the basal hypothalamus is excised. The tissue block is incubated at 4°C for 6 hr in 1 ml oxygenated Dulbecco's modified Eagle's medium containing 20 mM HEPES buffer (Flow Labs, Irvine, Scotland), 100 IU penicillin, and 0.1 mg streptomycin per milliliter. The block is fixed overnight in the usual fixative and the tissue processed as described above. PR immunoreactivity is well preserved under these conditions.

Back-filled neurons are strongly labeled with LY, which filled the cytoplasm and the nucleus (Fig. 3c). Although the nucleus is strongly fluorescent, detection of PR immunoreactivity does not suffer, and double labeling is observed (Fig. 3c,d).

3. Retrograde Axonal Tracing with Granular Blue

In our research, with the same purpose of labeling neurons projecting directly to the median eminence, administration of GB through the bloodstream has been used. It is known that intravenous or intraperitoneal administration of various fluorescent axonal tracers allows retrograde labeling of neurons projecting to areas devoid of blood–brain barrier, including the median eminence and the neurohypophysis (21). Obviously, incorporation of the tracer is achieved in terminals of intact neurons. With this method, parvocellular hypothalamic neurons which project to the median eminence have been identified. Some of these neurons contain PR immunoreactivity (10). Magnocellular neurons in the supraoptic and paraventricular nucleus of the hypothalamus which project to the neurohypophysis have also been identified (see Section IV,C).

FIG. 4 Photomicrographs of frontal sections through the anterior part of the supraoptic nucleus after triple labeling with GB. (a), Rabbit antiserum against oxytocin (b), and rat antiserum to ER (c). Triple labeled cells are indicated by arrows. OC, Optic chiasma. Bar, 50 μm.

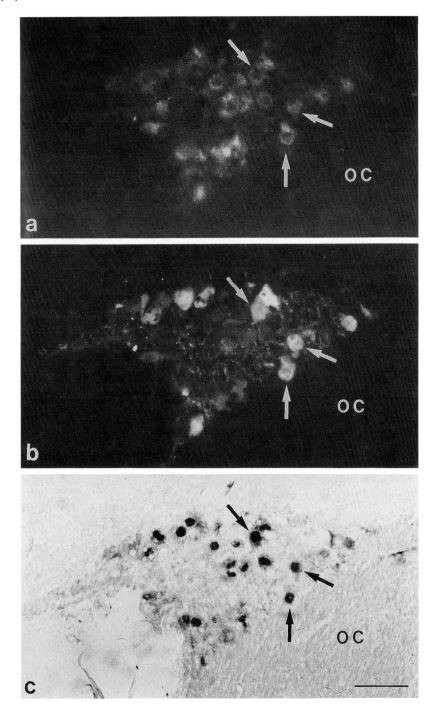

Small animals weighing 200 g have been chosen to save the tracer. Five milligrams GB is dissolved in 0.1 ml sterile water and sonicated, then injected intravenously through the jugular vein or intracardiacally. Injection does not result in death of the animal although the dye is not completely soluble. Long survival times (70–120 hr) result in an intense labeling of cell bodies after axonal retrograde transport and minimize labeling of neurons due to the spread of tracer into the arcuate nucleus. Animals are subjected to immunocytochemical analysis as described above.

Labeled cell bodies display silvery GB granules in the cytoplasm (Fig. 3e). Nucleoli appear as bright blue spots. No stain is present in the nucleus, thus facilitating observation of doubly labeled neurons (Fig. 3e,f). Production of GB was stopped since October 1991. This tracer can be successfully replaced with Fast Blue (same origin), injected under similar conditions of dilution and with similar survival times. A difference lies in the fact that the cytoplasm is homogeneously labeled with Fast Blue.

C. Combination of Retrograde Tracing and Sequential Demonstration of Neuropeptide and Steroid Receptor on the Same Tissue Section

The combination of retrograde fluorescent tracing with a sequential staining procedure for two antigens has been used in our experiments to demonstrate that ER-IR neurons in the supraoptic and paraventricular nuclei of the hypothalamus containing oxytocin immunoreactivity project to the neurohypophysis (22). It is essential that the two primary antisera be raised in different species. As the anti-ER antiserum is raised in rats, an antiserum directed against oxytocin raised in rabbits is used. Immunofluorescence and PAP methods are used on the same section from animals injected with GB. Sections are incubated in 1:2000 antioxytocin (U.C.B. Bioproducts, Braine-L'Alleud, Belgium) in 10 mM PBS containing 5% nonimmune donkey serum, overnight at 4°C. Oxytocin immunoreactivity is revealed by incubating the sections in 1:50 donkey anti-rabbit Ig conjugated to FITC (Amersham) for 1.5 hr at 4°C. The sections are wet-mounted and examined by fluorescence microscopy. The cells containing GB show blue fluorescence (Fig. 4a), whereas the cells containing oxytocin exhibit green fluorescence from FITC (Fig. 4b). After photography, the coverslips are removed, and sections are stained for ER immunoreactivity according to the PAP method as described above (Fig. 4c). Obviously, the method of photographing cells before the PAP procedure is long and expensive since it requires photographing many cells that subsequently may not stain for the antigen.

The application of combination techniques will provide important information on the interaction of steroid hormones with neuropeptidergic and neurotransmitter systems in the CNS. Studies on retrograde tracing of multiple antigen-containing neurons aid considerably in the neurochemical characterization of neuronal circuits.

Acknowledgments

We thank Professor E. Milgrom (INSERM, Unité 135, Bicêtre, France) for providing the monoclonal antibodies against the progesterone receptor used in our studies.

References

1. F. Lorenzo, A. Jolivet, H. Loosfelt, M. T. Vu Hai, S. Brailly, M. Perrot-Applanat, and E. Milgrom, *Eur. J. Biochem.* **176,** 53 (1988).
2. F. Logeat, M. T. Vu Hai, A. Fournier, P. Legrain, G. Butin, and E. Milgrom, *Proc. Natl. Acad. Sci. U.S.A.* **80,** 6456 (1983).
3. H. Loosfelt, F. Logeat, M. T. Vu Hai, and E. Milgrom, *J. Biol. Chem.* **259,** 14196 (1984).
4. F. Logeat, R. Pamphile, H. Loosfelt, A. Jolivet, A. Fournier, and E. Milgrom, *Biochemistry* **24,** 1029 (1985).
5. M. Perrot-Applanat, F. Logeat, M. T. Groyer-Picard, and E. Milgrom, *Endocrinology* (*Baltimore*) **116,** 1473 (1985).
6. M. Warembourg, F. Logeat, and E. Milgrom, *Brain Res.* **384,** 121 (1986).
7. M. T. Groyer-Picard, M. T. VuHai, A. Jolivet, E. Milgrom, and M. Perrot-Applanat, *Endocrinology* (*Baltimore*) **126,** 1485 (1990).
8. G. L. Greene, C. Nolan, J. P. Engler, and E. V. Jensen, *Proc. Natl. Acad. Sci. U.S.A.* **77,** 5115 (1980).
9. W. J. King and G. L. Greene, *Nature* (*London*) **307,** 745 (1984).
10. P. Poulain, M. Warembourg, and A. Jolivet, *J. Neurosci. Res.* **25,** 375 (1990).
11. L. L. Don Carlos, G. L. Greene, and J. I. Morrell, *Neuroendocrinology* **50,** 613 (1989).
12. M. Warembourg, A. Jolivet, and E. Milgrom, *Brain Res.* **480,** 1 (1989).
13. M. Warembourg, *J. Microsc. Biol. Cell.* **27,** 277 (1975).
14. M. Warembourg, *Brain Res.* **123,** 357 (1977).
15. M. Warembourg, *Neurosci. Lett.* **7,** 1 (1978).
16. M. Warembourg, *Mol. Cell. Endocrinol.* **2,** 67 (1978).
17. W. H. Oertel, M. L. Tappaz, A. Berod, and E. Mugnaini, *Brain Res. Bull.* **9,** 463 (1982).
18. M. Warembourg, P. Poulain, V. Jolivet, and A. Jolivet, *J. Neuroendocrinol.* **4,** 273, (1992).
19. V. A. Pieribone and G. Aster-Jones, *Brain Res.* **475,** 259 (1988).
20. L. C. Schmued and L. Heimer, *J. Histochem. Cytochem.* **38,** 721 (1990).
21. A. Van der Krans and P. V. Hoogland, *J. Neurosci. Methods,* **9,** 95 (1983).
22. M. Warembourg and P. Poulain, *Neuroscience* (*Oxford*) **40,** 41 (1991).

Section IV

Ligand Design

[23] N^α-Biotinylated Neuropeptide Y Analogs

A. Balasubramaniam, S. Sheriff, D. G. Ferguson,
S. A. Lewis Carl, and J. E. Fischer

Introduction

Neuropeptide Y (NPY), a 36-residue peptide amide originally isolated from
porcine brain (1), together with peptide YY (PYY) in the intestine (2) and
pancreatic polypeptide (PP) in the pancreas (3) constitute a family of homologous hormones. NPY is now regarded as a neurotransmitter or neuromodula-
tor widely distributed in the central and peripheral nervous systems. It is
also the most abundant peptide present in the mammalian brain and heart
(4). The properties of NPY have therefore been extensively investigated.
These studies have implicated NPY in a number of important regulatory
functions (see refs. 5 and 6 for reviews) and have led to the classification of
NPY among the most potent vasopressor and orexigenic peptides isolated
so far. PYY has also been shown to be equally effective in eliciting NPY-
like activities.

Although the pharmacological properties of NPY have received a great
deal of attention, there is much more to be done to understand the actions
of NPY at the receptor level. This aspect of investigation has been complicated due to the heterogeneity of NPY receptors (5, 6). In this regard, biotinyl-
NPY analogs may prove useful because the strong interaction between biotin
and the glycoprotein avidin has been exploited extensively to gain new
insights into ligand–receptor interactions. In this chapter, we describe the
strategies and methodologies that should be employed to prepare monobioti-
nylated analogs of NPY, **I** and **II** (Fig. 1), and illustrate how these analogs
could be used to visualize NPY receptors in rat cardiac ventricular tissues
and vascular smooth muscle cells. These analogs should also prove useful
in studying receptor distribution, down-regulation, purification, and receptor
subtypes.

General Strategies

1. Biotinylated ligands are, in most cases, prepared by coupling active esters
   of biotin with intact hormones. This strategy, however, produces many
   analogs. Therefore, the most desirable, fully active, biotinylated peptides

R - TYR - PRO - SER - LYS - PRO - ASP - ASN - PRO - GLY - GLU - ASP - ALA - PRO - ALA - GLU -
ASP - LEU - ALA - ARG - TYR - TYR - SER - ALA - LEU - ARG - HIS - TYR - ILE - ASN - LEU -
ILE - THR - ARG - GLN - ARG - TYR - NH$_2$

NPY, R = H

I BIOTIN-NPY, R =

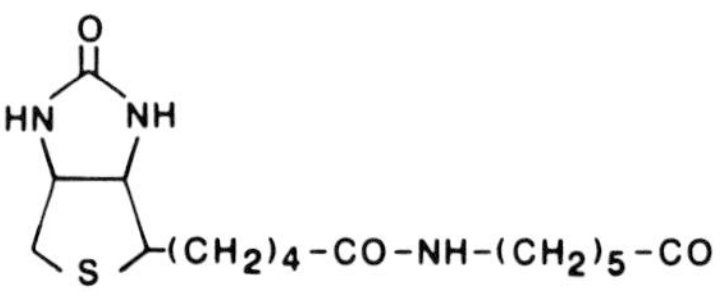

II BIOTIN-NPY, R =

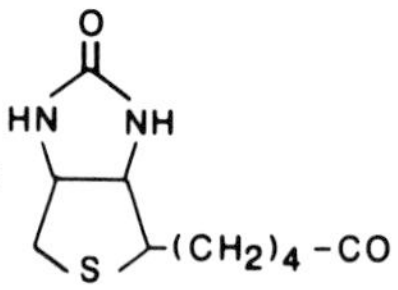

FIG. 1   Primary structures of NPY and N $^\alpha$-biotinylated NPY analogs. (Reprinted from Ref. 12 with permission from Pergamon Press.)

are obtained in low yields. These problems are avoided by incorporating the biotinylated group during the stepwise solid-phase peptide synthesis (SPPS) of intact peptide hormones.

2. The bulky biotinyl group can hinder the interaction of the ligand with its receptor. This is minimized by incorporating the biotinyl group at the N-terminal end, away from the receptor-recognizing C-terminal region of NPY. (If the N-terminal region is the active site, as in vasoactive intestinal polypeptide, the biotinyl group is introduced at the C-terminal end by starting the synthesis on N^α-Boc-N-$^\varepsilon$-biotinyl-Lys-resin).*

3. The usefulness of the biotinylated ligand depends on its ability to interact simultaneously with biotin antibodies/avidin and the peptide receptor (i.e., bifunctionality). This is facilitated by the introduction of a spacer group such as 6-aminohexanoic acid between the biotinyl group and the peptide.

4. The ability of the biotinylated hormones to exhibit comparable biological potency as the intact hormone as well as bifunctionality is first verified in a radioreceptor assay system before using the analogs in any investigation.

* IUPAC recommended abbreviations are used for amino acids and amino acid derivatives.

Materials

All the Boc-amino acid derivatives (Peninsula Labs, Palo Alto, CA), peptide synthesis reagents (Applied Biosystems, Foster City, CA), and solvents (American Scientific Products, McGraw Park, IL) are obtained commercially and used without further purification. N-Hydroxysuccinimide esters of biotin and (6-biotinylamido)hexanoic acids are purchased from Pierce Chemical Company (Rockford, IL). All the buffer reagents, proteolytic enzyme inhibitors, and histochemical reagents, unless otherwise stated, are obtained from Sigma Chemical Company (St. Louis, MO). Sprague-Dawley adult and neonatal rats are purchased from Charles River Laboratories (Wilmington, MA).

Preparation of Monobiotinylated Neuropeptide Y Analogs

Synthesis

Peptide synthesis is carried out on an Applied Biosystems Model 430A automated synthesizer utilizing the programs provided by the manufacturer for synthesis using p-α-tert-butyloxy-carbonylamino acid derivatives with the following side chain protecting groups: Tyr(2Br-Z), Lys(2Cl-Z), Ser(Bzl), Glu(OBzl), Asp(OBzl), Arg(Tos), His(Bom), and Thr(OBzl). p-Methylbenzhyldrylamine resin (0.45 mmol amino groups /g) is placed in the reaction vessel of the peptide synthesizer and the amino acids are coupled sequentially as the preformed symmetrical anhydrides (1.00 mmol, 2.2 equivalents). Boc-Arg(Tos), Boc-Gln, and Boc-Asn, however, are coupled as the preformed 1-hydroxybenzotriazole esters (HOBT) (2 mmol, 4.4 equivalents) to avoid dehydration and/or lactam formation. All amino acids are coupled twice. The symmetrical anhydrides and the HOBT esters are generated automatically by the synthesizer. The synthesis is terminated using the end-nh2 function which deprotects the N^α-Boc group, neutralizes, and air dries the side chain protected peptide resin. The partially protected resin is dried under reduced pressure overnight and divided into three equal batches (~0.15 mmol each).

Biotinylation

Biotinylation is carried out manually in a two-neck round-bottomed flask (50 ml) fitted with a drying tube. Peptide resin with free α-NH$_2$ groups (~0.15 mmol) is suspended in dimethylformamide (DMF) (20 ml) in the round-bottomed flask and stirred using a magnetic bar for 20 min before adding diisopropylethyl amine (0.15 mmol), HOBT (0.30 mmol) and *N*-hydroxysuccinimidyl esters of biotin, or (6-biotinylamido)hexanoic acid (0.30 mmol) (Fig. 1). Biotinylation is generally complete within 18 hr as judged by a qualitative ninhydrin test (7). The peptide resin is transferred to a fritted filter funnel, washed repeatedly with DMF, CH$_2$Cl$_2$, and ethanol, and dried overnight *in vacuo* at room temperature.

Hydrogen Fluoride Cleavage

HF is a corrosive gas, and therefore all operations are handled in a Teflon HF apparatus (Toso Kasei; Protein Research Foundation, Osaka, Japan) housed in a well-ventilated hood, taking all the precautions described by Stewart and Young (8). The HF method, if carried out properly, gives superior quality peptides compared to other cleavage methods. Biotinylated peptide resin (~0.15 mmol), *p*-cresol (~0.5 g), *p*-thiocresol (~0.25 g) (required to avoid oxidation of the thioether group in the biotin moiety), and a magnetic bar are placed in the reaction vessel of the HF apparatus. The HF line is evacuated, and approximately 10 ml of HF is collected by cooling the reaction vessel with a dry ice–acetone mixture. The vessel is then allowed to warm up to −2 to 0°C and stirred at this temperature for 1 hr. HF is evaporated at −2 to 0°C and the residue transferred to a fritted filter funnel using diethyl ether (2 × 30 ml) and washed with diethyl ether (2 × 30 ml) to remove the organic scavengers. The crude peptide product is then extracted with 30% acetic acid (2 × 15 ml), diluted to 10%, and lyophilized. Free NPY is obtained similarly using a mixture of HF (~10 ml) and *p*-cresol (~0.5 g).

Purification

Purification is accomplished by reversed-phase chromatography on a Waters (Milford, MA) Model 600 high-performance liquid chromatography (HPLC) system in conjunction with the U6K injector, Model 481 Spectrophotometer, and Baseline 810 Data Collection system installed in an IBM-XT computer. The crude peptide (20 mg/run) is dissolved in 6 *M* guanidine hydrochloride (2 ml) and passed through a 0.45-µm filter (Micron Separations, Inc., Westboro, MA) to remove particulates. The peptide solution is then loaded onto a Vydac (Hesparia, CA) semipreparative column (C$_{18}$, 250 × 10 mm, 10 µm particle size, 300 Å pore size) and purified using 1% triethylammonium phosphate, pH 2.25 (solvent A), and 60% acetonitrile in 1% triethylammonium phosphate, pH 2.25 (solvent B) (Fig. 2A). The fractions (4.7 ml each)

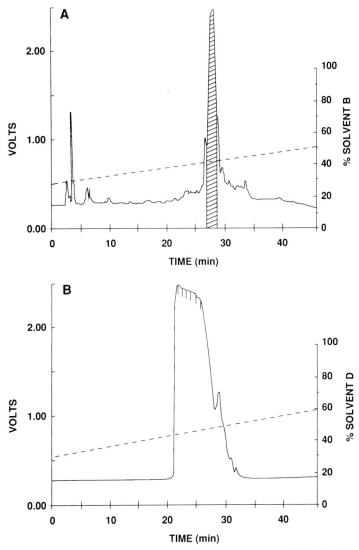

FIG. 2   (A) Semipreparative reversed-phase chromatography of biotin-NPY analog
I (~20 mg) obtained by HF cleavage. Fractions corresponding to the shaded region
are collected. (B) Desalting of the combined peptide fractions obtained in five semipreparative purification runs as shown in (A). Fractions indicated by the vertical
lines are collected. Conditions are described in the text.

corresponding to the major peak obtained in five such runs are combined, diluted with 2 volumes of water, and pumped onto the same column. The column is flushed with 0.1% aqueous trifluoroacetic acid (solvent C) for 15 min to remove the salts and is then subjected to a gradient generated using solvent systems C and D (0.1% trifluoroacetic acid, 19.9% water, and 80% acetonitrile) (Fig. 2B). The homogeneity of the fractions obtained in the desalting steps is verified on a Vydac analytical column (C_{18}, 250 × 4.6 mm, 5 μm particle size, 300 Å pore size) using solvent systems C and D (e. g., Fig. 3). The homogeneous fractions are concentrated in a Speed-Vac centrifuge (Savant, Farmingdale, NY), pooled with an equal volume of water, and lyophilized.

Characterization

The purified peptides are characterized in our laboratory by analytical reversed-phase chromatography and amino acid analysis (Waters Pico-Tag system). Mass spectral analyses are performed at Beckman Research Institute (City of Hope, CA). The peptides had the expected amino acid compositions. Furthermore, the observed and theoretical masses of NPY (4253.8

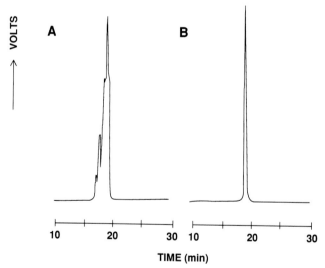

FIG. 3   Verification of the purity of the fractions obtained in the desalting step (Fig. 2B) by analytical reversed-phase chromatography. Gradient: 40–100% D in C in 55 min. Chromatograms of fractions 1 (A) and 2 (B) are shown as examples. See text for conditions.

versus 4254.7), analog **I** (4593.5 versus 4593.2), and analog **II** (4481.4 versus 4481.1) are in close agreement.

Verification of Applicability of Biotinylated Neuropeptide Y Analogs to Probe Cardiac Receptors

Binding Studies

Radioreceptor assays (9) are performed by incubating rat cardiac ventricular membranes (200 μg protein) with ^{125}I-labeled NPY (40 pM) in the presence of various concentrations of NPY or biotinylated NPY in a total volume of 0.25 ml of 20 mM HEPES assay buffer, pH 7.6, containing 2% bovine serum albumin (BSA), 100 μM phenylmethylsulfonyl fluoride (PMSF), 4 μg/ml leupeptin, 4 μg/ml chymostatin, 5 kIU/ml aprotinin, and 0.1% bacitracin for 2 hr at 18°C in a shaking water bath. The incubation mixture is then mixed on a vortex mixer, and 150-μl aliquots are transferred to polypropylene tubes containing 250 μl ice-cold assay buffer. The ^{125}I-labeled NPY bound to the membranes is separated by centrifugation at 10,000 g for 10 min followed by aspiration of the supernatant. The radioactivity in the pellets is determined in a Micromedic γ counter.

Comment

^{125}I-labeled NPY tagged at only the N-terminal tyrosine exhibits superior binding to cardiac membranes compared to other NPY tracers, including those obtained from a number of commercial sources (9). The radiolabeled NPY is prepared by the standard chloramine-T method, and the iodination mixture is purified on a Vydac C$_4$ column (250 × 4.6 mm, 5 μm particle size, 300 Å pore size) under isocratic conditions of 32% acetonitrile containing 0.1% trifluoroacetic acid. The first of the two major peaks obtained contains the NPY labeled at N-terminal tyrosine.

Adenylate Cyclase Activity

Assay of adenylate cyclase activity is performed according to our published procedure (10). Each experiment is carried out in a total volume of 200 μl containing 30 mM Tris-HCl, pH 7.4, 150 mM NaCl, 8.25 mM MgCl$_2$, 0.75 mM EGTA, 1.5 mM theophylline, 20 μg/ml aprotinin, 100 μg/ml bacitracin, 1 mg/ml BSA, 1 mM ATP, 20 mM creatine phosphate, 1 mg/ml phosphocreatine kinase, 10 μM isoproterenol, 10 μM GTP, and various concentra-

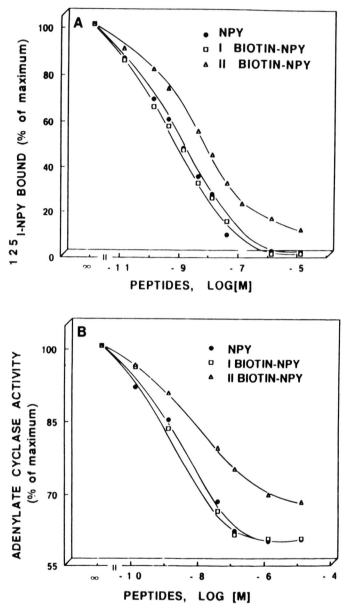

FIG. 4   (A) Effects of increasing concentrations of biotin-NPY analogs and NPY on
[125]I-labeled NPY binding to rat cardiac ventricular membranes. Results are expressed
as the percentage of ^{125}I-labeled NPY binding obtained in the absence of competing
peptides. (B) Effects of increasing concentrations of biotinylated NPY analogs and

tions of peptides (0–10 μM). The reaction is initiated by the addition of 50 μg (50 μl) of membrane proteins. After incubation for 10 min at 35°C in a shaking water bath, the reaction is arrested by adding 100 μM EDTA and boiling for 3 min. The cAMP formed is quantitated using a radioimmunoassay kit (Du Pont–NEN, Boston, MA, NEK033), according to the instructions provided by the manufacturer.

Bifunctionality

Simultaneous binding of the biotinylated NPY analogs to cardiac NPY receptors and to avidin is examined under the same conditions used in the binding experiments, but with biotinylated NPY analogs instead of ^{125}I-labeled NPY. After incubating analogs I or II with the membranes in the presence of 10 μM NPY, the unbound analog is separated by centrifugation at 10,000 g for 10 min followed by aspiration of the supernatant. The pellet with the bound analog is suspended in a total volume of 250 μl assay buffer and incubated with ^{125}I-labeled avidin (350,000 cpm/tube) (Amersham, Arlington Heights, IL) for 30 min at 18°C in a shaking water bath. The unbound ^{125}I-labeled avidin is separated, and the radioactivity remaining in the pellet is determined as described above.

Inference

Although NPY and both analogs I and II inhibit ^{125}I-labeled NPY binding and adenylate cyclase activity in rat cardiac ventricular membranes in a dose-dependent manner, only analog I with the 6-aminohexanoic acid spacer arm exhibits comparable potency to NPY (Fig. 4). Furthermore, analog I exhibits superior bifunctional properties than analog II (Fig. 5). These observations suggest that the spacer arm is crucial to facilitate the NPY–receptor and biotin–avidin interactions with minimal hindrance. Analog I is therefore chosen to probe NPY receptors.

NPY on the isoproterenol-stimulated adenylate cyclase activity of rat cardiac ventricular membranes. Results are expressed as the percentage of isoproterenol-stimulated activity in the absence of peptides. (Reprinted from Ref. 12 with permission from Pergamon Press.)

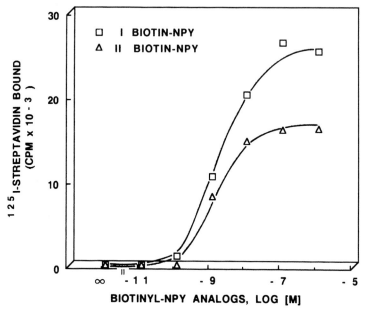

FIG. 5   Binding of streptavidin to biotin-NPY analogs bound to rat cardiac ventricular membranes. Biotin-NPY analogs are bound to rat cardiac ventricular membranes in the absence and presence of 10 μM NPY for 90 min. The membranes are then washed and incubated with ^{125}I-labeled streptavidin for 30 min and the bound radioactivity determined. The presence of 10 μM NPY completely inhibited the binding of ^{125}I-labeled streptavidin (not shown).

Immunofluorescence Labeling of Neuropeptide Y Receptors

Cardiac Receptors

Rats (250–300 g) under anesthesia are sacrificed by cervical dislocation. The heart is rapidly excised and the ventricles separated by dissection. Small samples of fresh ventricular tissue (~2 mm in diameter and 5–10 mm long) are attached to flat cork specimen mounts using gum traganth and are immediately snap-frozen in isopentane cooled in liquid nitrogen. The cork mounts are attached to precooled specimen holders using Tissue Tek O.C.T. compound, and frozen sections, 4–6 μm thick, are cut using a cryostat microtome. The sections are mounted on cooled glass slides precoated with 0.5% gelatin. The slides are warmed to room temperature and incubated for 45 min with phosphate-buffered saline (PBS) containing BSA (150 mM NaCl, 20 mM Na_2HPO_4, and 1% BSA, pH 7.2) to block nonspecific binding sites. Excess

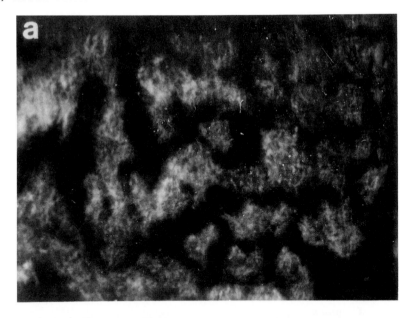

FIG. 6   Cryostat sections of rat cardiac ventricular muscles treated with (a) analog
I followed by antibiotin–FITC or (b) a slight excess of NPY followed by analog
I and then antibiotin–FITC. Magnification: ×1300. (Reprinted from Ref. 12 with
permission from Pergamon Press.)

blocking solution is removed using a fine-tipped pipette, and the sections are then incubated with 20 μg biotinylated NPY analog **I** for 1 hr in 100 μl PBS/BSA solution containing 100 μM PMSF and 0.1% bacitracin. The incubation solution is removed, and the sections are thoroughly washed with three 20-min changes of PBS/BSA.

The NPY receptors are visualized via fluorescein isothiocyanate (FITC) fluorescence by incubating the sections in a 100-μl aliquot of affinity-purified goat antibiotin–FITC secondary immunoprobe (Sigma Immunochemicals) for 1 hr at room temperature. The excess secondary label is removed with three 20-min changes of PBS/BSA. The sections are mounted using Citifluor (Citifluor Ltd., London, England), a glycerol-based mounting medium. The coverslips are sealed using nail polish, and the specimens are examined immediately following mounting. Fluorescence micrographs are obtained using a Nikon Labophot microscope. Experiments are repeated by incubating tissues with biotinylated NPY solutions containing a 12-fold excess of NPY. Additional control experiments are performed in which sections are treated only with goat antibiotin–FITC secondary immunoprobe.

Comment

Best results are obtained when sections are stained immediately after collection, but it is possible to store the sections at −20°C for several days before staining. Also, Citifluor gives better results than other mounting media.

Vascular Smooth Muscle Receptors

Study is performed using sympathetic neuron–mesenteric vascular muscle coculture. The mesenteric blood vessels and the superior cervical ganglion dissected from neonatal rats are enzymatically dispersed and cultured together in Dulbecco's modified Eagle's medium supplemented with 10% fetal calf serum and nerve growth factor (50 ng/ml) in collagen-coated culture dishes (35 mm) according to the procedure described by Ferguson *et al.* (11). Culture medium is removed from the dishes after 4 days, and the cells are fixed in 4% formaldehyde for 10 min. The cells are washed with three 10-min changes of PBS and then permeablized by incubating in 0.5% Triton

---

FIG. 7   (a) Mesenteric smooth muscle cells cocultured in the presence of neurons from the superior cervical sympathetic ganglion treated with analog **I** followed by antibiotin–FITC. (b) Mesenteric smooth muscle cells cultured in the absence of sympathetic neurons treated with analog **I** followed by antibiotin–FITC. (c) Cocultures treated with only antibiotin–FITC. Magnification: ×1300.

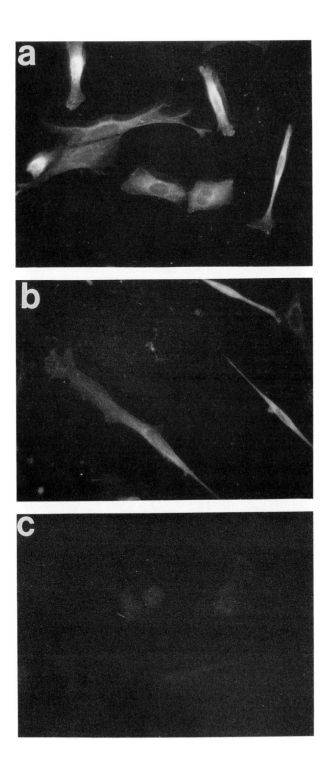

X-100 for 15 min. The cells are washed with three 10-min changes of PBS to remove excess detergent. Nonspecific binding sites are blocked by incubating the cells for 45 min in PBS/BSA. Excess blocking solution is removed, and the cells are incubated for 30 min in 800 μl PBS/BSA solution containing 200 μg biotinylated NPY analog **I**. The cells are washed with three 10-min changes of PBS/BSA to remove biotin-NPY. A 1-ml aliquot of antibiotin–FITC (1 : 200 dilution in PBS/BSA) is then applied, and the cells are incubated for 1 hr at room temperature. The excess secondary label is removed by washing for 10 min with PBS/BSA and then with two 10-min changes of PBS. The cells are coverslipped using Citifluor and examined as before.

Inference

Localization experiments show that analog **I** can label the NPY receptors in both cardiac ventricular membranes and mesenteric smooth muscle cells (Figs. 6 and 7). These labelings are specific since intact NPY blocked the staining. Furthermore, it appears that neurogenic influences may prolong the retention of the smooth muscle phenotype because mesenteric smooth muscle cells cultured alone are less brightly stained than those cocultured with sympathetic neurons (Fig. 7).

Acknowledgments

This work was supported in part by grants from National Institutes of Health [GM38601 (A.B.) and HL34779 (D.G.F.).] The mesenteric smooth muscle–sympathetic neuron cocultures were generously provided by Dr. Ray Pun, Department of Physiology and Biophysics, University of Cincinnati Medical Center, and this work was supported in part by an American Heart Association Ohio affiliate Grant-in-Aid (SW-91-18).

References

1. K. Tatemoto, *Proc. Natl. Acad. Sci. U.S.A.* **79,** 5485 (1982).
2. K. Tatemoto, *Proc. Natl. Acad. Sci. U.S.A.* **79,** 2514 (1982).
3. J. R. Kimmel, L. J. Hayden, and H. G. Pollock, *J. Biol. Chem.* **250,** 9369 (1975).
4. J. K. McDonald, *Crit. Rev. Neurobiol.* **4,** 97 (1988).
5. C. Wahlestedt, L. Grundemar, R. Håkanson, M. Heilig, G. H. Shen, Z. Zukowska-Grojec, and D. J. Reis, *Ann. N.Y. Acad. Sci.* **611,** 7 (1990).
6. S. Sheriff and A. Balasubramaniam, *J. Biol. Chem.* **267,** 4680–4685 (1992).

7. E. Kaiser, R. L. Colescott, C. D. Bossinger, and P. I. Cook, *Anal. Biochem.* **34,** 595 (1970).
8. J. M. Stewart and J. D. Young, "Solid Phase Peptide Synthesis." Pierce Chemical Co., Rockford, Illinois, 1984.
9. A. Balasubramaniam, S. Sheriff, D. F. Rigel, and J. E. Fischer, *Peptides* **11,** 545 (1990).
10. A. Balasubramaniam and S. Sheriff, *J. Biol. Chem.* **265,** 14724 (1990).
11. D. G. Ferguson, S. A. Lewis and R. Y. K. Pun, *Adv. Exp. Med. Biol.* **304,** 507 (1991).
12. A. Balasubramaniam, S. Sheriff, D. G. Ferguson, M. Stein, and D. F. Rigel, *Peptides* **11,** 1151 (1990).

[24]   Designing Selective Analogs for Peptide Receptors: Neuropeptide Y as an Example

John L. Krstenansky

Introduction

Both agonist and antagonist analogs having receptor selectivity are invaluable tools in dissecting the physiological and pharmacological importance of a receptor system. Often the value of a particular system as a pharmaceutical target can only be assessed when the appropriate analogs have been discovered. Therefore, there is often a considerable synthetic effort given to a particular hormone or neurotransmitter when it is discovered, even when its utility in treating a disease state is unproven. Recent examples of this are atrial natriuretic factor (ANF), endothelin, and amylin. Neuropeptide Y (NPY) could also fall into this category except that the level of synthetic effort which has been given to it is far less than that given to ANF and endothelin. Although NPY may have physiological roles in the regulation of appetite, memory, blood pressure, gastrointestinal secretion, and kidney function, it is not certain that selective agonists or antagonists would be useful in the treatment of human disease. However, the most definitive way of determining the therapeutic utility of receptor-selective NPY agonists and antagonists is to discover or design them and test them in the appropriate models. This is the approach that we, as well as a number of other groups, pursued. At least two groups, Pfizer and Heumann Pharma (1), utilized a screening protocol, and Tatemoto used a soluble peptide library approach (2), but most groups applied systematic and/or rational design schemes (3–43).

How one approaches the design of selective ligands for peptide receptors can vary depending on the nature of the natural peptide ligand. Short peptides (<15 amino acids) are typically treated somewhat differently from the medium to large peptides. Reviews of design strategies for smaller peptides are more numerous than those which are applicable to the larger peptides (45–49). This chapter discusses approaches to the design of larger peptides using neuropeptide Y as an example.

The Problem

Peptide hormones and neurotransmitters consisting of 20–50 amino acids present a unique challenge to the medicinal chemist. Smaller peptides often assume conformational states involving reverse turns and can be dealt with

Methods in Neurosciences, Volume 12

by focusing on primary and secondary structure. Larger peptides are more complex and in some cases may utilize tertiary structure in their receptor-bound conformations.

Avian pancreatic polypeptide has been described by Blundell and co-workers as a small globular protein (50–53). Many larger peptides contain amphipathic α-helical regions (54), such as members of the glucagon/secretin family (vasoactive intestinal peptide, glucagon-like peptide, helodermin), members of the pancreatic polypeptide family (avian pancreatic polypeptide, neuropeptide Y, peptide YY), members of the calcitonin family (calcitonin gene-related peptide, amylin), and corticotropin-releasing factor. However, at this time no general rules are evident on how these larger hormones interact with their receptors. For example, the function of the amphipathic α helix in many of these hormones is not known. Is it there as a lipid-binding moiety so that the membrane lipid can act as a "catalyst" of the peptide–receptor interaction (55, 56)? Does it bind to an amphipathic site on the receptor itself? Does it serve a structural role in the stabilization and orientation of other portions of the peptide so that they are properly poised for receptor interaction? Is the role of this structural feature different in each peptide, or is there a universal theme present here?

Therefore, at present we must analyze our target individually, being careful about assumptions based on data from related peptides. It is also essential to develop as deep an understanding of the nature of the receptor as possible, because utilizing assay systems that consist of multiple receptor subtypes only complicates the interpretation of the data obtained. Ideally, systems having predominantly one receptor subtype yield the most useful information for compound design and for understanding the function of the individual receptor subtypes. Unfortunately, the receptor pharmacology for a given peptide often becomes clear only when the selective analogs are already available.

Neuropeptide Y is an interesting example to study. It is not a "solved" problem, but work has been performed by several groups that have yielded important insights.

Analysis of the Ligand

One of the most useful starting points in examining new peptide ligands is to study what is known about homologous peptides. Often the new peptide is a member of a family of peptides for which there may be considerable information on one or more of the other members. For neuropeptide Y this was the case. It is a member of a family of peptides that contain the pancreatic polypeptides (PP) and peptide YY (PYY) (57). Figure 1 gives the sequences of this family of peptides. It was recognized early that the pancreatic polypeptides and NPY utilized separate receptor systems (65). The "Y" peptides, NPY and PYY, could either potently bind to the receptors of one another, or they utilized the same receptors and their subtypes (66–68).

With this information, the homology to other family members gives some instant structure–activity relationships (SAR) from which one can gain insights into regions of importance for potency and receptor selectivity. In 1987, based on the sequences that were known at the time, we noted that some residues were common to the entire family (residues 2, 5, 8, 9, 12, 13 or 14, 15, 27, 29, 32, 33, 35, and 36), some were common only to the "Y"

		10	20	30	34	ref. #
NPY						
sheep	YPSKPDNPGD	DA	PAEDLARY	YSALRHYINL	ITRQRY#	62
porcine	YPSKPDNPGE	DA	PAEDLARY	YSALRHYINL	ITRQRY#	57
ox	YPSKPDNPGE	DA	PAEDLARY	YSALRHYINL	ITRQRY#	62
rabbit	YPSKPDNPGE	DA	PAEDMARY	YSALRHYINL	ITRQRY#	58
rat	YPSKPDNPGE	DA	PAEDMARY	YSALRHYINL	ITRQRY#	58
guinea pig	YPSKPDNPGE	DA	PAEDMARY	YSALRHYINL	ITRQRY#	58
human	YPSKPDNPGE	DA	PAEDMARY	YSALRHYINL	ITRQRY#	58
PYY						
lamprey	MPPKPDNPSP	DASP	EELSKY	MLAVRNYINL	ITRQRY#	59
anglerfish	YPPKPETPGS	NASP	EDWASY	QAAVRHYINL	ITRQRYG	60
sculpin	YPPQPESPGG	NASP	EDWAKY	HAAVRHYVNL	ITRQRY#	60
eel	YPPKPENPGE	DASP	EEQAKY	YTALRHYINL	ITRQRY#	60
skate	YPPKPENPGD	DAAP	EELAKY	YSALRHYINL	ITRQRY#	60
bowfin	YPPKPENPGE	DAPP	EELARY	YSALRHYINL	ITRQRY#	60
gar	YPPKPENPGE	DAPP	EELAKY	YSALRHYINL	ITRQRY#	60
salmon	YPPKPENPGE	DAPP	EELAKY	YTALTHYINL	ITRQRY#	60
porcine	YPAKPEAPGE	DASP	EELSRY	YASLTHYLNL	VTRQRY#	57
rat	YPAKPEAPGE	DASP	EELSRY	YASLTHYLNL	VTRQRY#	61
human	YPIKPEAPGE	DASP	EELNRY	YASLTHYLNL	VTRQRY#	61
PP						
ostrich	GPAQPTYPGD	DA	PVEDLVRF	YDNLQQYLNV	VTRHRY#	63
chicken	GPSQPTYPGD	DA	PVEDLIRF	YDNLQQYLNV	VTRHRY#	63
turkey	GPSQPTYPGD	DA	PVEDL?RF	YNDLQQYLNV	VTRHRY#	63
goose	GPSQPTYPGN	DA	PVEDLIRF	YDNLQQYRLV	VFRHRY#	63
alligator	TPLQPKYPGD	GA	PVEDLIQF	YNDLQQYLNV	VTRPRF#	63
bullfrog	APSEPHHPGD	QATP	DQLAQY	YSDLYQYITF	ITRPRF#	14
ovine	ASLEPEYPGD	NATP	EQMAQY	AAELRRYINM	LTRPRY#	63
bovine	APLEPEYPGD	NATP	EQMAQY	AAELRRYINM	LTRPRY#	63
rat	APLEPMYPGD	YATH	EQRAQY	ETQLRRYINT	LTRPRY#	63
guinea pig	APLEPVYPGD	DATP	QQMAQY	AAEMRRYINM	LTRPRY#	64
porcine	APLEPVYPGD	DATP	EQMAQY	AAELRRYINM	LTRPRY#	63
canine	APLEPVYPGD	DATP	EQMAQY	AAELRRYINM	LTRPRY#	63
feline	APLEPVYPGD	NATP	EQMAQY	AAELRRYINM	LTRPRY#	14
human	APLEPVYPGD	NATP	EQMAQY	AADLRRYINM	LTRPRY#	63

FIG. 1 Sequences of the pancreatic polypeptide family in standard one-letter code. The references given are simply references that cite the sequences and do not in all cases represent the original paper in which the sequence was determined. #, a C-terminal amide.

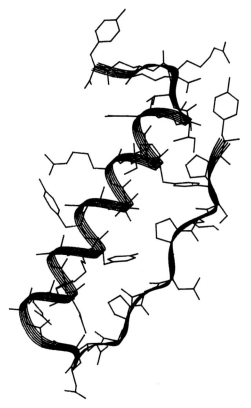

FIG. 2 NPY model based on the crystal structure of APP (4,50–53).

peptides (residues 1, 4, 24, and 34), and one was specific for the pancreatic polypeptides (residue 7) (3). Now with the expanded list of sequences shown in Fig. 1 the three lists would be narrowed somewhat: the entire family (residues 2, 5, 8, 12, 27, 33, and 35), the "Y" peptides (residue 34), and the pancreatic polypeptides (residue 7). We concluded that "some or all of the residues at positions 1, 4, 7, 26, and 34 may be responsible for receptor selectivity in these hormones" (3). The highly conserved residues common to the entire peptide family were thought to be important for promoting a favorable conformation and/or for direct receptor interaction and activation. Schwartz and co-workers also made similar observations and eventually made use of them (10, 15, 16). These studies are discussed later.

One very useful aspect of the observed homology among all of the peptides of this family was that at least some of the highly conserved residues were serving a conformational role. For example, residues 2, 5, and 8 are all proline residues, and they are spaced three residues apart. This pattern of proline residues is characteristic of a polyproline helix. Indeed, the N-terminal portion of avian pancreatic polypeptide (APP; chicken) adopts a left-handed polyproline helical conformation in the crystal (50–53). Residues 14–31 adopt an α-helical conformation that is amphipathic in nature. The polyproline helix and the α helix run antiparallel to one another, and their lipophilic faces participate in a number of intramolecular van der Waals interactions. Thus, APP is an intramolecularly stabilized, folded structure that has been described as a "small globular hormone" (52, 53). By analogy then, owing to the high homology of this family of peptides at positions that would be considered to be important for the conformation seen for APP in its crystal structure, one might infer that NPY assumes a similar intramolecularly stabilized, folded conformation. Indeed, most groups working with NPY utilized this assumption, and a number of models of NPY quickly appeared in the literature (4, 10, 14, 18, 24, 25, 42, 69–71). Figure 2 depicts one such model of NPY. Circular dichroic (CD) spectroscopy and analytical ultracentrifugation gave evidence that porcine NPY appeared to have similar physical properties to those reported for APP (3). These data added some measure of confidence to the use of these hypothetical models of NPY based on the crystal structure of APP.

Systematic Approaches

The earliest work done on NPY compared it to various fragments of the peptide in a number of bioassays (32–34, 41, 72). This work showed that in some systems the entire peptide was required for potent activity, whereas, in other systems, fragments of NPY consisting of its C-terminal region [e.g., NPY(2–36), NPY(13–36), and NPY(19–36)] retained reasonable potency. These data were quite invaluable as they gave the first evidence that NPY may utilize at least two receptors (or receptor subtypes). The C-terminal fragment peptides were, in fact, the first receptor-selective analogs of NPY. The designations of Y_1 and Y_2 receptors were coined to distinguish the two NPY receptor subtypes (72). NPY binds to the Y_1 receptor with high picomolar to low nanomolar affinity, but NPY(13–36) does not bind to this receptor even at much higher concentrations. NPY(13–36) exhibits approximately one-tenth the affinity of NPY for Y_2 receptors (68). For both receptors the C-terminal amide functionality was essential (72).

The C-terminal pentapeptides of APP and NPY have been shown to exhibit specific binding to ^{125}I-labeled APP binding sites in chick cerebellar membrane preparations in the micromolar range (73). Systematic modification of the N- and C-terminal regions with D-amino acids or Ala again pointed to the critical importance of residues 33–36 for receptor recognition (19, 22, 26, 27, 29, 35). The N-terminal region was comparatively tolerant of D-amino acid substitutions, especially in position 1. Although the aromatic side chain was important for potent Y_1-type receptor affinity, it can be acylated or alkylated, the amino group can be removed, or it can be replaced with the D isomer and still retain full potency (19, 22). Together, these data suggest the primary importance of the C-terminal tail of NPY for receptor recognition and activation. The role of the majority of the peptide was not made entirely clear by these studies.

In an attempt to study the similarities and differentiation of members of the "Y" peptide and pancreatic polypeptide (or "PP") family, Schwartz and co-workers engaged in synthetic studies which hybridized the two classes of family members. By doing this they sought to differentiate those regions of the peptides which played a common structural role throughout the "Y" and "PP" subfamilies from those which gave receptor recognition. As noted above, homology exists at several levels within this family of peptides. Schwartz and co-workers focused on exchanging residues that were specific to either the "Y" or "PP" subfamily to peptides of the opposite subfamily. For example, when Pro^{34} of porcine pancreatic polypeptide was replaced with Gln from the "Y" peptide subfamily, the resulting peptide exhibited high affinity for the Y_2 receptor (14). Similarly, $[Ile^{31}, Gln^{34}]APP$ bound to Y_2, but not Y_1, receptors with high affinity (15, 18). The pancreatic polypeptides do not bind to either the Y_1- or Y_2-type receptors (14, 15). The analog which represents the converse experiment, $[Leu^{31}, Pro^{34}]NPY$, as expected had greatly decreased affinity for the Y_2 receptor and increased affinity for the PP receptor, relative to NPY (15). The unexpected result was that affinity for the Y_1 receptor was unaffected (15, 16). Thus, the first highly potent and selective Y_1 receptor agonist had been discovered.

Concurrently, we were engaged in a similar study. Based on homology observations (3), we looked at Pro or His substitutions at position 34 in full-length and fragment analogs of NPY, which derive from the mammalian and avian pancreatic polypeptides, respectively (8). We found that both Pro and His substitutions gave a greater decrease in Y_2 receptor affinity than in Y_1 receptor affinity; however, the Pro modification was much more effective than His in achieving selectivity and retaining Y_1 receptor affinity (8). These data and further studies by Schwartz and co-workers (10) showed that the Pro modification alone is needed for the Y_1 receptor selectivity and that the additional Leu^{31} modification of the earlier analog was not required for the selectivity.

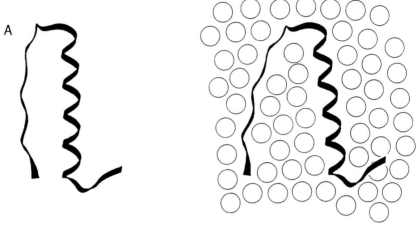

IN SOLUTION   **BOUND TO MEMBRANE**

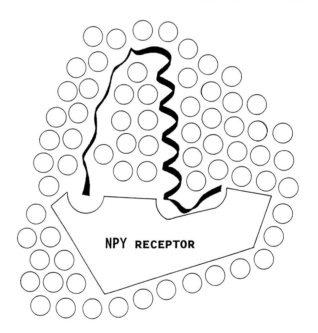

BOUND TO RECEPTOR ON THE MEMBRANE

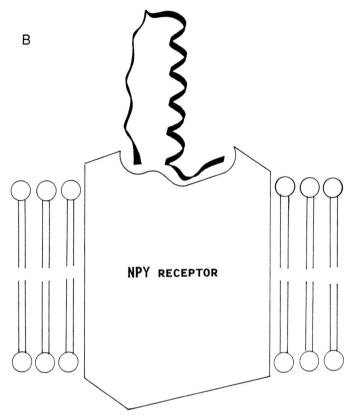

Fɪɢ. 3   Hypothetical models of the interaction of NPY with its receptors. NPY exists in solution as an equilibrium of folded and unfolded structures. (A) NPY might partially open up to expose its amphipathic structures which can then interact with membranes. (The view shown is looking directly down onto the membrane surface. The circles represent phospholipid head groups). The membrane then serves to facilitate the interaction of NPY with the receptor. (B) NPY interacts directly as a folded structure with its receptor. There is no involvement of the lipid membrane.

Conformational Approaches

Some of the work that is more distinctive of larger peptide hormones and neurotransmitters focuses on the conformational aspects of the peptide. As mentioned above, although much of the conformational work with shorter peptides deals with reverse turns and side chain conformations (46–48), longer peptides present different problems. For instance, which is more valuable, results from CD studies, NMR studies, or X-ray crystallography? The availability of the APP crystal structure clearly influenced all conformational approaches to NPY, but there is no guarantee that NPY will adopt the same conformation. Nor is there any assurance that the crystal structure resembles either the solution conformation or the receptor-bound conformation. One NMR study of NPY failed to confirm the presence of a folded conformation like that of the APP crystal structure (74). Thus, since we cannot decide which data are most relevant to the receptor-bound structure, we must devise ways of testing various models of receptor-bound conformations that are as unambiguous as possible. For NPY some approaches were to analyze, then optimize, components of the secondary structure.

The presence of the amphipathic α-helical region near the C terminus of NPY presented a natural avenue to explore. The history of the importance of this structural feature in other longer peptides (54–56) implied that it might play an important role in NPY receptor affinity. In the APP crystal structure, the lipophilic face of the α helix is involved in van der Waals interactions with the N-terminal polyproline helix (50–53). On viewing this, one might conclude that the amphipathic α helix plays primarily a structural role in maintaining a folded conformation, which may be important for receptor interaction. However, amphipathic α helices are known to have an affinity for membrane surfaces that has been proposed as being important for receptor recognition in these types of peptides (55, 56). Indeed, NPY, presumably via its amphipathic helical domains, interacts with lipid quite well (6, 7). Thus, two hypothetical models are shown in Fig. 3. Clearly, other variations are possible, but these serve as radically different alternatives that could be tested with the appropriate strategies.

Schwartz and co-workers conducted CD studies on selected full-length and fragment analogs of NPY. NPY and NPY(8–36) gave CD spectra that indicated similar α-helical contents, and these values were close to the α-helical content seen in the APP crystal structure (18). The weighted difference spectra of NPY and NPY(8–36) resembled the CD spectra obtained from calf skin collagen, which has a polyproline II helix structure. This gives indirect evidence of the presence of a polyproline helical conformation for NPY(1–7). Although this is the expected result from the APP crystal structure and agrees with the structure that was observed for bovine pancreatic polypeptide in solution (77), the polyproline helix and the folded structure were not observed in the NMR study of NPY by Saudek and Pelton (74). The introduction of Pro at position 20 for the Tyr residue in NPY yielded an analog with a markedly different conformation, as observed by CD, and this analog now bound only to Y_2 receptors, not Y_1 receptors, with an affinity as high as that of NPY(19–36) (Ref. 18). Thus, the disruption of the C-terminal α helix through the introduction of the Pro residue, and thereby the disruption of the folded structure, yielded the equivalent of a C-terminal fragment analog of NPY. Combined, these results strengthened the hypothesis that the APP-like folded conformation may be important for receptor interaction, especially at the Y_1-type NPY receptors. One interesting observation that was made in conjunction with these studies was that the folded conformation of NPY yields a structure that has a significant charge dipole, where the apposed termini have a generally positive charge and the turn region between the helices has an overall negative charge. The importance of this dipolar nature for receptor interaction is unknown.

Studies were done by Taylor and co-workers investigating replacement of the C-terminal amphipathic α helix of NPY with modified amphipathic α-helical segments, as had been done by the Kaiser group for a variety of larger peptides (42, 54), and stabilizing the α helix via i to $i + 4$ side chain cyclization (43). In the first study, several analogs were made that had multiple modifications introduced into the 13–32 region of NPY. The analogs retained the amphipathic α-helical nature of this region, and subsequently their ability to interact with lipid, but all were less potent than NPY in inhibiting electrically stimulated contraction of the rat vas deferens (42). This latter fact makes it difficult to draw any conclusions concerning the individual importance of any residue in the helical region since the drop in potency could be due to the loss of a functional group or a change in conformation. The data do suggest that it is the general amphipathic nature of regions of NPY that account for its membrane affinity and not particular residues in those amphipathic regions. Stabilization of the α-helical structure in NPY through i to $i + 4$ side chain cyclization gave an analog, cyclo(18,22)-[Lys^{18},Asp^{22}]NPY, that was equipotent to NPY in inhibiting electrically stimulated contractions in the rat vas deferens and had slightly decreased receptor binding affinity in the same tissue (43).

Studies involving a large leap of faith in favor of the folded conformational model were performed separately by Jung and co-workers (24, 25, 27) and our group (4, 5). These involved removal of the central "structurally important" region and linkage of the "functional" N- and C-terminal regions with a spacer residue of appropriate length to allow proper apposition of the N- and C-terminal regions. Jung and co-workers found that linking NPY(1–4) to NPY(25–36) with a 6-aminohexanoyl residue yielded an analog with a potency midway between NPY and NPY(19–36) for a presynaptic (or Y_2-like) activity (suppression of noradrenaline release) (27). On a postsynaptic (or Y_1-like) activity (blood pressure increase) the analog has a potency similar to that of NPY(19–36) (Ref. 27). Thus, this analog appears to possess Y_2 selectivity.

Our approach was to produce a series of centrally truncated analogs that bridged the N and C termini with an 8-aminooctanoyl residue (Aoc) at points of closest approach between the two amphipathic helices (Fig. 4). As the object of our study was to also attempt to differentiate between the proposed models of receptor interaction for NPY that are depicted in Fig. 3, we introduced a disulfide bond to link the N and C termini. The function of the disulfide bond was 2-fold. One, the truncation of the structure meant that less van der Waals stabilization of the conformation would be available; therefore, to compensate for this the disulfide would, theoretically, make up for the loss of van der Waals interaction. Two, covalent linkage of the N and C termini would prevent the opening of the folded structure and thereby interfere with the ability of the analog to interact with lipid via its amphipathic structures. The three analogs that resulted were [D-Cys^7,Aoc$^{8-17}$,Cys^{20}]NPY, [Cys^5,Aoc$^{7-20}$,D-Cys^{24}]NPY, and [Cys^2,Aoc$^{5-24}$,D-Cys^{27}]NPY which are abbreviated as C7-NPY, C5-NPY, and C2-NPY, respectively (4–6).

C7-NPY exhibited good affinity for the mouse brain receptors (Y$_1$-like), whereas C5-NPY had only weak affinity and C2-NPY possessed extremely poor affinity (4–6). On porcine spleen (Y$_2$-like) all of the analogs had good affinity relative to NPY (6); however, on other Y$_2$ systems only C2-NPY gave consistently high potency (10, 11). This is interesting from the standpoint that both C7-NPY and C5-NPY contain significantly more of the C-terminal region than C2-NPY and therefore would be expected to be at least as good as the shorter analog. This implies that the conformation present in the C7-NPY and C5-NPY analogs is not compatible with certain Y$_2$ receptor systems. It should be noted that C2-NPY is very similar to the linear analog of Jung and co-workers mentioned above which also appears to be Y$_2$ selective as are the C-terminal fragment analogs of NPY.

In regard to differentiating between the proposed receptor models in Fig. 3, the ability of these centrally truncated and cyclized analogs to interact

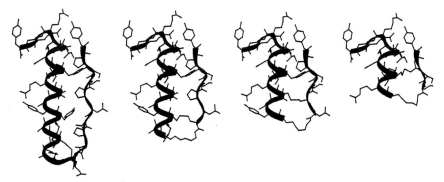

FIG. 4   Folded model of NPY (left) and (left to right) the series of systematically centrally truncated structures C7-NPY, C5-NPY, and C2-NPY. (Figure reprinted with the permission of ESCOM Science Publishers from Ref. 5.)

with lipid was examined using CD, fluorescence spectroscopy, differential scanning calorimetry, and the clearing of liposomes (6). C7-NPY retained the ability to clear liposomes, but its binding to them was much reduced relative to NPY. Neither C5-NPY nor C2-NPY were able to interact with lipid to any significant extent. In comparing these data to the receptor affinities of these analogs (especially C2-NPY), it would appear that lipid interaction is not required for interaction with Y_2 receptors. The attenuated lipid affinity yet good mouse brain receptor affinity for C7-NPY suggests that lipid interaction may not be essential for Y_1 receptor interaction as well.

More recently, both Jung and co-workers (28, 75) and Rivier and co-workers (76) have synthesized centrally truncated analogs of NPY that are cyclized via an amide side chain linkage. The analog cyclo(2,27)-[Glu^2,Pro^5,Gly^6,des-AA^{7-24},D-2,3-diaminopropionyl^{27}]NPY has been shown to be a potent Y_2 selective analog (76). It is hoped that structural studies on these constrained analogs will yield more specific information on the conformational requirements for receptor interaction.

Discussion

The interpretations that one can derive from all of the work on NPY can vary depending on one's perspective, but our present understanding (and current hypotheses) can be summarized as follows:

1. The primary determinants of NPY binding to its receptor reside in its C-terminal tail, and the functionality and/or conformation of this region can determine receptor affinity and selectivity. The conformational requirements for receptor interaction of the C-terminal tail are still unknown. It is not clear which residues, if any, other than those in the very C terminus, interact directly with the NPY receptors.
2. The amphipathic helical regions play a role in the intramolecular structural stabilization of NPY, which may be important for receptor recognition. The ability of NPY to interact with lipid is a consequence of the amphipathic nature of NPY, but it does not appear that this lipid interaction plays a role in the interaction of NPY with its receptor. The importance that the dipolar nature of the folded NPY molecule may have in NPY–receptor interaction is not clear.
3. The role of the residue at position 1 for Y_1 receptor interaction may be to promote a favorable conformation in the C-terminal tail region due to its apposition to that region in the folded structure.

Thus, while many questions remain, significant advances have been made in understanding NPY and the relationship of its properties to NPY–receptor interactions. The approaches to studying NPY that have been described have also been applied to other peptide hormones and neurotransmitters, with varying levels of success. They represent the systematic dissection of the molecule and can certainly be applied to new peptides and peptide families as they are discovered. NPY is perhaps unique in that it appears to have been better "behaved" than other peptides in that its structure and relatively well-defined C-terminal functional domain have aided evaluation. Some other hormones (glucagon, for example) seem to be more resistant to evaluation despite substantial effort and progress in analog design. The troublesome peptides, the ones which show our hypotheses about them to be wrong and defy our most clever analyses, will be the ones that force us to devise new ways of thinking about them.

Acknowledgments

I wish to thank Dr. Walt Lovenberg who in 1986 first suggested that I take a look at neuropeptide Y and see if there was anything that could be done with it. I also wish to thank all of my collaborators within and outside of the Marion Merrell Dow Research Institute (Cincinnati, OH) who have made invaluable contributions and made working in this area a very enjoyable experience.

References

1. M. C. Michel and H. J. Motulsky, *Ann. N.Y. Acad. Sci.* **611,** 392 (1990).
2. K. Tatemoto, *Ann. N.Y. Acad. Sci.* **611,** 1 (1990).
3. J. L. Krstenansky and S. H. Buck, *Neuropeptides* **10,** 77 (1987).
4. J. L. Krstenansky, T. J. Owen, S. H. Buck, K. A. Hagaman, and L. R. McLean, *Proc. Natl. Acad. Sci. U.S.A.* **86,** 4377 (1989).
5. J. L. Krstenansky, T. J. Owen, S. H. Buck, K. A. Hagaman, and L. R. McLean, *in* "Peptides: Chemistry, Structure and Biology: Proceedings of the Eleventh American Peptide Symposium" (J. E. Rivier and G. R. Marshall, eds.), p. 319. ESCOM, Leiden, The Netherlands, 1990.
6. L. R. McLean, S. H. Buck, and J. L. Krstenansky, *Biochemistry* **29,** 2016 (1990).
7. L. R. McLean, B. M. Baron, S. H. Buck, and J. L. Krstenansky, *Biochim. Biophys. Acta* **1024,** 1 (1990).
8. J. L. Krstenansky, T. J. Owen, M. H., Payne, S. A. Shatzer, and S. H. Buck, *Neuropeptides* **17,** 117 (1990).
9. E. A. Gordon, J. L. Krstenansky, and P. H. Fishman, *Neurosci. Lett.* **119,** 187 (1990).
10. T. W. Schwartz, J. Fuhlendorff, L. L. Kjems, M. S. Kristensen, M. Vervelde, M. O'Hare, J. L. Krstenansky, and B. Bjornholm, *Ann. N.Y. Acad. Sci.* **611,** 35 (1990).

11. H. M. Cox and J. L. Krstenansky, *Peptides* **12**, 323 (1991).
12. M. M. T. O'Hare, S. Tenmoku, L. Aakerlund, L. Hilsted, A. Johnsen, and T. Schwartz, *Regul. Pept.* **20**, 293 (1988).
13. S. P. Sheikh, M. M. T. O'Hare, O. Tortora, and T. W. Schwartz, *J. Biol. Chem.* **264**, 6648 (1989).
14. T. W. Schwartz, J. Fuhlendorff, N. Langeland, H. Thorgersen, J. C. Jorgensen, and S. P. Sheikh, *in* "Neuropeptide Y: XIV Nobel Symposium" (V. Mutt, T. Hokfelt, K. Fuxe, and J. M. Lundberg, eds.), p. 143. Raven, New York, 1989.
15. J. C. Jorgensen, J. Fuhlendorff, and T. W. Schwartz, *Eur. J. Pharmacol.* **186**, 105 (1990).
16. J. Fuhlendorff, U. Gether, L. Aakerlund, N. Langeland-Johansen, H. Thorgersen, S. G. Melberg, U. B. Olsen, O. Thastrup, and T. W. Schwartz, *Proc. Natl. Acad. Sci. U.S.A.* **87**, 182 (1990).
17. L. Aakerlund, U. Gether, J. Fuhlendorff, T. W. Schwartz, and O. Thastrup, *FEBS Lett.* **260**, 73 (1990).
18. J. Fuhlendorff, N. Langeland-Johansen, S. G. Melberg, H. Thorgersen, and T. W. Schwartz, *J. Biol. Chem.* **265**, 11706 (1990).
19. J. H. Boublik, N. A. Scott, M. R. Brown, and J. E. Rivier, *J. Med. Chem.* **32**, 597 (1989).
20. J. H. Boublik, N. A. Scott, J. Taulane, M. R. Brown, M. J. Goodman, and J. E. Rivier, *Int. J. Pept. Protein Res.* **33**, 11 (1989).
21. J. H. Boublik, N. A. Scott, R. D. Feinstein, M. Goodman, M. R. Brown, and J. E. Rivier, *in* "Peptides: Chemistry, Structure and Biology: Proceedings of the Eleventh American Peptide Symposium" (J. E. Rivier and G. R. Marshall, eds.), p. 317. ESCOM, Leiden, The Netherlands, 1990.
22. J. H. Boublik, M. A. Spicer, N. A. Scott, M. R. Brown, and J. E. Rivier, *Ann. N.Y. Acad. Sci.* **611**, 27 (1990).
23. M. A. Spicer, J. H. Boublik, R. D. Feinstein, M. Goodman, M. Brown, and J. Rivier, *Ann. N.Y. Acad. Sci.* **611**, 359 (1990).
24. A. Beck, G. Jung, G. Schnorrenberg, H. Koppen, W. Gaida, and R. Lang, *in* "Peptides 1988, Proceedings of the 20th European Peptide Symposium" (G. Jung and E. Bayer, eds.), p. 586. de Gruyter, Berlin, 1989.
25. A. Beck, G. Jung, W. Gaida, H. Koppen, R. Lang, and G. Schnorrenberg, *FEBS Lett.* **244**, 119 (1989).
26. A. Beck-Sickinger, W. Gaida, G. Schnorrenberg, R. Lang, and G. Jung, *Int. J. Pept. Protein Res.* **36**, 522 (1990).
27. A. G. Beck-Sickinger, G. Jung, W. Gaida, H. Koppen, G. Schnorrenberg, and R. Lang, *Eur. J. Biochem.* **194**, 449 (1990).
28. H. Durr, E. Hoffmann, A. G. Beck-Sickinger, and G. Jung, *in* "Peptides 1990: Proceedings of the 21st European Peptide Symposium" (E. Giralt and D. Andreu, eds.), p. 216. ESCOM, Leiden, The Netherlands, 1991.
29. A. G. Beck-Sickinger, W. Gaida, G. Schnorrenberg, and G. Jung, *in* "Peptides 1990: Proceedings of the 21st European Peptide Symposium" (E. Giralt and D. Andreu, eds.), p. 646. ESCOM, Leiden, The Netherlands, 1991.
30. C. R. Baeza and A. Unden, *FEBS Lett.* **277**, 23 (1990).

31. C. R. Baeza and A. Unden, *in* "Peptides 1990: Proceedings of the 21st European Peptide Symposium" (E. Giralt and D. Andreu, eds.), p. 649. ESCOM, Leiden, The Netherlands, 1991.

32. F. Rioux, H. Bachelard, J. C. Martel, and S. St-Pierre, *Peptides* **7**, 27 (1986).

33. J. M. Danger, M. C. Tonon, M. Lamacz, J. C. Martel, S. Saint-Pierre, G. Pelletier, and H. Vaudry, *Life Sci.* **40**, 1875 (1987).

34. V. Donoso, M. Silva, S. St-Pierre, and J. P. Huidoboro-Toro, *Peptides* **9**, 545 (1988).

35. M. Forest, J. C. Martel, S. St-Pierre, R. Quirion, and A. Fournier, *J. Med. Chem.* **33**, 1615 (1990).

36. R. Quirion, J. C. Martel, Y. Dumont, A. Cadieux, F. Jolicoeur, S. St-Pierre, and A. Fournier, *Ann. N.Y. Acad. Sci.* **611**, 58 (1990).

37. A. Cadieux, M. T-Benchenkroun, A. Fournier, and S. St-Pierre, *Ann. N.Y. Acad. Sci.* **611**, 369 (1990).

38. A. Balasubramanian, L. Grupp, M. A. Matlib, R. Benza, R. L. Jackson, J. E. Fischer, and G. Grupp, *Regul. Pept.* **21**, 289 (1988).

39. A. Balasubramanian, V. Renugopalakrishnan, D. F. Rigel, M. S. Nussbaum, R. S. Rapaka, J. C. Dobbs, L. A. Carreira, and J. E. Fischer, *Biochim. Biophys. Acta* **997**, 176 (1989).

40. A. Balasubramanian, S. Sheriff, V. Renugopalakrishnan, S. G. Huang, M. Prabhakaran, W. T. Chance, and J. E. Fischer, *Ann. N.Y. Acad. Sci.* **611**, 355 (1990).

41. W. Danho, J. Triscari, G. Vincent, T. Nakajima, J. Taylor, and E. T. Kaiser, *Int. J. Pept. Protein Res.* **32**, 496 (1988).

42. H. Minakata, J. W. Taylor, M. W. Walker, R. J. Miller, and E. T. Kaiser, *J. Biol. Chem.* **264**, 7907 (1989).

43. M. Bouvier, J. W. Taylor, and G. Osapay, *in* "Peptides 1990: Proceedings of the 21st European Peptide Symposium" (E. Giralt and D. Andreu, eds.), p. 652. ESCOM, Leiden, The Netherlands, 1991.

44. M. O. Perlman, J. M. Perlman, M. L. Adamo, R. L. Hazelwood, and D. F. Dyckes, *Int. J. Pept. Protein Res.* **30**, 153 (1987).

45. W. F. DeGrado, *Adv. Protein Chem.* **39**, 51 (1988).

46. V. J. Hruby and H. I. Mosberg, *Peptides* **3**, 329 (1982).

47. V. J. Hruby, *Trends Pharmacol. Sci.* **6**, 259 (1985).

48. V. J. Hruby, F. Al-Obeidi, and W. Kazmierski, *Biochem. J.* **268**, 249 (1990).

49. M. Rosenblatt, *N. Engl. J. Med.* **315**, 1004 (1986).

50. S. P. Wood, J. E. Pitts, T. L. Blundell, I. J. Tickle, and J. A. Jenkins, *Eur. J. Biochem.* **78**, 119 (1977).

51. T. L. Blundell, J. E. Pitts, I. J. Tickle, S. P. Wood, and C.-W. Wu, *Proc. Natl. Acad. Sci. U.S.A.* **78**, 4175 (1981).

52. I. Glover, I. Haneef, J. Pitts, S. Wood, D. Moss, I. Tickle, and T. Blundell, *Biopolymers* **22**, 293 (1983).

53. I. D. Glover, D. J. Barlow, J. E. Pitts, S. P. Wood, I. J. Tickle, T. L. Blundell, K. Tatemoto, J. R. Kimmel, A. Wollmer, W. Strassberger, and Y.-S. Zhang, *Eur. J. Biochem.* **142**, 379 (1985).

54. E. T. Kaiser and F. J. Kezdy, *Science* **223**, 249 (1984).

55. R. Schwyzer, *in* "Peptides: Structure and Function. Proceedings of the Ninth American Peptide Symposium" (C. M. Deber, V. J. Hruby, and K. D. Kopple, eds.), p. 3. Pierce Chemical Co., Rockford, Illinois, 1985.

56. D. F. Sargent and R. Schwyzer, *Proc. Natl. Acad. Sci. U.S.A.* **83,** 5774 (1986).

57. K. Tatemoto, *Proc. Natl. Acad. Sci. U.S.A.* **79,** 5485 (1982).

58. M. M. T. O'Hare, S. Tenmoku, L. Aakerlund, L. Hilsted, A. Johnsen, and T. W. Schwartz, *Regul. Pept.* **20,** 293 (1988).

59. J. M. Conlon, B. Bjornholm, F. S. Jorgensen, J. H. Youson, and T. W. Schwartz, *Eur. J. Biochem.* **199,** 293 (1991).

60. J. M. Conlon, C. Bjenning, T. W. Moon, J. H. Youson, and L. Thim, *Peptides,* **12,** 221 (1991).

61. G. A. Eberlein, V. E. Eysselein, M. Schaeffer, P. Layer, D. Grandt, H. Goebell, W. Niebel, M. Davis, T. D. Lee, J. E. Shively, and J. R. Reeve, Jr., *Peptides* **10,** 797 (1989).

62. R. Sillard, B. Agerberth, V. Mutt, and H. Jornvall, *FEBS Lett.* **258,** 263 (1989).

63. D. Litthauer and W. Oelofsen, *Int. J. Pept. Protein Res.* **29,** 739 (1987).

64. J. Eng, C.-G. Huang, Y.-C.E. Pan, J. D. Hulmes, and R. S. Yalow, *Peptides* **8,** 165 (1987).

65. M. J. Kilborn, E. K. Potter, and D. I. McKloskey, *Regul. Pept.* **12,** 155 (1985).

66. M. W. Walker and R. J. Miller, *Mol. Pharmacol.* **34,** 779 (1988).

67. A. Inui, M. Okita, T. Inoue, N. Sakatani, M. Oya, H. Morioka, K. Shii, K. Yokono, N. Mizuno, and S. Baba, *Endocrinology (Baltimore)* **124,** 402 (1989).

68. S. P. Sheikh, R. Hakanson, and T. W. Schwartz, *FEBS Lett.* **245,** 209 (1989).

69. J. Allen, J. Novotny, J. Martin, and G. Heinrich, *Proc. Natl. Acad. Sci. U.S.A.* **84,** 2532 (1987).

70. A. D. MacKerell, Jr., *J. Comput.-Aided Mol. Des.* **2,** 55 (1988).

71. A. D. MacKerell, Jr., A. Hemsen, J. S. Lacroix, and J. M. Lundberg, *Regul. Pept.* **25,** 295 (1989).

72. C. Wahlestedt, N. Yanaihara, and R. Hakanson, *Regul. Pept.* **13,** 307 (1986).

73. M. O. Perlman, J. M. Perlman, M. L. Adamo, R. L. Hazelwood, and D. F. Dyckes, *Int. J. Pept. Protein Res.* **30,** 153 (1987).

74. V. Saudek and J. T. Pelton, *Biochemistry* **29,** 4509 (1990).

75. A. Beck-Sickinger, W. Gaida, H. Durr, E. Hoffmann, G. Schnorrenberg, and G. Jung, *in* "Proceedings of the Twelfth American Peptide Symposium" (J. A. Smith and J. E. Rivier, eds.), p. 17. ESCOM, Leiden, The Netherlands, 1991.

76. D. A. Kirby, R. D. Feinstein, S. C. Koerber, M. R. Brown, and J. E. Rivier, *in* "Proceedings of the Twelfth American Peptide Symposium" (J. A. Smith and J. E. Rivier, eds.), p. 480. ESCOM, Leiden, The Netherlands, 1991.

77. X. Li, M. J. Sutcliffe, T. W. Schwartz, and C. M. Dobson, *Biochemistry* **31,** 1245 (1992).

Index